JANE'S NBC PROTECTION EQUIPMENT

SECOND EDITION

EDITED BY
TERRY J GANDER

1989-90

ISBN 0 7106 0894 2
JANE'S DEFENCE DATA
"Jane's" is a registered trade mark

In the USA and its dependencies
Jane's Information Group Inc, 1340 Braddock Place, Suite 300, PO Box 1436, Alexandria, Virginia 22313-2036, USA
Printed in the United Kingdom

Contents

Alphabetical list of advertisers

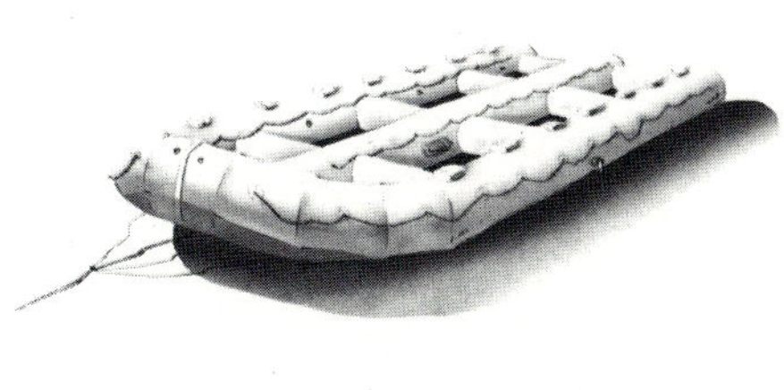

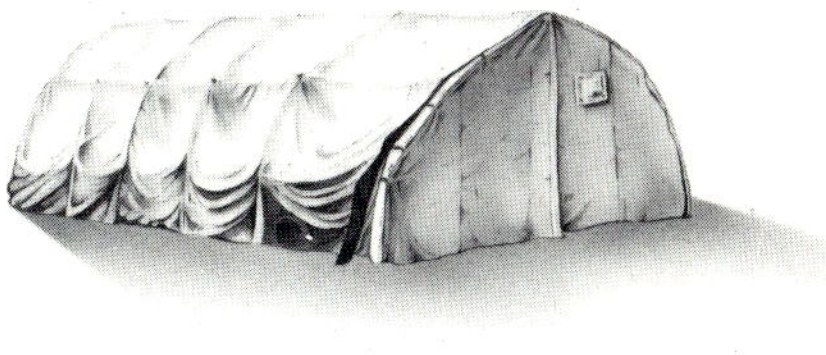

Classified list of advertisers

The following companies have informed us that they are involved in the fields of manufacture listed below:

Chemical and biological detection and identification equipment
Duphar
Environics
Environmental Technologies
Graseby Ionics
Jasmin Simtec
J & S Franklin
Plessey Controls

Chemical training maintenance and other equipment
Graseby Ionics

Collective protection equipment and associated equipment
Booth Engineering
Cristanini
Graseby Ionics
Howden Aircontrol
J & S Franklin
Sekur

Consulting engineers
Tyréns

Design and planning of NBC protected shelters and other installations
Tyréns

Decontamination kits and equipment
Cristanini
Environmental Technologies
Hughes Safety Showers
J & S Franklin

Dosimeters and charging units
Alnor OY
Plessey Controls

Filter canisters
Sekur

Medical and life support systems
Duphar
Howden Aircontrol

NBC casualty bags
J & S Franklin
Lantor

NBC facelets
J & S Franklin
Lantor
Sekur

NBC gloves
J & S Franklin
Sekur

NBC hazard prediction computer
Zengrange

NBC overboots
Avon Industrial Polymers
J & S Franklin
Sekur

NBC protective clothing
Avon Industrial Polymers
J & S Franklin
Graseby Ionics
New Pac Systems
Lantor
Sekur

Personnel protection equipment
Howden Aircontrol

Radiation survey meters and detection equipment
Alnor UY
Jasmin Simtec
Plessey Controls

Respirators
Avon Industrial Polymers
Sekur

NBC Equipment

NBC RESPIRATOR

The J.&S. Franklin Group of Companies manufacture and supply NBC Equipment to meet all requirements. Franklin specialise in Individual and Collective Protective Systems, Advanced Gas Detectors and Alarms for remote or personal use and Decontamination Equipment. Franklin also supply Associated Simulation Equipment and Training.

CASUALTY BAGS

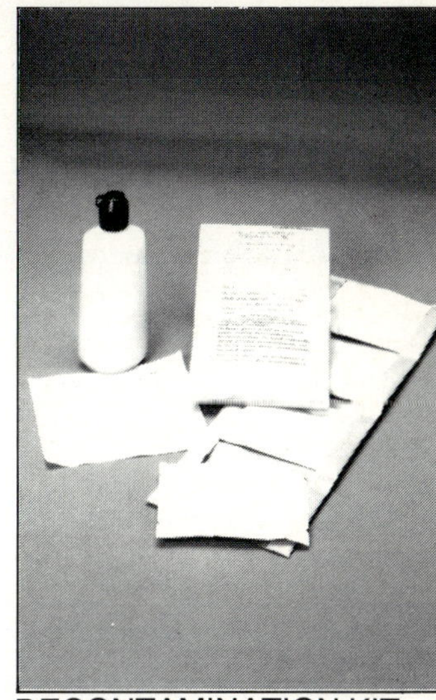

DECONTAMINATION KIT

PERSONAL PROTECTION

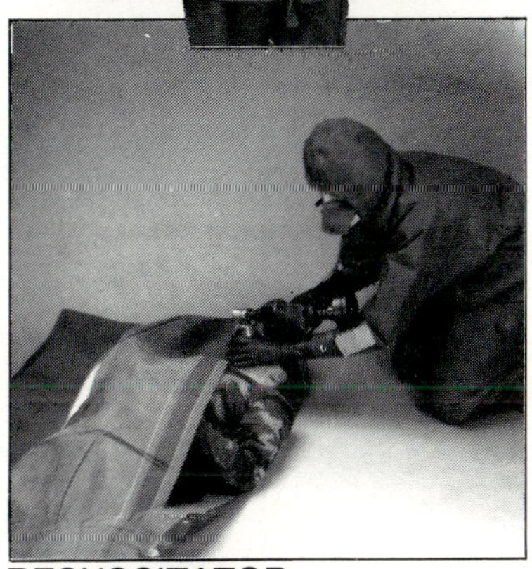

RESUSCITATOR

COLLECTIVE PROTECTION

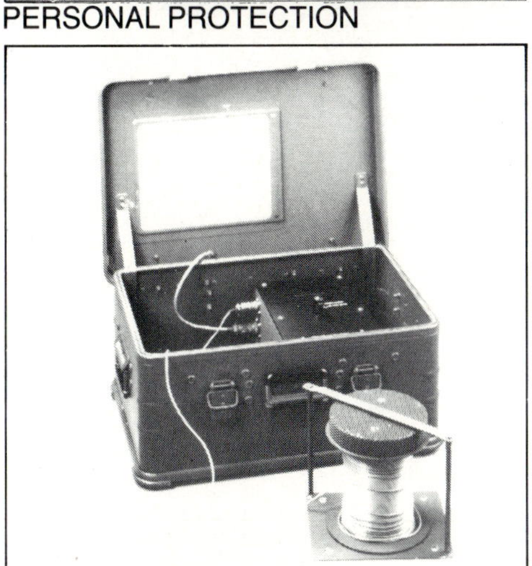

REMOTE DETECTORS

INDIVIDUAL DETECTOR

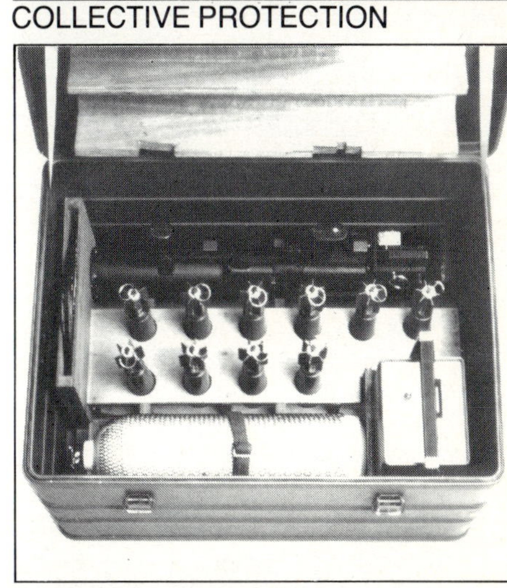

CHEMICAL DETECTORS

J&S
FRANKLIN EQUIPMENT

J.&S. FRANKLIN LTD.

Franklin House, 151 Strand, London WC2R 1HL
Telex: 897222 FRNKLN-G Fax: 01-836 2784 **Telephone: 01-836 5746**
Manufacturers of Military & Civil Equipment

New Pac Disposable C-Cover S/89

Description

In close co-operation with the Swedish Defence Material Administration (FMV), NEW PAC has developed a disposable plastic co-overall for NBC warfare protection.

A complete C-Cover S/89 consists of two parcels. One Body Cover and a pair of Foot Covers, both tightly rolled into a compact unit. One size fits everybody! The C-Cover S/89 is especially designed to meet the FMV requirements:

1. Compact dimensions
2. Minimum weight (560 gram)
3. Get covered within 10 seconds
4. C-proof at least 5 hours
5. Abrasion proof at least for 5 km walking

A patented Zig-Zag folding of the Body Cover provides easy donning even when lying on the ground. The Body Cover also has a patented sleeve construction, allowing the arms to "swim" easily into the sleeves through one common sleeve opening.

The C-Cover S/89 is carefully tested by FMV and thus granted for army, marine and air force services.

Status

In production since 1988. 80,000 C-Covers in service with the Swedish Air Force. 500,000 purchased by FMV for delivery during 1989/1990.

Manufacturer

NEW PAC

New Pac SYSTEMS AB

NEW PAC SYSTEMS AB
P.O. Box 15
564 01 Bankeryd, Sweden

PORTAflex 300
PORTABLE, LIGHTWEIGHT, MULTI-PURPOSE DECONTAMINATION SHOWER

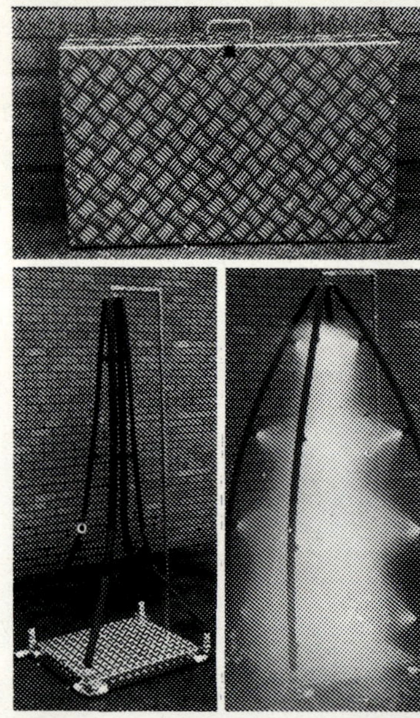

PATENT No. 2179549

Developed for:-
- The Armed Services
- Civil & Military Fire Brigades
- Civil Defence Units

COMPACT
Foldaway unit with dual purpose foot platform/ carrying case - all of extremely rugged construction. Storage size: 750 x 450 x 170mm.

LIGHTWEIGHT
Easily carried by one man. Total weight 26 Kg.

QUICK ASSEMBLY
Time for one man to assemble unit - 30 seconds

The increasing threat of future warfare involving dangerous chemicals or nuclear fallout has led to the development for military use of the Portaflex 300.

Speedily erected and connected to a water supply (either fire hydrant or pumped from reservoir, pond, etc.), the hoses become rigid under pressure, adopting the curved form as seen in the illustration. Enough water is delivered through the fog-producing nozzles to dilute any chemical contamination suffered by personnel quickly and efficiently.

HUGHES
SAFETY SHOWERS

WHITEFIELD ROAD,
BREDBURY,
STOCKPORT, CHESHIRE
SK6 2SS. ENGLAND.
Telephone: 061-430 6618
Telex: 668958
Facsimile: 061-430 7928

FROM A HAS

AND AWACS HANGARS TO UNDERGROUND SHELTERS

BOOTH
MAKE THE DOORS

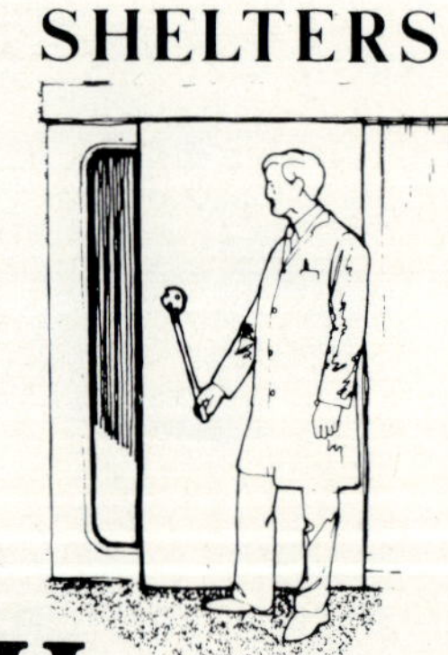

BOOTH
Booth Engineering Ltd.

PO Box 50
Hulton Steelworks
Bolton BL3 3ST

Telephone: 0204 61191
Telex: 635679 FYRDOR G
Telefax: 0204 64646

A MEMBER OF BOOTH INDUSTRIES PLC

Foreword

This is the second edition of *Jane's NBC Protection Equipment*. The first edition arrived on the scene just as the Chemical Weapons Convention was getting under way in Geneva and it seemed at that time that it would not be long before chemical and biological weapons would be the subject of some form of international treaty or international agreement banning their use and production. The Convention is still in progress and would appear to be set for a long haul ahead.

As ever in the sphere of international agreements, things are not quite as straightforward as they might at first appear. Every thinking person would dearly love to live in a world where chemical and biological warfare had not even been thought of, but it does exist and recent years have seen the re-introduction of those weapons on a large scale. At the same time the means to produce chemical weapons are proliferating to the extent where international conferences have taken place to discuss possible ways of preventing their further spread. But as always, just because common sense and even a small sense of humanitarian feelings urges the acceptance of international treaties and courses of action, it does not necessarily follow that the path to such understandings will be smooth or easy.

The first edition of this Yearbook was written in the aftermath of the well documented nerve gas attack against the hapless and unprotected citizens of Halamjeh. The world was appalled at the resultant sights. The sense of universal revulsion was such that it seemed at the time there would be an acceleration of efforts to rectify matters and put the chemical warfare genie back into the bottle. Despite the efforts of many that has not yet happened.

The reasons for this seeming lack of international progress to a sensible end are many and it must be stated here and now that the Geneva Convention has made considerable efforts and progress towards some form of ban on chemical weapons. The trouble is that the further one goes into the subject of producing a chemical weapons ban that would stick, the more the difficulties working against such an objective would seem to multiply.

Verification

One of the first stumbling blocks to any proposed ban is the positive verification of any treaty or convention that might be produced. Chemical (and to a large extent biological) weapons are unlike any others for several reasons. One is that they can be produced using industrial plant and facilities that have many peaceful commercial and other uses. Nuclear weapons have to be manufactured and stored in highly specialised facilities that can have few, if any, other uses and the same applies to the production of most conventional weapons and their associated systems. Few chemical weapons require such specialised and easily identified production facilities. To police and supervise every industrial facility that could be turned to weapon production would be a vast and almost impossible undertaking. Even in these days of satellite surveillance it would be very difficult to keep a permanently watchful eye on every potential chemical agent production facility but without such supervision and verification any form of treaty would rapidly become meaningless.

One large problem involved is that the means to produce chemical warfare agents are not all that difficult to come by. On a small scale it could even be possible to produce lethal agents in backroom workshops or even domestic facilities, and it would take only a few small scale uses of the resultant chemical weapons to run a railroad through any treaty. As yet none of the numerous terrorist groups that infest the modern world have turned their attention to the possible uses of chemical warfare but the potential for them to do so has been around for a very long time. If a cynical eye were turned on that subject it could be stated that the actual event of such an attack taking place can only be a matter of by whom, when and where.

However, the prospect of chemical agent-involved terrorism is only a small aspect of the proposed international understanding. The subject of verification is likely to be a major stumbling block but it is to be earnestly hoped that it is an obstacle that can be overcome. Just because something is seen to be difficult there is no reason why attempts should not be made to overcome those difficulties. There is too much at stake.

Evidence

It seems very likely that the resultant treaty involving the outlawing of chemical weapons will involve some form of international court to sift evidence of any treaty violation and impose some form of sanction against offenders. Such activities will involve a great deal of legal procedure that will mean the imposition of strict rules of evidence. The last two decades have seen numerous accusations of the use of chemical warfare in many parts of the world but few, if any, have actually been proved to have taken place (the horrors of Halamjeh and the Iran-Iraq conflict apart). Too many of the accusations have been based on hearsay evidence and partial or inarticulate witnesses. Forensic-type evidence has been lacking as well, and if it has been forthcoming, has been proved to be either inconsequential or too unreliable to act as hard proof of any occurrence.

Thus, any future treaty will have to involve the introduction of some form of international and impartial body that will supervise or undertake the collection and preparation of evidence of any treaty transgressions to the extent that their endeavours will be acceptable and usable to the necessary legal proceedings. Here we are treading in murky waters, for the legal world has a way of either ignoring common sense or raising objections that can overturn the most obvious of factual evidence. Precise and incontrovertible facts have to be firmly established to the extent of actual time and places, exactly where each item of forensic evidence was collected, the path the evidence took to reach the court, the influence of collecting methods and materials upon the evidence and so on, to the extent that even the very nature of truth becomes the subject of controversy. Such is the legalistic world in which specialist technicians will have to tread.

So once again there would seem to be a major obstacle in the way of any chemical warfare ban, but thankfully the legal area is not one that imposes insurmountable barriers. Modern electronics can be readily modified to identify places and times accurately, even to the point of involving satellite navigation and position finding, and sealed videos could provide hard evidence of collecting and test procedures and all the other necessary legal niceties. The legal requirements of any future ban are only one aspect of the much larger subject but are mentioned here to provide part of the reason why it will take a long time to reach a meaningful international agreement banning the use of chemical weapons.

PFIB

To add one further aspect to the problems besetting the delegates at Geneva, a paper was submitted to the Convention in April 1989. The paper originated from the United Kingdom's Chemical Defence Establishment at Porton Down (Wiltshire) and drew the attention of the delegates to the subject of PFIB.

PFIB is the abbreviation for perfluoroisobutene, a substance that has been known about for some time and which seems to have a potential use as a chemical warfare agent. PFIB is a clear and odourless gas that is highly toxic, causing death by the production of pulmonary oedema, or 'dry land drowning', in which the lungs fill with fluid to the extend that normal breathing cannot take place. Even non-lethal doses can cause severe medical problems such as pneumonia. Even minute doses measured in parts per million can cause disablement when inhaled over a period.

PFIB poses particular problems for any treaty organisers for two reasons. One is that it is produced when polytetrafluoroethene, or PTFE as it is more commonly known, is pyrolised, that is burned at high temperatures. Considering the widespread use of PTFE for both commercial and domestic purposes, it requires little thought as to how the substance could be readily produced with only a minimum of knowledge to create considerable potential harm.

The second aspect relating to PFIB is not primarily a concern of the negotiators at Geneva but is one of protection. The amounts of PFIB that are potentially harmful are so small that many existing protection systems may well be less than fully effective against the substance. This is a matter that can no doubt be technically overcome but in the meantime some protection and filtration systems may be found wanting if ever they are called upon to protect against PFIB.

Good News

To provide at least a partial counter to the above outline of doom and difficulty, there has been considerable progress in the continuing efforts

to rid the world of NBC weapons. The SALT treaties are progressing well and at least a few nuclear delivery devices have been destroyed in satisfying and spectacular operations. More such operations are on the way and at the end of the current proceedings the world will be rid of a small portion of its nuclear arsenal.

The dangers of NBC proliferation have been taken firmly by the horns and, as mentioned above, conferences have taken place to discuss the subject and discover ways of preventing the spread of chemical and biological weapons. If Halamjeh and the Iran-Iraq conflict can possibly be said to have had a positive outcome it is this realisation of the dangers that await, should chemical warfare ever again become established as an accepted weapon of war. Actual results have so far been few but they are on the way and we can only welcome them when they arrive.

Hardware

This edition contains a few new national names. For the first time mention is made of nations such as Greece, Iraq and Romania in the production of NBC protection equipment. Some of it has no doubt been around for a while but the spread of such military hardware to nations that at one time would seem to have no need of such material is an indication of the increasing importance that the possible occurrence of NBC warfare now plays in military planning all over the globe.

Following the recent Third International Symposium on Protection against Chemical Warfare Agents held at Umea, Sweden in June 1989, it is possible to detect a few trends relating to hardware design and manufacture. One is that more and more consideration is being provided to making NBC protective clothing more 'user friendly' than has been thought necessary in the past. Items such as cooling garments, ventilation systems, flash proofing and so on are now becoming established military requirements. No doubt the perspiring soldiers who have had to endure the current series of protective garments would say that it is about time too, but it does make military sense to ensure that a serviceman wearing NBC clothing can carry out his normal duties in as unimpaired a manner as possible. In any military system the most important and expensive single item is always the individual soldier so any extra money spent in providing that extra cooling or ease of movement will be money well spent.

The second noticeable trend is towards ease of use of detection instruments. Too many in-service radiac and chemical agent detectors resemble too closely the laboratory instruments they were originally developed from and are thus covered in too many small, twiddly controls, and confusing display scales. More recent instruments have only a few controls and those provided are intended to be used when wearing NBC protection gloves, while displays are simple to the point where they can be understood after a minimum of training or use. Such matters are no doubt obvious to many potential users but are all too often overlooked by designers who never leave their laboratories.

Biological Warfare

The foreword for the first edition of *Jane's NBC Protection Equipment* ended with a mention of biological warfare and this edition will do the same. Over the past year there have been too many indications that the subject is becoming a major consideration when the general use of NBC warfare is discussed. Until recently biological warfare was looked upon as a possible strategic weapon of uncertain effectiveness that could only be considered in the long term. Recent innovations in genetic and related technology have changed all that.

Biological warfare is now considered by many forward planners to be a distinct possibility in the very near future. Several major powers have openly renounced all possible use of such weapons but the danger still lurks that some nations would be only too happy to employ biological weapons to further their own ends in some future conflict. The moralities involved and other such considerations have few meanings to some current forms of political and religious philosophies so mere pronouncements of good intention will be unable to prevent the proliferation of biological warfare in the foreseeable future. The time has come to give the possibility of biological warfare a much greater degree of consideration than it has received in the past.

The first edition of *Jane's NBC Protection Equipment* received a favourable response from its intended users but although it has been produced using the long arm of the Jane's Information Group, the Editor feels that there are still gaps that should be filled. Future editions will be amended and enlarged to meet the changing requirements of the market so any comments, suggestions and extra information would be gladly received; the address is provided below. This edition is appreciably larger and more comprehensive than the first and the intention is to make this Yearbok as encyclopaedic and usable as possible, so any assistance would be most welcome.

Considerable assistance in the compilation of this edition has been forthcoming from many sources, not least from the ever-helpful corps of Jane's Yearbook editors; my old friend Christopher F Foss has been especially forthcoming in guidance and advice. Many others within the Jane's Information Group have contributed and have been more than helpful, especially Tony Currow and his team. They have to be placed alongside the many members of the NBC defence industry and associated establishments around the world who have given so freely of their time and help. There are too many of them for the Editor to give more than his collective thanks.

A Yearbook editor would be hard pressed to complete his work without the able and willing assistance of the Jane's Yearbook publishing staff. My thanks are due to them all and especially to Kevin Borras who has had the main task of converting paper to published pages. My thanks are once again due to the unrelenting and cheerful efforts of Butler and Tanner who have produced a Yearbook worthy of standing alongside the many other Jane's publications produced over the decades.

July 1989 Terry J Gander

Note

As mentioned above, any additions, comments and alterations regarding the information contained in this Yearbook should be sent to the Editor, Jane's NBC Protection Equipment, Jane's Defence Data, Sentinel House, 163 Brighton Road, Coulsdon, Surrey CR3 2NX, United Kingdom.

PLESSEY

Radiation Alarm Systems For Ships and Submarines

The installed Radiac 22NRS systems, for use in naval surface vessels, monitor gamma radiation in the atmosphere and the sea resulting from fallout from a nuclear weapon burst. The equipment is shown in the illustration.

The strategic placement of the seven sensor units provides a continuous picture of the radiation profiles throughout the vessel enabling the command to initiate appropriate nuclear, biological and chemical defence (NBCD) counter-measures.

Indicators are normally mounted in the damage control headquarters (DCHQ) and in the operations room. The DCHQ indicator continuously displays dose rates within the ship's environment, and total accumulated doses. A visual and audible alarm is activated when a preset level is exceeded. The operations room indicator repeats the visual and audible alarms, and the continuous digital display of the upper deck high-level sensor output.

Simulator programs are available for various fallout profiles and pre-wetting effects can be simulated.

The Radiac 23NRS is designed for use in submarines and smaller surface vessels. The system is based on two sensors, plus an indicating unit which continuously displays dose rates measured by the two sensors and the total dose as indicated by the high level sensor.

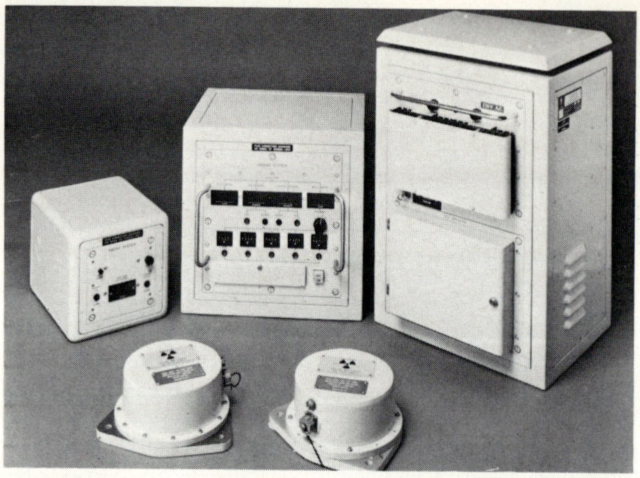

The indicator also has a visual alarm and facilities for an audible alarm activated when a preset level is exceeded.

In submarines, by placing the low level sensor adjacent to the snort tube, an early indication of contamination is given, so that snorting can be controlled. This sensor also monitors external activity through the hull. A simple simulator permits the manual input of fallout profiles.

The equipment can be used for installation in selected shore establishments.

Both systems have been developed for the British Ministry of Defence (Navy) by Plessey Controls Limited.

Personal Dosimetry System for Radiation Dose Assessment

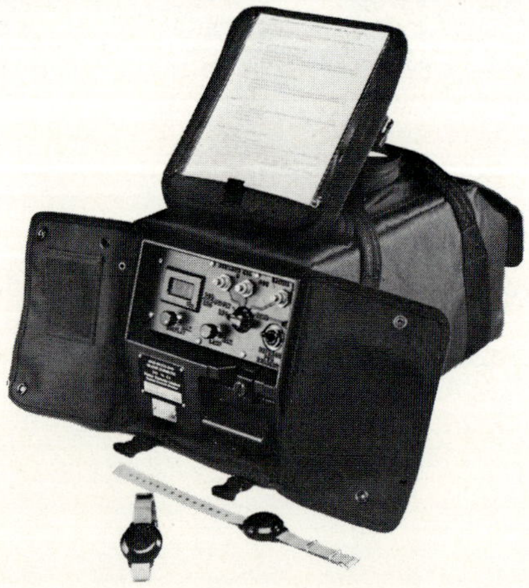

Exposure to gamma and neutron radiation can quickly reduce combat effectiveness of troops. The personal dosimetry system allows the radiation exposure state (RES) of a fighting unit to be calculated. With this information the military commander can rapidly assess and quantify the effects of radiation on his troops, and plan accordingly.

The system comprises a dosimeter locket, worn by each combatant, and a reader unit. Both the instantaneous and residual gamma and neutron radiation dose received are recorded. To establish dose, each dosimeter is placed in the reader and interrogated. The system is fast and simple to operate, requiring no technical or medical skills; maintenance has been reduced to a minimum.

The British and US Armies are both major customers for this equipment.

Portable Dose Rate Meter

The Plessey range of dose rate meters are rugged, waterproof, lightweight instruments designed to satisfy United Kingdom Government requirements. The portable device is intended for post incident nuclear monitoring in civil defence and military applications. The fixed post version allows remote monitoring from a protected bunker or shelter. A decontamination monitor incorporating an audible indication enables checks to be made on personnel, equipment and vehicles following the cleansing process. The battery powered instruments feature large LCD displays with rising and falling dose rate indication, self checking and fault indication.

The low cost Plessey instruments present the opportunity to deploy many units throughout the field, providing the command with a strategic advantage in operating in potentially contaminated areas.

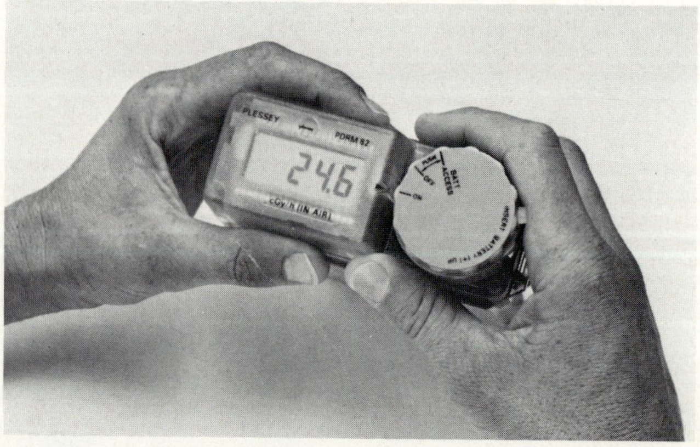

Plessey Controls Limited
Sopers Lane, Poole, Dorset BH17 7ER
Telephone: (0202) 782000 Telex: 41272 Fax: (0202) 782331
International Telephone: 44 202 782000

GLOSSARY

ABC	Atomic, biological and chemical
AC	Hydrogen cyanide
ACADA	Automatic Chemical Agent Detector Alarm
ACAS	Airfield Chemical Alarm System
AFU	Air filtration unit
ALAD	Automatic Liquid Agent Detector
AMCOMM	Armament, Munitions and Chemical Command
ANP	Appareil normal de protection
APC	Armoured personnel carrier
ARRV	Armoured repair and recovery vehicle
ARS	Aerial Radiac System
AVM	Airborne Vapour Monitor
BBC	Bromobenzlycyanide
BITE	Built-in test equipment
BZ	Benactyzine
BZS	Bundesamtes fur Zivilschutz
CADS	Chemical Agent Decontamination Simulant
CAM	Chemical Agent Monitor
CAS	Casualty Assessment System
CATM	Chemical Agent Training Mixture
CAU	Central alarm unit
CAVA	Chemical Agent Vapour Alarm
CB	Chemical-biological
CCA	Contamination control area
CDE	Chemical Defence Establishment
CG	Phosgene
CK	Cyanogen chloride
CN	Chloroacetophone
CNB	Code name for tear agent
CNC	Code name for tear agent
CNS	Code name for tear agent
CP	Command post
cpm	counts per minute
CPR	Cardio-pulmonary resuscitation
CRAM	Chemical Agent Remote Monitor System
CRDEC	Chemical Research, Development and Engineering Center
CS	O-chlorobenzamalonitrile
CTC	Combat Training Center
CVR(T)	Combat Vehicle Reconnaissance (Tracked)
CX	Phosgene oxime
DA	Diphenylchloroarsine
DADS	Direct Application Decontamination System
DAP	Decontamination Apparatus, Portable
DC	Diphenylcyanoarsine
DEDAS	Decontamination Emulsion Direct Application System
DM	Adamsite
DP	Diphosgene
DU	Detector unit
ECAS	Electronic Chemical Agent Alarm System
EC/NBC	Environmental Control/Nuclear, Biological and Chemical
ED	Ethyldichloroarsine
EMP	Electro-magnetic pulse
ERADCOM	Electronics Research and Development Command
FAT	First article testing
FMGC	Filtre mixte de grande capacité
G	Nerve agent
GA	Tabun
GB	Sarin
GD	Soman
GRD	Gruppe fur Rustungdienste
H	Mustard; Levenstein mustard
HCN	Hydrogen cyanide
HCPE	Hybrid Collective Protection Equipment
HD	Distilled mustard
HL	Mustard-Lewisite mixture
HMMWV	High mobility medium wheeled vehicle
HN	Nitrogen mustard
HVM	High velocity missile
IA	Irrespirable atmosphere
IFV	Infantry fighting vehicle
IOC	Initial operational capability
L	Lewisite
LCD	Liquid crystal display
LED	Light-emitting diode
MBT	Main battle tank
MCCU	Micro-Climate Cooling Unit
MD	Methyldichloroarsine
MILES	Multiple Integrated Laser Engagement System
MPDS	Multi-Purpose Decontamination System
MPS	Motor pump set
MR	Simulated mustard
NAIAD	Nerve Agent Immobilised enzyme Alarm and Detection
NATO	North Atlantic Treaty Organisation
NBC	Nuclear, biological and chemical
NBCF	Nuclear, biological, chemical and fire
NBCRS	Nuclear, biological and chemical reconnaissance system
NCO	Non-commissioned officer
OFM	Otto Fuel Monitor
OP	Observation post
PCAS	Persistent Chemical Agent Simulant
PD	Phenyldichloroarsine
PDRM	Portable Dose Rate Meter
PEG	Polyethelene glycol
PS	Chloropicrin
RAP	Regimental aid post
RAS	Remote Active Spectrometer
RDA	Remote display/alarm
RF	Radio frequency
RFP	Request for proposals
RPV	Remotely-piloted vehicle
SA	Arsine
SAWE	Simulated Area Weapons Effects
SCAS	Site Chemical Agent Detection System
SCPS	Survivable Collective Protection System
SICAS	Ship Installed Chemical Alarm System
SICS	Ship Installed Chemical System
SPAL	Simulator Projectile Airburst Liquid
T	Code name for sulphur and chlorine compound
TDCC	Trousse de Detection Chimique de Controle
TECOM	Test and Evaluation Command
TEG	Thermo-electric generator
TFA	Toxic free area
USAMRDC	United States Army Medical Research and Development Command
VX	Code name for nerve agent

JANE'S
NBC PROTECTION EQUIPMENT 1990-91

☐ **YES!** Please send me next year's edition of JANE'S NBC PROTECTION EQUIPMENT 1990-91.

NAME

TITLE/RANK

ORGANISATION

ADDRESS

CITY

STATE/ZIP/COUNTRY

If you sign and return this order card, we will register your subscription to next year's edition, and invoice you prior to publication so that the book can be sent to you without delay.

However, we will hold the price of next year's edition at the current level if you send a remittance with the order card, and we receive it by 31 January 1990.

PLEASE COMPLETE BOTH SIDES OF ORDER FORM

JANE'S
NBC PROTECTION EQUIPMENT 1989-90

☐ **YES!** Please send me _____ additional copy(ies)of this edition of JANE'S NBC PROTECTION EQUIPMENT 1989-90.

NAME

TITLE/RANK

ORGANISATION

ADDRESS

CITY

STATE/ZIP/COUNTRY

US PRICE
$135.00

UK &
REST OF WORLD
PRICE
£80

PLEASE COMPLETE BOTH SIDES OF ORDER FORM

OTHER PRODUCTS FROM JANE'S INFORMATION GROUP

*	Title	Price US$	Price £	Qty
	JANE'S DEFENCE AND AEROSPACE YEARBOOKS 1989			
	Jane's Military Communications 1989	$140	£80	
	Jane's Military Logistics 1989	$190	£100	
	Jane's Fighting Ships 1989-90	$170	£100	
	Jane's Land-Based Air Defence 1989	$120	£70	
	Jane's Radar and EW Systems 1989-90	$140	£80	
	Jane's Infantry Weapons 1989-90	$170	£95	
	Jane's Avionics 1989-90	$170	£95	
	Jane's Underwater Warfare Systems 1989-90	$140	£80	
	Jane's Battlefield Surveillance 1989-90	$120	£70	
	Jane's C³I Systems 1989-90	$140	£80	
	Jane's Armour and Artillery 1989-90	$145	£85	
	Jane's AFV Systems 1989-90	$145	£85	
	Jane's All The World's Aircraft 1989-90	$170	£100	
	Jane's Security & Co-In Equipment 1989-90	$110	£65	
	Jane's Military Training Systems 1989-90	$135	£80	
	Jane's NATO Handbook 1989-90	$110	£65	

*	Title	Price US$	Price £	Qty
	JANE'S DEFENCE AND AEROSPACE BINDER PRODUCTS 1989			
	Jane's Strategic Weapon Systems	$300	£175	
	Jane's Defence Appointments (Pacific Rim)	$1050	£600	
	Jane's Defence Appointments (NATO Europe)	$1050	£600	
	Jane's Air-Launched Weapons	$300	£175	
	Soviet High Command	$935	£550	
	Jane's Naval Weapon Systems	$300	£175	
	Warsaw Pact High Command	$825	£500	
	INTERAVIA DIRECTORIES 1989			
	Interavia ABC Aerospace Directory 1989	$295	£175	
	Interavia Space Directory 1989-90	$140	£80	
	International Defense Directory 1990	$365	£230	

☐ Please send me a complete list of all JANE'S INFORMATION GROUP products.

The publisher reserves the right to alter prices without prior notification.

*Please indicate products required ☑

PLEASE COMPLETE BOTH SIDES OF ORDER FORM

☐ I enclose a cheque payable to JANE'S INFORMATION GROUP

☐ Please invoice my Company/Organisation
Purchase Order No. ______________________

____
SIGNATURE

Send US orders to:

JANE'S INFORMATION GROUP
Department Yearbooks
1340 Braddock Place, Suite 300
P.O. Box 1436
Alexandria, VA 22313-2036
(703) 683-3700
800 243-3852

Send UK & Rest of the World orders to:

JANE'S INFORMATION GROUP
Department DSM
Sentinel House
163 Brighton Road
Coulsdon
Surrey CR3 2NX
United Kingdom

International (+441) 763 1030
de la France 19..05.90.83.98 (appel gratuit)
aus der B.D.R. 0130-81 01 53 (Nahtarif)
in UK 0800 282 455 (toll free)
or UK 01-763 1030 (not toll free)

☐ I enclose a cheque payable to JANE'S INFORMATION GROUP

☐ Please invoice my Company/Organisation
Purchase Order No. ______________________

____
SIGNATURE

Send US orders to:

JANE'S INFORMATION GROUP
Department Yearbooks
1340 Braddock Place, Suite 300
P.O. Box 1436
Alexandria, VA 22313-2036
(703) 683-3700
800 243-3852

Send UK & Rest of the World orders to:

JANE'S INFORMATION GROUP
Department DSM
Sentinel House
163 Brighton Road
Coulsdon
Surrey CR3 2NX
United Kingdom

International (+441) 763 1030
de la France 19..05.90.83.98 (appel gratuit)
aus der B.D.R. 0130-81 01 53 (Nahtarif)
in UK 0800 282 455 (toll free)
or UK 01-763 1030 (not toll free)

DELIVERY DETAILS

____
NAME

____
TITLE/RANK

____
ORGANISATION

____
ADDRESS

____
CITY

____
STATE/ZIP/COUNTRY

Send US orders to:

JANE'S INFORMATION GROUP
Department Yearbooks
1340 Braddock Place, Suite 300
P.O. Box 1436
Alexandria, VA 22313-2036
(703) 683-3700
800 243-3852

PAYMENT METHOD

☐ I enclose a cheque payable to JANE'S INFORMATION GROUP

☐ Please invoice my Company/Organisation
Purchase Order No. ______________________
(All prices include shipping by surface mail. Air mail rates on application. Individual updates will be sent airmail.)

____
SIGNATURE

Send UK & Rest of the World orders to:

JANE'S INFORMATION GROUP
Department DSM
Sentinel House
163 Brighton Road
Coulsdon
Surrey CR3 2NX
United Kingdom

International (+441) 763 1030
de la France 19..05.90.83.98 (appel gratuit)
aus der B.D.R. 0130-81 01 53 (Nahtarif)
in UK 0800 282 455 (toll free)
or UK 01-763 1030 (not toll free)

Nuclear Weapons and their Effects

Nuclear weapons fall into two main categories, fission and fusion. To the layman and the media in general these categories correspond to atomic and hydrogen weapons. In practical terms the main difference between the two is one of degree with the fusion weapons being by far the most powerful. The term powerful is a relative one for fission weapons are still more powerful by far than even the most powerful conventional chemical energy-based weapons.

Nuclear fission and fusion weapons make use of the latent power contained within the smallest items of matter, namely atomic particles. The potential energy content of these miniscule particles was only realised following extensive scientific and theoretical research into the fundamental properties of the building blocks of matter that began in earnest around the beginning of the 20th Century. Following a halcyon period of pure research that culminated during the 1930s, the military possibilities of the massive amounts of power that could be unleashed from the atom began to dominate investigations. The point was eventually reached where pure research turned into practical development with the establishment of the Manhattan Project in the United States from 1942 onwards.

The first end product of the Manhattan Project was the detonation of the world's first nuclear device, a fission device set off in the Alamogordo Desert at 0530 on 16th July 1945. The first military use of the new weapon was made when a uranium-based fission bomb was dropped over Hiroshima on 7th August 1945, soon to be followed by a second nuclear strike on Nagasaki on 9th August 1945.

Those two weapons, the only nuclear devices that have ever been used to actually attack a target in war, initiated an arms race that continues to this day and led to the state of mutually-assured destruction that overshadows all life on this planet. Thousands of nuclear devices now stand ready to be used at a few seconds notice and are installed as warheads on inter-continental and tactical rockets delivered from the land, sea and air, on artillery shells, on torpedoes and depth charges, on aircraft bombs, buried underground as mines and placed into back-packs for use as super-powerful demolition charges. Plans are being made to base nuclear weapons in space.

Nuclear fission devices depend for their power upon the splitting of atoms of heavy elements such as uranium or plutonium. By contrast fusion devices depend on the fusing together of separate atoms of a light element, usually hydrogen. While the latter process releases enormous amounts of energy the actual fusion demands extremely high temperatures to make the process possible. The usual method is to produce such heat by detonating a fission device; in effect the fission device acts merely as a fuze. Thus it can be seen that in relative terms the power of fission and fusion weapons is widely disparate. While a fission weapon can have devastating effects on any target area the power unleashed by a fusion weapon is much, much greater. Hence the somewhat surprising division of nuclear weapons. Fission weapons are now widely regarded in military terms as tactical weapons while fusion weapons are strategic.

The ordinary mind finds it difficult to encompass the notions relating to the amounts of energy that nuclear devices can produce. Ordinary physical measuring yardsticks convey only a vague idea of the potential destructive power of such weapons but in order to provide some form of measurement, no matter how vague and indefinite in concept, the power (or yield) of a nuclear device is stated in the equivalent power of so many thousand tons, or kilotons (usually written as kt), of a conventional high explosive, namely TNT. Thus a fission device may have a power in the region of from 0.5 to 100 kt; the Hiroshima bomb was a 19 to 20 kt weapon. By contrast a fusion device can reach levels of millions of tons of TNT, so such weapons have yields of megatons, or mt. A typical fusion device could have a power of 10 mt, while 50 mt monsters have been produced. All nuclear devices, whatever their size or form, produce three major effects when they are detonated. They are heat, blast and radiation, with a fourth effect being a by-product in the form of a powerful electro-magnetic pulse, or EMP. All these effects can be traced to the massive amounts of energy released as a nuclear device is detonated.

The detonation of any form of nuclear device releases massive amounts of energy within an extremely short period of time (about one millionth of a second) and within the confines of a limited quantity of matter. The immediate result is an extremely rapid rise in temperature that can reach a peak of several tens of millions of degrees Centigrade. The immediate result of this immense rise in temperature is the vapourisation of the device itself and anything else in the immediate vicinity. However these dramatic effects are secondary to the effects of the rise in temperature produced by the energy release. A great deal of the heat is radiated as intense light but more energy is released in the form of pure energy, or radiation, of which more anon. The amount of heat produced will vary according to the power of the device detonated.

The immediate physical effect of the heat release is a fireball around the point of detonation. Due to the temperatures involved the fireball is not static and expands outwards and starts to rise immediately it is formed. The actual surface temperature of the fireball does not vary greatly according to the size of the device involved; instead the size and duration of the fireball vary.

To the observer the degree of light produced by a nuclear device detonation is far brighter than anything previously encountered. Even in daylight the nuclear flash will appear to be infinitely brighter than the sun at noon. The light may be visible many miles away, even in broad daylight, and can be so intense it causes temporary or permanent blinding. At night these effects are considerably magnified.

The light is soon followed by a wave of heat. At close ranges the heat can produce immediate conflagrations of imflammable materials while at greater distances charring and scorching will result. The effects on the human body will also vary with range from the point of detonation, ranging from immediate death by burning to reddening of the skin.

The heat and light are followed by blast, which is really another effect of the heat production. As a device detonates the rapid release of energy and rapid rise in temperature causes an enormous expansion of the air close to the production of the fireball. The expanding air around the fireball moves outwards at high speed and thus creates a blast wave (often referred to as a shock wave) that moves rapidly away from the point of detonation. Any point passed by the blast wave will experience a very rapid and intense rise in pressure, pressure so intense that even robust structures can be knocked down or damaged. As the blast wave passes a point will then experience a period of relatively low pressure in the wake of the blast wave. The law of physics will then take effect after a short time period and air will rush inwards to fill the partial vacuum. This forms a secondary blast wave to add to, or complete, the destructive effects of the main blast wave. Despite the long term effects of nuclear radiation, heat (and light) and blast are the main destructive effects produced by nuclear devices. Throughout this summary no attempt has been made to quantify the actual results of these destructive effects for they can vary enormously and are dependent on many factors. One factor is the size of the device involved. As has already been mentioned, fusion devices are much more powerful than fission devices and so can be much more destructive. Then there is the matter of the height at which the device is detonated. If a device is detonated under, on, or close to ground (or sea) level a great deal of the heat and blast will be absorbed by the surrounding terrain (or sea) and the long range effects of the heat and blast will be reduced. Such a burst is known as a ground burst. By contrast a device detonated in the air above the surface can have its destructive effects considerably enhanced and spread over a wide area.

For most purposes a nuclear device detonated in the air at such a height that no part of the resultant fireball touches the Earth's surface is known as an air burst. The height above which no part of the fireball touches the ground is known as the critical height and varies with the power of the device involved. The point below the centre of a burst is known as the ground zero.

The difference between an air and ground burst can make a considerable difference to the third effect of a nuclear detonation, namely radiation. Radiation produced by a nuclear burst can be produced in two forms, immediate and residual.

Immediate radiation is that formed at the actual instant of detonation. It is formed by the vapourisation of the device involved and anything else in the immediate vicinity. Not all the energy released by a detonation is released as heat and light. Some pure energy is released, mainly in the form of gamma and beta rays and neutrons. This immediate radiation can be extremely harmful to life but the ranges and duration of the radiation involved is so short that any recipients will be casualties many times over as a result of the heat and blast effects produced at the same time.

However the immediate radiation has the effect of irradiating all material in its immediate vicinity. As the nuclear fireball rises it takes all manner of debris, dust and material with it. If the device involved has been a ground burst the amounts of material could be considerable. The radiation produces numerous radioactive isotopes and compounds that in their turn produce radiation. They produce radiation for various periods before they decay or turn into other isotopes and compounds, some with lives or durations of a few milliseconds and others that can remain radioactive for years.

As a nuclear fireball ascends into the upper reaches of the atmosphere it cools and gradually loses its thermal energy. Gradually the debris and particles accumulated by the fireball will fall back to earth in the form of 'fallout'. Prevailing winds will scatter the fallout far and wide, depending on the wind strengths and directions at various altitudes and thus spread the residual radiation from the nuclear device. The radioactivity strengths, its extent and duration will depend upon many factors, from the size of the device to the altitude at which the bomb was detonated. Residual radiation from a high altitude air burst could be negligible in human risk terms while a ground burst could result in massive amounts of potentially harmful fallout.

The effects of nuclear radiation on life can vary from the minute to the drastic. All radiation, in whatever form it is produced, can have some effect or another on life but in excess it produces all manner of physiological disturbances from preventing myosis (cell division) to the creation of dangerous cancers. If large amounts of nuclear radiation are received within a short time, death can result after only a short period. Equally large radiation doses absorbed over a long time period can have few discernable effects although adverse effects may appear in the long term. There are no medical resources to deal with or arrest the effects of radiation on the human body, although various palliatives can be dispensed.

One side-effect of a nuclear detonation that was mainly overlooked until comparatively recently is the electro-magnetic pulse, or EMP. It is another by-product of the huge energy release produced at the instant of a nuclear detonation when some of the energy produced has the effect of altering the electrical properties of electrons in the nearby atmosphere. The result is intense electrical and magnetic fields of variable duration that can extend for considerable distances from the point of detonation. These fields can establish eddy currents of considerable strength that by themselves can have significant effects on electronic-based equipments and processes, usually by producing larger eddy currents than the equipments are rated for and thus causing breakdowns. They can also create interference across many radio frequencies. The effects of the EMP can travel further than those for the other more destructive hazards, especially where air bursts are involved, and may last for a significant period after the other immediate nuclear hazards have subsided.

Despite the well-broadcast message that there is no defence against nuclear weapons, some defensive measures can be taken. In short they are summed up in one word – shelter. Even a modest covering of earth or hard cover provided by structures can provide some measure of defence against the heat and blast effects of nuclear devices. The same protective measures can be used against radiation, including residual radiation. Examples of a few of the more formalised of these protective shelters can be found in these pages.

Defence against EMP is also possible but it has to take the form of careful and thorough screening of all electrical and electronic equipment likely to be affected. Such screening is expensive in cost and materials but without such protection few items of electronic equipment will be able to function correctly in the aftermath of a nuclear burst.

The above summary is a very short and simplified outline of the effects of nuclear weapons. There are many variables involved and some of the more involved explanations of some of the processes that take place when a nuclear device is detonated have not even been touched upon. The explanation for this over-simplification is that to do the subject justice would fill this Yearbook several times over and would be of little utility or interest to most readers of this work. If students of the subject wish to investigate the topic further there are many erudite and learned works already published that delve as deeply as the most interested student would wish.

For the average reader one work that has proved extremely useful is *The Effects of Nuclear Weapons*, published by the United States Department of Defense, 1977. It is only one of many similar works that can be recommended.

Biological Warfare

Compared to nuclear and chemical warfare, biological warfare has a pedigree almost as long as that of warfare itself. Ever since mankind started to fight with his neighbour attempts have been made to inflict hurt on an enemy by somehow introducing illness to the squabble. Biological warfare involves the use of living organisms (such as bacteria) to attack the normal functioning of the human body to render an opponent ineffective. Primitive men (and their later generations), even though they had no idea of what bacteria or similar organisms could be, often sought to harm their enemies by introducing illness to an enemy's community. Having observed the way in which disease and epidemics could lay low whole communities the simplest and most straightforward method of biological attack usually employed was leaving the corpses of infected animals or men where they would be encountered by an enemy, or forcibly hurling infected cadavers or organs into an enemy compound or fortress, or introducing similar noxious substances to water supplies. An even simpler ploy was to release infected prisoners back into their own community or territory.

The drawbacks of waging what we now know as biological warfare were as apparent to primitive man as they are to his present-day counterpart. Leaving aside for the moment the fact that it can be a double-edged weapon, biological warfare is rarely an instant casualty producer. Attacking the way the body operates using living organisms takes time. The process cannot be hurried and results are often uncertain. Another drawback is that it is rare for any biological warfare attack or even a virulent natural epidemic to completely wipe out or incapacitate an entire population or garrison; some effectives always seem to escape any attack or epidemic and remain operational in military and practical terms.

For these reasons biological warfare has never been used in isolation. It cannot be deemed to be a tactical weapon but is instead regarded rather more as a strategic weapon which could produce long term results, even when modern technology is involved. The fact is that modern technology has been brought to bear on the prospect of biological warfare to a remarkable degree. Despite earlier tentative experimentation, it is only recently that the technology has existed to manufacture what can be regarded as bacteriological warfare weapons in anything like militarily useful quantities.

The point has now been reached where virtually any disease known to man can be manufactured, not only in laboratories but often on an industrial scale. Most of the diseases involved have been known to man since the beginning of history but modern technology has produced the ability to manufacture new diseases or produce new and more virulent strains of existing diseases. Genetic engineering has opened up new prospects for the proponents of biological warfare to an alarming degree.

Technology has also produced the counters to many diseases. Inoculation is one obvious counter but inoculations are usually only effective against a known disease or strain of disease. Any entirely new disease or variant will require an entirely new counter and the rate of introduction of new potential biological warfare weapons makes any 'blanket' inoculation of a population or armed force virtually impossible.

Manufacturing a disease for military purposes is one thing. Spreading it to a significant number of personnel in a target population is another. When the organisms involved are spores or something similar the task can usually be left to wind currents. However, many diseases require some living organism to carry them. These carriers are known as vectors and many long-established diseases rely on vectors such as mosquitos, rats, lice and other similar unpleasant living things. For military purposes such vectors have to meet a testing specification. They have to be bred in sufficient numbers and have to have a survival life long enough to actually spread the required disease. They also have to be hardy enough to endure long periods of storage and be dispensed from either high altitude or in extremes of climate. Insects such as mosquitos can meet such requirements and have been bred in huge numbers for various military biological warfare campaigns. Fortunately none of these campaigns has been initiated on anything like a large scale – yet.

Disease can also be spread using viruses of varying types. Anyone alive has been affected by some form of virus or another at some stage of their life, usually in the form of some strain of influenza. Every time a new strain of influenza appears it seems to spread like wildfire through populations spread far apart around the world. Thankfully some viruses remain more geographically localised in their effects. The introduction of a 'designer virus' based on, say, Yellow Fever could mean the arrival of a highly efficient biological warfare weapon. Viruses therefore have their own section under this heading.

Toxins are also included in this section. Toxins are inorganic substances produced by living organisms that can adversely affect the functioning of the human body. The number of substances in this category are legion and many are well known. Once again genetic engineering and other forms of modern technology can produce new toxic substances hitherto unknown and for which no counter may be readily available. However, how such toxins could be spread is a different matter. Water supply is always a viable method but most modern societies take stringent precautions to keep their water supplies safe from such hazards.

Biological warfare agents may be dispensed in many ways. The age-old ploy of simply introducing infected persons or animals into a population is still a distinct possibility but other methods could include the dispensing of viruses in aerosol form from aircraft. The spreading of spores (such as Anthrax) from aircraft bombs is another and vectors such as insects could be spread far and wide by releasing them into the efflux from a jet engine, either on the ground or from an aircraft.

It should not be forgotten that biological warfare can be conducted on animals and plants as well as humans. Diseases such as Swine Fever can decimate animal herds, and populations dependent on a staple crop such as rice or maize could be rendered helpless by the introduction of previously unknown specific plant diseases. Insects such as the Colorado beetle can destroy potato crops almost overnight and locusts can still reduce whole areas to virtual deserts within hours.

As has been outlined above, the following list is only an incomplete outline of diseases known to have been considered for use in war. New diseases,or virulent variants of otherwise harmless ailments, could be introduced at almost any time. It should not be forgotten, however, that biological warfare is an uncertain weapon. It takes time to take effect, is not always 100 per cent effective and has a nasty tendency to be harmful to perpetrators as well as victims.

RICKETTSIAE

Rickettsia Prowazeki

The micro-organism is a nonmotile, minute, coccoid or rod-shaped rickettsia, occurring sometimes in pairs or chains, with a diameter of about 0.3 micron and with varied sizes and shapes.

The disease is known as a classic epidemic (human or louse-borne) typhus. It is an acute infectious disease of man and is characterised by severe headache, sustained high fever, general pains, and a skin rash.

Infections are acquired from persons with the disease.

Transmission is mainly by body lice which have fed upon infected persons. The individual becomes infected usually by scratching or rubbing louse faeces into a wound made by the louse bite; by crushing an infected louse on the skin; by rubbing louse faeces into the eyes. It is possible that louse faeces from dirty clothing may be transmitted through the air to the respiratory tract.

Incubation is from 6 to 15 days, averaging 12 days.

All peoples are susceptible. One attack confers immunity, which is not always permanent.

The disease is widely distributed in peoples living under crowded and unhygienic conditions (famine, war), particularly in Europe and Asia. Cases occur throughout the year, with an increase in occurence during the cold months.

Mortality is from 10 to 80 per cent, varying in different epidemics and with the age of individuals.

Some vaccines confer considerable protection of uncertain duration; immunisation should be repeated every 4 months where the danger of typhus is present. The vaccine reduces the risk of infection, modifies the course of the disease, and lowers the mortality.

The course of the disease can be shortened by use of appropriate chemotherapy and antibiotics (chloromycetin, aureomycin, terramycin). Supportive treatment and prevention of secondary infections are essential.

The disease is not contagious or communicable from man to man. Epidemics usually occur in winter under crowded and unsanitary conditions, particularly during famine and war, and when the population is heavily infested with body lice. Louse eradication is used to control epidemics.

The organism is destroyed by heat at 62°C, in 15 to 30 minutes and inactivated by use of 0.1 per cent formalin and 0.5 per cent phenol.

Rickettsia Mooseri (Rickettsia Typhi)

This rickettsia is similar to Rickettsia Prowazeki, but with less variation in appearance.

Murine (endemic, rat, or flea-borne) typhus is similar to classic epidemic typhus except that the disease is milder and has a slower onset.

Infected rodents, particularly rats, are sources of infection.

The disease is transmitted from rodents to man by the bite of the rat flea.

The incubation period ranges from 6 to 14 days, mostly 12 days.

Susceptibility is general. One attack confers immunity, which sometimes is not permanent.

The disease is widely distributed in temperate, sub-tropical, and tropical countries. It is transmissible to man throughout the year, with an increase during summer and autumn.

Case fatality rate is about 2 per cent, increasing in older people (over 50 years of age).

Vaccine similar to that for classic epidemic typhus is probably equally effective.

Treatment is similar to that for classic epidemic typhus. Delayed recognition of the disease may result in the optimum period of treatment being passed.

The disease is not contagious. Rodent and flea eradication are the most successful control measures.

Stability is the same as for classic epidemic typhus.

Rickettsia Tsutsugamushi

This rickettsia appears as small diplococcus-like structures or short rods which are 0.3 to 0.5 micron in length and 0.2 to 0.4 micron in width.

Scrub typhus (tsutsugamushi disease or mite-borne typhus), an acute infection in man, is characterised by sudden onset with chills and fever and headache which may increase in intensity. A dull red eruption appears on the trunk of the body on the fifth to eighth day, and may extend to the arms and legs. Coughing and other signs of pneumonia are frequently present.

The rickettsia is perpetuated in infected mites, where it is passed on from generation to generation and is maintained by the mites feeding on susceptible wild rodents, particularly mice and rats of different species.

Transmission to man is made by the bite of infected larval mites.

The incubation period is from 7 to 10 days, sometimes it may be as long as 14 days.

Susceptibility is general. One attack confers immunity, which is not always permanent.

The disease occurs in south-eastern Asia, including India, Burma, Malaysia, Vietnam, islands of the West and South Pacific, Japan, Taiwan, Indonesia, New Guinea, and Australia. Transmission is limited to the summer months in Japan but may occur throughout the year near the Equator.

Mortality varies with age and locality and may go as high as 50 per cent.

There is no effective vaccine available for general use. The immunity produced in man (by scrub typhus) is not as permanent as that produced by other rickettsiae.

Prompt use of an appropriate antibiotic (chloromycetin) is highly effective.

The disease is not communicable from man to man. Control is by the elimination of rodents and eradication of mites.

Stability is the same as for typhus rickettsiae.

Rickettsia Rickettsii

The rickettsia is a diplococcus-like micro-organism, with the distal ends tapered so that the pairs resemble minute pneumococci. The average is 1 micron in length and 0.2 to 0.3 micron in width.

Rocky Mountain Spotted Fever (Sao Paulo Fever in South America), an acute infectious disease, is characterised by fever and joint and muscular pains. A skin rash usually appears on the third or fourth day, rapidly spreading from the ankles and wrists to the legs, arms, and chest.

Infected ticks such as the common dog tick in the eastern and southern United States, the wood tick in the north-western United States, and occasionally, the lone star tick in the south-west are sources of infection. In other countries the vectors may be other species of ticks.The infection is passed from generation to generation in ticks and is probably maintained by larvae feeding on susceptible wild rodents.

The disease is transmitted to man by the bite of an infected tick, by contamination of abraded skin with infected tick tissues or faeces, or by contact of tick tissues with unbroken skin.

The incubation period is from 3 to 10 days.

Susceptibility is probably general; recovery from an attack is followed by immunity which may or may not be permanent.

The disease occurs throughout North America and in some parts of South America. Predominant occurrence is in the spring and early summer, when adult ticks appear.

The fatality rate may vary from 20 to 60 per cent, depending on the locality.

An effective egg-yolk vaccine is available, which lessens the chance of infection and lowers the mortality rate. Yearly booster doses are needed to attain maximum protection.

Appropriate antibiotics (chloromycetin, aureomycin, and terramycin) are effective in reducing fatality and shortening the course of the disease. Supportive treatment also is indicated.

The disease is not communicable from man to man. Tick eradication and preventive immunisation are the best control measures.

The rickettsia can be killed by exposure to a temperature of 62°C for 10 minutes and by drying 10 hours. It is inactivated by 0.1 per cent formalin and 0.5 per cent phenol.

Coxiella Burneti (Rickettsia Burneti)

This rickettsia is a bacterium-like, gram-negative organism which may vary in size and shape from a lanceolate rod 0.25 by 0.5 micron to a diplobacillus 0.25 by 1.5 microns. Because the small forms will pass through bacterial filters, they are considered by some to be viruses.

Q Fever, also known as Nine Mile Fever and (North) Queensland fever, is an influenza-like disease that is moderately incapacitating, and is characterised by an acute fever of sudden onset, headache, chills, weakness, and severe perspiration. Pulmonary involvement occurs in the majority of cases, accompanied by mild cough, scanty expectoration, and chest pains. Q Fever is distinguished from other rickettsial diseases in its failure to cause a skin rash.

Cows, sheep, goats, ticks, and wild animals appear to be natural reservoirs, with numerous ticks transmitting the disease among them. The organism may be found in the milk and mammary glands of infected cows, in the milk of sheep and goats, and in the dust-laden air of dairy cattle barns and goat pens which harbour infected animals.

In man the disease appears to be transmitted by the inhalation of infected dust from one source or another as well as by ingestion. Raw milk from cows and goats, dried milk, raw wool, hides, infected meat, goat hair, and tick faeces, as well as cultures of infected tissues, have been involved in infections.

The incubation period is from 14 to 26 days with a mean of 19 days.

Susceptibility is general. An attack confers immunity of long duration.

The disease has been reported in many parts of the United States, particularly in California; it has also been reported in Australia, the Mediterranean area, Western Europe, the Near East, the Balkans, Spain, England, Panama, and scattered areas of Africa.

Fatalities are rare, but may be up to 4 per cent during epidemics.

Vaccines have been effective when used by laboratory personnel and slaughterhouse and stockyard workers.

Appropriate antibiotics (aureomycin, chloromycetin, and terramycin) may be effective, although they have not been completely evaluated. Supportive treatment is indicated.

The disease is relatively non-contagious. Outbreaks have occurred among slaughterhouse and stockyard workers in the United States.

The micro-organism is resistant to temperature changes from +40 to −52°C, is resistant to 0.5 per cent phenol, and is relatively resistant to desiccation. It is killed by 0.5 per cent formalin. It probably persists on surfaces from 5 to 60 days.

Bacillus Anthracis

Bacillus Anthracis is a rod-shaped, gram-positive, aerobic sporulating micro-organism, the spores constituting the usual infective form.

Anthrax may appear in three forms in man; cutaneous, pulmonary, and intestinal. The cutaneous or skin form is also referred to as malignant pustule, occurring most frequently upon the hands and forearms of persons working with infected livestock, and is characterised by carbuncles and swelling at the site of infection. Sometimes this local infection will develop into systemic infection. The pulmonary form, known also as Woolsorters Disease, is an infection of the lungs contracted by inhalation of the spores; it occurs mainly among workers handling infected hides, wool, and furs. The intestinal form, which is rare in man, is contracted by the ingestion of insufficiently cooked meat from infected animals.

Cattle, sheep, and horses are the chief animal hosts, but other animals may be infected. The disease may be contracted by the handling of contaminated hair, wool, hides, flesh, blood, and excreta of infected animals, and from manufactured products such as bone meal.

Transmission is made through scratches or abrasions of the skin, wounds, inhalation of spores, eating uncooked infected meat, or by flies.

Incubation is from 1 to 7 days. It is usually less than 4 days and may be less than 24 hours in pulmonary cases.

All human populations are susceptible. Recovery from an attack of the disease may be followed by immunity.

Anthrax is rare in man and is associated only with animal infections or the handling of infected hides or furs.

In man, the mortality of untreated cutaneous anthrax ranges up to 25 per cent; in pulmonary cases, it is almost 100 per cent, while the rare intestinal cases are usually fatal.

Artificial active immunisation measures have been developed and are presently being evaluated for man. Anti-anthrax serum confers some passive immunity.

Cutaneous anthrax can be treated effectively with some antibiotics, including penicillin, aureomycin, terramycin, and chloromycetin; sulfadiazine; and immune serum. Similar treatment for respiratory and intestinal infections may be useful in the very early stages, particularly if combined with the use of immune serum, but is of uncertain value after the disease is well established.

The disease is not epidemic in man. Control of the disease is accomplished by the disposal of carcasses (by burning or deep burial) and by the decontamination of animal products.

The spores are very stable and may remain alive for many years in soil and water. They will resist sunlight for several days. Steam under pressure or exposure to dry heat above 159°C for an hour are necessary to kill spores. Effective decontamination also can be accomplished by boiling contaminated articles in water for 30 minutes or by using some of the common disinfectants. Iodine and chlorine are most effective in destroying spores and vegetative cells.

Shigella Dysenteriae

Shigella dysenteriae is a rod-shaped, gram-negative, non-motile, nonsporulating bacterium.

Bacillary dysentery, an infectious disease of man, is characterised by mild or severe irritation of the lower gastro-intestinal tract; it is usually accompanied by fever, abdominal pain, diarrhoea, weakness or prostration, and ulceration of the mucous membranes of the intestine.

Faeces of infected human patients and carriers are sources of infection.

Transmission is made by the ingestion of contaminated food, water, or milk; by hand-to-mouth transfer of contaminated material soiled with faeces of a patient or carrier; or by flies.

Incubation is from 1 to 7 days, usually less than 4 days.

Most persons are susceptible, but the disease is more common and severe in children than in adults. Recovery from the disease is followed by a relative, transitory immunity. Washing, ordinary sterilisation methods and use of some of the common disinfectants are effective decontamination measures.

The disease is endemic throughout the world, and epidemics or sporadic outbreaks occur where sanitation is lacking or inadequately applied or enforced, particularly in relation to sewage disposal, food handling and preparation, and infant hygiene. Outbreaks are most common during the summer months and occur frequently in large institutions.

Mortality is variable, ranging from 2 to 20 per cent in untreated cases, depending on the particular strain of organism.

Beneficial results may be obtained from treatment with sulfadiazine, terramycin, chloromycetin, and aureomycin.

The disease is highly contagious, particularly under unsanitary conditions. Control measures include rigid sanitation (careful handling of excreta and adequate sewage disposal) and fly control.

The dysentary organisms remain viable for considerable periods in water, ice, and mucous discharges but are readily killed by sunlight. Sterilization by steam and common disinfectants are effective decontaminants.

Brucella Group

In this group are included three closely related organisms, Brucella melitensis, Brucella abortus, and Brucella suis. All are non-motile, non-sporulating, gram-negative, rod-shaped bacilli.

Brucellosis or Undulant Fever in man, a general infection, is characterised by irregular prolonged fever, profuse sweating, chills, pain in joints and muscles, and fatigue. The illness lasts for months, sometimes for years, and may be caused by any one of the three related organisms. Brucella abortus is a parasite of milk cows, producing contagious abortion in cattle; the organism has also been reported in mares, sheep, rabbits, and guinea pigs. Brucella melitensis is primarily a strict parasite of goats and sheep; Brucella suis is a parasite of swine. Br. melitensis and Br. suis are more virulent for man than Br. abortus.

Brucella organisms are found in the tissues, milk, and dairy products of infected goats, cattle and swine.

These diseases are transmitted to man by the ingestion of contaminated milk and other dairy products, pickled meats, and uncooked foods and water contaminated by the excretions of infected animals; and by direct contact with infected animals or animal products. Infection has also occurred by inhalation and by accidental inoculation among laboratory workers.

Incubation is from 6 to 60 days or more, averaging 14 days.

Most individuals have some degree of resistance or acquired partial immunity to the abortus strains of the organism, probably from the ingestion of small doses. The susceptibility of man to Br. melitensis infection ranges from 50 to 80 per cent although it may range from 75 to 80 per cent; susceptibility to Br. suis appears to be approximately equal to that of Br. melitensis; susceptibility to Br. abortus appears to be not more than 50 per cent.

Brucellosis is prevalent in most areas where cattle, goats, and swine are raised. The infection of humans occurs more often in males than in females, particularly among persons working with cows, hogs, goats, and dairy products, or among those using unpasteurised milk from cows or goats.

Mortality of untreated infections is said to average 2 to 3 per cent with Br. abortus and 3 to 6 per cent with Br. suis and Br. melitensis.

Immunisation methods are unsatisfactory for man. The immunisation of calves is effective as a control measure.

The course of the disease may be shortened by appropriate treatment with antibiotics, particularly by a combination of streptomycin and terramycin. However, some cases are resistant to all forms of therapy.

The disease is not communicable from man to man. Epidemics could result from the wide-scale consumption of contaminated, unpasteurised dairy products.

Brucella organisms will remain alive for weeks in water, unpasteurised dairy products, and soil, and are very resistant to low temperatures. Contaminated materials are easily sterilised or disinfected by common methods. Pasteurisation is effective for contaminated dairy products.

Vibrio Comma

This micro-organism is a short, slightly bent, motile, gram-negative, non-sporulating rod.

Cholera, an acute infectious gastrointestinal disease of man, is characterised by sudden onset with nausea, vomiting, profuse watery diarrhoea with 'rice-water' appearance, the rapid loss of body fluids, toxemia, and frequent collapse.

Faeces and vomitus of patients, faeces of convalescents, and temporary carriers are sources of infection.

Transmission is made through direct or indirect faecal contamination of water or foods, by soiled hands or utensils, or by flies.

Incubation is from 1 to 5 days, usually 3 days.

All populations are susceptible, while natural resistance to infection is variable. Recovery from an attack is followed by a temporary immunity which may furnish some protection for years.

Endemic centres exist in India and South-East Asia, from where the disease may spread along human communication lines to more remote countries and cause epidemics. It is normally absent from the Western Hemisphere.

Mortality ranges from about 3 to 30 per cent in treated cases to 50 per cent in untreated cases.

Artificial immunisation with vaccines is of variable degree and uncertain duration (6 to 12 months). Acquired immunity lasts for many years.

The first consideration in the treatment of cholera is to replenish the fluid and mineral losses of the body. Drug therapy has little or no effect upon the clinical course of the disease. However, chloromycetin, aureomycin, and terramycin, given by mouth, cause rapid disappearance of the vibrio organisms, thus reducing the spread of the disease.

Epidemicity is very high under unsanitary conditions, especially those concerned with water supplies, foods, and fly control.

The organism is easily killed by drying. It is not viable in pure water, but will survive up to 24 hours in sewage, and as long as 6 weeks in certain types of relatively impure water containing salts and organic matter. It can withstand freezing for 3 to 4 days. It is readily killed by dry heat at 117°C, by steam and boiling, by short exposure to ordinary disinfectants, and by chlorination of water.

Corynebacterium Diptheriae

This bacterium (a slender, often slighly curved rod) is gram-positive, non-motile, non-sporulating, and non-acid fast. It varies in size from 2 to 7 microns in length and from 0.5 to 1 micron in diameter. The rodlike forms are usually arranged in palisades and often exhibit club-shaped terminal swellings. They stain irregularly, displaying bars or granules as a result of irregular distribution of protoplasm within the cell. Although the organism is normally aerobic, it is often capable of anaerobic cultivation. It produces a highly potent exotoxin, both in the body and in culture.

Diptheria, an acute febrile disease, is generally characterised by local infection, usually involving the air passages. The systemic manifestations are due to absorption of the soluble toxin into the bloodstream. The bacteria multiply rapidly in the tonsils, nose and throat, where grayish membraneous patches appear on the mucous membranes, causing sore throat swelling and stoppage of air passages. Skin and wound infections are not uncommon in tropical and sub-tropical climates. Early diptheria is usually a surprisingly mild disease unless symptoms of obstruction develop. During the first few days of infection, the throat is not particularly sore, there is only slight fever, and there are no severe constitutional symptoms. This lack of obvious symptoms is characteristic of diptheria in the adult and is especially dangerous when infection occurs in the nasal passages, because the infection is not recognised or treatment is not begun until sufficient exotoxin has been absorbed to cause irreparable damage to other parts of the body.

Discharges from the nose and throat of infected persons and healthy carriers or from skin lesions are sources of infection.

The disease is contracted by direct contact with patients or carriers, by droplet infection, or through articles freshly contaminated with nose and throat discharges of infected individuals.

The incubation period is usually from 2 to 5 days but occasionally may be longer.

Susceptibility is general in the absence of previous contact with the organism or its toxin. In the past there was a very high percentage of immunity in the adult population because of repeated contact with usually unrecognised sources of infection. Due to the widespread practice of diptheria vaccination during infancy, there is less frequent opportunity for natural exposure to doses sufficient to produce immunity. Therefore, the adult population is now more susceptible to the disease than in the past. This susceptibility can be accurately measured by means of the Schick test. Recovery from the disease does not necessarily result in immunity.

The disease is endemic and epidemic around the world. It is more common in temperate zones than elsewhere and during the autumn and winter. Age distribution of cases and deaths depends largely upon childhood immunisation practices.

The fatality rate is variable, depending upon the virulence of the infecting strain; among untreated cases it may range from 10 to 50 per cent. In cases receiving antitoxin treatment, this rate is lowered to 2 to 8 per cent.

Diptheria toxoid is extremely effective. Permanent immunity may be maintained by means of booster inoculations at regular intervals.

Diptheria antitoxin is effective when given promptly and in adequate dosage. Penicillin as a supplementary treatment suppresses secondary invaders, shortens the period of illness, and reduces the number of convalescent carriers.

Epidemicity is high, depending on the immunity status of the population and degree of exposure to the disease. A large proportion of the cases occur in children under 5 years of age.

The diptheria organism is more resistant to light, drying, and freezing than are most non-sporulating bacilli, remaining viable for a long time in air and dust. It is capable of surviving many hours on a cotton swab and has been cultured from dried bits of diptheritic pseudomembrane after 14 weeks. It is destroyed by ordinary antiseptics and by being boiled for 12 minutes or by being heated to 75°C for 10 minutes.

Bacterium Tularense (Pasteurella Tularensis)

This is a small, aerobic, gram-negative cocco-bacillus, often varying in size and shape. It is non-motile and non-sporulating.

Tularemia is also known as Rabbit Fever and Deer Fly Fever. It is a fatal septicemic (blood poisoning) disease of wild rodents, accidentally communicable to man, in whom it is characterised by sudden onset with chills, fever, and prostration and by a tendency to pneumonic complications. In man it is an acute, severe, weakening disease, later becoming chronic, and may be accompanied by enlargement of the regional lymph glands with or without a lesion at the site of infection, or by typhoid-like symptoms with no local lesion or enlargement of the local lymph glands.

Wild rabbits or hares, deer flies, ticks, and many other animals (including the woodchuck, coyote, opossum, tree squirrel, skunk, cat, deer, fox, hog, sage hen, and some snakes) are sources of infection.

Transmission is made by infection through the skin, eyes, or lungs from handling infected animals, as in skinning or dressing the animals or performing autopsies; by bites of infected flies and ticks; by eating insufficiently cooked rabbit meat; or by drinking contaminated water. Laboratory infections are not infrequent.

Incubation is from 1 to 10 days, usually about 3 days.

All ages are susceptible, and recovery from an attack is followed by permanent immunity. The infectivity rate is from 90 to 100 per cent.

The disease is present throughout North America and in many parts of continental Europe and Japan. It occurs in every month of the year in the United States.

Untreated cases have a death rate of 4 to 8 per cent averaging 5 per cent.

Vaccination greatly reduces the severity of the disease and may prevent infection in some cases.

The antibiotics, particularly streptomycin, aureomycin, and chloromycetin, are effective.

The disease is essentially sporadic, but may be epidemic when modes of transmission (see above) are prevalent. It is not transmitted directly from man to man.

The organism remains viable for weeks in water, soil, carcasses, and hides, and for years in frozen rabbit meat. It is resistant for months to temperatures of freezing and below. It is rather easily killed by heat at 38°C or above for a few minutes and by 0.5 per cent phenol in 15 minutes.

Pasteurella Pestis

Pasteurella Pestis is a rod-shaped, non-motile, non-sporulating, gram-negative, aerobic bacterium.

Plague, or Black Death, occurs as three clinical types in man; bubonic, pneumonic and septicemic. Another type of plague, Sylvatic Plague, is an infectious disease of wild rodents; it is transmissible to man by flea bites. In general, Plague is characterised by a rapid clinical course with high fever, extreme weakness, glandular swelling, pneumonia, and/or haemorrhages in the skin and mucous membranes.

(1) Bubonic Plague, the most common, is transmitted to man by the bite of an infected flea, the disease being perpetuated by the rat-flea-rat transmission cycle. The flea bites are usually on the lower extremities where the bacilli spread rapidly through the lymphatic system, enlarging the lymph nodes (buboes) in the groin. The bacilli escape from the nodes, invade the bloodstream and produce a generalised infection. Other parts of the body that are affected are the spleen, lungs, and meninges.

(2) Pneumonic Plague, transmitted by inhalation, spreads rapidly until the entire lung is involved in a haemorrhagic, pneumonic process. The disease is usually fatal, the patient dying of suffocation and/or general toxaemia.

(3) Septicemic Plague occurs as the result of gross invasion of the blood stream by Plague bacilli, which cause small haemorrhages in tbe skin and mucous membranes. Death occurs before buboes or pulmonic manifestations appear.

(4) Sylvatic Plague, transmitted by wild rodent fleas, is somewhat different from the Plague transmitted by the rat flea (Bubonic). The flea bites usually occur in the upper extremities and buboes originate

in the armpits rather than in the groin.It frequently changes over to the pneumonic type.

Infected rodents and human patients with Pneumonic Plague are sources of infection. The primary source of the disease is Plague of wild rodents, including the ground squirrel, pack rats and harvest mice of the United States, and various species of wild rodents in other parts of the world. Infection may reach man from these sources or more often through the medium of the domestic rat.

Pneumonic Plague is usually transmitted directly from man to man by droplet infection. Bubonic Plague is generally transmitted to man by the bites of fleas from infected rats and other rodents.

Incubation is from 1 to 7 days for Pneumonic Plague, 4 to 7 days for Bubonic Plague.

Susceptibility is general, particularly to the Pneumonic form. Recovery is followed by temporary, relative immunity.

The disease is rare in North America and island possessions of the United States. Occasional cases of the bubonic type occur west of the Mississippi River from bites of fleas from infected wild rodents. The disease has foci of infection in various parts of the world, particularly in Asia.

Untreated Bubonic Plague has a mortality of 30 to 60 per cent, while untreated Pneumonic Plague kills from 90 to 100 per cent of its victims.

Anti-Plague serum produces an artificial passive immunity of 2 weeks duration. Active immunisation with killed bacterial vaccines is protective for some months when adminstered in 2 or 3 doses at weekly intervals, repeated stimulating doses being necessary. According to some authorities, vaccines prepared from living avirulent strains confer a better and longer immunity than do vaccines from other sources.

Prompt treatment with sulfonamides and streptomycin (chloromycetin and aureomycin may be used if resistance to streptomycin develops) and combined with serum therapy is essential and is effective if used early. Supportive treatment for pneumonic and septicemic forms is required. These measures shorten the duration of the disease and reduce its mortality.

Bubonic Plague is not directly communicable from person to person, but Pneumonic Plague is intensely communicable during the acute period. Strict area quarantine and sanitation, in addition to other measures such as rat flea extermination (DDT), are essential to control outbreaks.

The organism probably will remain viable in water from 2 to 30 days and in moist meal and grain for about 2 weeks. At near freezing temperatures, it will remain alive from months to years but is killed by 15 minutes exposure to 72°C. It also remains viable for some time in dry sputum, flea faeces, and buried bodies but is killed by 3 to 5 hours exposure to sunlight. Decontamination is effected by boiling, use of dry heat above 72°C or steam, and treatment with lysol or chloride of lime.

Malleomyces Mallei

This organism is a slender, non-motile, non-sporulating, gram-negative, aerobic, rod-shaped bacterium.

Glanders, an infection occasionally communicated to man, is characterised by nodular, ulcerative lesions of the skin, mucous membranes. and viscera. It is an acute or chronic disease mainly of horses, mules and asses, communicable to dogs, goats, and sheep. The acute form is limited to the nasal mucosa and upper respiratory tract; the chronic form, called Farcy, is characterised by Farcy buds, ulcers, and pus-forming lesions in the joints and muscles.

Infected horses, mules, and asses are sources of infection.

Transmission is usually made by droplet infection (inhalation) or through breaks in the skin; it is sometimes made through the gastro-intestinal tract.

Incubation is from 3 to 5 days.

Man is highly susceptible to Glanders, and the disease does not confer immunity against a second attack.

Glanders is prevalent among horses, mules, and asses in the Balkans, Russia, South-East Asia and India. It is uncommon elsewhere.

In untreated cases, the acute form has a mortality of nearly 100 per cent, while mortality of the chronic form ranges from 50 to 70 per cent.

Satisfactory immunisation procedures have not been developed.

Prompt and radical surgery is the most effective treatment for the chronic form. Sulfadiazine and streptomycin are reported to be effective. The value of newer antibiotic drugs is not yet clearly established.

Although the disease is contagious, an epidemic spread in man is improbable.

The organism resists drying for 2 or 3 weeks but is killed by direct sunlight in a few hours. It may remain alive in decaying matter for 2 to 3 weeks. It is easily killed by the common disinfectants and by being heated at 72°C for 10 minutes.

Malleomyces Pseudomallei

This bacterium is motile, non-sporulating, gram-negative, aerobic, rod-shaped, and small (1 to 2 microns long and 0.5 microns wide). It is often marked by bipolar staining. It closely resembles Malleomyces Mallei.

Melioidosis, also known as Whitmore's Disease, is a Glanders-like disease primarily of rodents but occasionally found in man. It tends to run a more rapid course than Glanders does and in man is almost always acute and rapidly fatal, death occurring usually in 3 to 4 weeks, often within 10 days. Reports of chronic infections involving lungs and lymph glands, bones, joints, and legs have appeared. The disease is characterised by sudden onset with severe chills, high fever, rapid prostration, headache, muscle and joint pains, cough, laboured breathing, nausea and vomiting. In a short time, numerous small abscesses form in the skin, bones, lymph nodes, lungs, and other internal organs.

Probable sources of infection are food or other materials contaminated with rodent excreta and possibly rat fleas.

Transmission takes place apparently by the ingestion of food contaminated with excreta from infected rats and by rat flea bites.

Although not accurately known, the incubation period is probably only a few days.

Susceptibility is general but the organism apparently is not extremely infective under natural conditions, considering the low prevalence of human disease in endemic areas. However, man appears to have little or no resistance to it once infection is established.

Cases of the disease have been found chiefly in Malayasia, Indo-China and Sri Lanka. They have also been reported in Guam, the Philippine Islands and the United States.

Acute Melioidosis is usually fatal.

No vaccine has been developed. Little is known of any immunity acquired through infection.

Chloromycetin alone or in combination with terramycin or aureomycin has been used with marked success. Sulfadiazine is also effective in the acute stage.

The disease is not normally contagious. Spread by droplet infection might occur in a cold climate, which is more suitable to this type of transmission.

The organism is extremely resistant to drying and may survive a month or more in dried soil, in excreta, and in water. It is easily killed in 10 minutes by 1 per cent phenol or 0.5 per cent formalin and by moist heat at 74°C.

Salmonella Typhosa

This organism is a rod-shaped, motile, nonsporulating, gram-negative bacterium. It is also known as S. typhi.

Typhoid Fever is a systemic infection characterised by continued fever, lymphoid tissue involvement, ulceration of the intestines, enlargement of the spleen, rose-coloured spots on the skin, diarrhoea and constitutional disturbances.

The faeces and urine of infected individuals and carriers are sources of infection.

Transfer of organisms is made through the alimentary tract by direct or indirect contact with a typhoid patient or a chronic carrier, by consumption of contaminated water, food, milk, or shellfish; and by flies.

Incubation is from 3 to 38 days, usually 7 to 14 days.

Susceptibility is general, except that some adults have an acquired immunity from unrecognised infection. Recovery is usually followed by permanent immunity.

The disease is widespread throughout the world. Once endemic in most large cities of North America, it has been steadily falling in incidence, particularly in areas supplied with safe water and pasteurised milk, and where modern sewage disposal facilities are used. It is still endemic in some rural areas of the United States, usually as sporadic cases or in small carrier or contact epidemics.

The mortality in untreated cases ranges from 0 to 10 per cent.

Inoculation with typhoid vaccine produces an artificial active immunity of about 2 years' duration. High protection lasting for about a year can be maintained by an annual booster injection of vaccine.

Prompt use of appropriate antibiotics (chloromycetin, aureomycin) shortens the period of communicability and rapidly cures the disease.

Epidemicity is high in the presence of carriers; in the absence of sanitary control for water, food, and milk supplies; and where individuals are not protected by immunisation.

The organism remains viable for 2 to 3 weeks in water, up to 3 months in ice and snow, and for 1 to 2 months in faecal material. Pasteurisation, exposure to 73°C for 20 minutes, exposure to 5 per cent phenol or 1:500 bichloride of mercury for 5 minutes, cooking, and boiling are effective decontamination measures.

Salmonella Paratyphi and Salmonella Schottmuelleri

The organisms are short, plump, rod-shaped, motile, non-sporulating, gram-negative bacteria. S.paratyphi is known at type A of the group; S. Schottmuelleri is also known as S. paratyphi B.

Paratyphoid Fever (an acute, febrile, generalised infection) is very similar to Typhoid Fever (indistinguishable clinically), but its symptoms are usually milder. It is characterised by continued fever, severe diarrhoea, and abdominal pain, with involvement of the lymphoid tissues of the intestines, enlargement of the spleen, and sometimes rose-coloured spots on the trunk. S. Schottmuelleri (type B) is responsible for more cases of the disease than type A and may also produce gastro-enteritis. (Salmonella Hirschfeldii (S. Paratyphi C) may also produce Paratyphoid Fever.)

The contaminated faeces and urine of patients and carriers are sources of infection.

Transfer of organisms is the same as for Typhoid.

Incubation is variable, from 1 to 10 days, depending on strain of organism but averaging less than a week.

Susceptibility is general, and recovery is followed by permanent immunity.

The disease is world-wide, but incidence has declined with that of Typhoid. Outbreaks are sporadic or limited and are due to contact or to the consumption of contaminated foods such as milk or water. There are probably many unrecognised cases. Paratyphoid Fever caused by S. Hirschfeldii has been found relatively frequent in parts of Asia, Africa, and south-east Europe but is almost unknown in the United States.

Fatalities are low, perhaps between 1 and 2 per cent.

Paratyphoid vaccine for types A and B is usually incorporated with typhoid vaccine.

Chemotherapy with 'sulfa' drugs and use of antibiotics, such as chloromycetin, aureomycin, and streptomycin, shorten the period of communicability and hasten cure of the disease.

The epidemicity is similar to that for Typhoid, depending on presence of carriers; inadequate sanitary controls for water, food, and milk supplies; and absence of immunisation.

Stability is the same as for S. typhosa. De-contamination measures include chlorination, pasteurisation, boiling, and cooking.

Salmonella Typhimurium

This bacterium is a short plump rod which occurs singly and measures 0.5 micron in width and from 1 to 1.5 microns in length. It is gram-negative, non-sporulating, and motile.

Salmonella food poisoning (gastro-enteritis) is most frequently caused by S. typhimurium in man. The onset of the infection is nearly always sudden, characterised by headache, chills, and usually by abdominal pains. This is followed by nausea, vomiting, severe diarrhoea with a rise in temperature and prostration. Recovery is usually complete within 2 to 4 days.

The sources of infection are usually rodents, especially rats and mice, human carriers who handle food, eggs, and meat from diseased animals.

The disease is usually obtained by the ingestion of contaminated food (particularly meat), water, or milk. It may also be obtained by direct contact with infected persons or carriers, by direct contact with articles contaminated by discharges (faeces, urine, and vomitus) of infected persons or carriers, or from flies.

Food poisoning occurs after an incubation period ranging from 6 to 24 hours but seldom after more than 48 hours. The short interval suggests that large numbers of the organisms are usually ingested.

Susceptibility is general. The disease is more severe in infants and young children than in adults. Natural immunity is believed to exist in some persons, while acquired immunity is usually permanent after recovery from the disease. (The bacterium is primarily pathogenic for animals; in mice it produces a Typhoid-like disease with high mortality.)

The disease is widely distributed geographically and occurs in almost all warm-blooded animals.

Fatalities range from 1 to 2 per cent in epidemics.

Vaccination is not practical.

Treatment is mainly physiologic. A saline purge is indicated if the infected food has not been eliminated by nature. The restoration of fluid balance is most important, particularly in the very young and old. Streptomycin, chloromycetin, terramycin, and aureomycin reduce the number of organisms in the intestinal tract.

The infection is contagious. Epidemics usually occur when the mass consumption of contaminated food occurs. Explosive epidemics occur in animals, particularly rodents, and many surviving animals become chronic carriers. Spread of the infection can be halted by the elimination of carriers as food handlers, proper sanitation, pasteurisation of milk, elimination of rodents and flies where food is prepared, and careful handling and adequate cooking of food.

Mycobacterium Tuberculosis

Tubercle bacilli are slender straight or slightly curved rods with rounded ends. They vary from 0.2 to 0.5 micron in width and from 1 to 4 microns in length. They are acid-fast, non-motile, and gram-positive and are strictly aerobic.

Pulmonary Tuberculosis is characterised by severe lung involvement accompanied by coughing, fever, fatigue, and loss of weight. This form of the disease is the chief cause of morbidity and mortality. The primary type is acute, healing or progressing in a relatively short time, and is most commonly seen in infants and children and occasionally in adults who have escaped childhood infection. Adults acquiring their first infection may manifest the post-primary type, passing through the primary phase inconspicuously because of its rapid development. The post-primary (re-infection) type is more stable and more chronic than the primary type and is associated with a significant, but inadequate, degree of resistance. Tuberculosis infection in the bones, joints, skin, or other tissues is usually caused by the bovine variety of M. Tuberculosis, although this type may also invade the lungs.

Infection is acquired from persons with draining lung cavities. Tuberculous cattle, and particularly their raw milk, is the source of the bovine variety.

Transmission usually occurs through the discharges of the respiratory tract, by direct or indirect personal contact. Primary infection is almost always a result of inhalation of the bacilli in droplet form. The post-primary type of Tuberculosis may be caused either by organisms which have survived in primary lesions or by newly inhaled bacilli. The bovine type is acquired by contact with tuberculous cattle or ingestion of their raw milk but can be transmitted also by the same route as the human type, from person to person. Natural infection usually requires continued and intimate exposure.

This period is variable, depending on dosage, age, and other factors, but probably is not less than one month. The period may be reduced considerably by exposure to heavy concentrations of the organism.

Man is very susceptible to tuberculous infection but remarkably resistant to tuberculous disease. Susceptibility to the disease is dependent upon age, race, family characteristics, and previous exposure to the organism. It is lowest in persons from 3 to 12 years of age and is greater in the undernourished and fatigued and among peoples who have not previously been exposed to the disease. In the United States the rapidly progressive primary type is seen more often in young adult Negroes than in Caucasian adults. Recovery from the disease leaves no solid immunity, but resistance is altered so that re-infection is not as acute as the previous infection. The chance of contracting progressive clinical Tuberculosis is higher in tuberculin-negative than in tuberculin-positive individuals.

Tuberculosis is one of the most common of the infectious diseases of man. It occurs in all parts of the world, although it has never appeared in some isolated groups of people. Infection is widespread in the urban population, as is shown by the 50 to 95 per cent tuberculin reactor rate, but the progressive disease develops in only a small proportion of those infected.

The fatality rate is high, especially among infants, adult males up to old age, and adolescent and young adult females. Despite the many improvements in case-finding and the newer therapeutic procedures, Tuberculosis is still among the ten leading causes of death in countries such as the United States.

Vaccination of tuberculin-negative persons with living strains of attenuated tubercle bacilli of Calmette and Geurin (BCG) confers some protection against naturally acquired tuberculous disease (primary and post-primary). However, prevention and control of Tuberculosis depends largely upon detection and isolation of carriers and the general improvement of environmental and economic conditions.

Streptomycin, particularly when combined with para-aminosalicylic acid (PAS) or isonicotinic acid hydrazine (INH), is valuable in arresting the disease, particularly in its acute manifestations, but this treatment in itself is not curative. Success in treatment depends upon supportive therapy, including rest, good food (supplementing the diet with vitamins A, C, and D), and fresh air. Local rest of the lungs is promoted by several methods. Surgery is required in some cases.

The spread of Tuberculosis occurs in large part through continued family or household case association, the disease being transmitted slowly from one generation to the next. Under favourable conditions, epidemic outbreaks may take place. Occurrence is influenced by occupation, as by continued exposure to mineral (silica) dust, which predisposes to infection.

When organisms are exposed to direct sunlight, in artificial culture they are killed in 2 hours but in sputum under the same conditions they may survive 20 to 30 hours. When organisms are protected from the sun, they will live in putrefying sputum for weeks and in dried sputum for as long as 6 to 8 months. They are resistant to the usual chemical disinfectants, 24 hours being required for the decontamination of sputum by 5 per cent phenol, but possess no greater resistance to moist heat than other bacteria do, being killed in 15 to 20 minutes at 78°C. Pasteurisation is effective in destroying the organisms in milk.

BACTERIA

Bacillus Anthracis

This organism and the infection it produces in man have been described under anti-personnel agents. The discussion which follows will be limited to considerations of its pathogenicity for animals.

Anthrax, also known as charbon or splenic fever, is principally a disease of herbivorous animals (cattle, sheep, horses and mules), although occasionally it affects pigs, dogs, cats and some wild animals. Sheep are the most susceptible domestic animals, followed by cattle. In animals, anthrax occurs as an acute febrile septicaemic infection characterised by depression; weakness; difficult breathing; inflamed mucous membranes; bloody diarrhoea; and swellings in the neck, chest, flank and lumbar regions. The spleen becomes enlarged and may in some cases be so distended that its capsule ruptures; there is also marked liver damage. The intestinal tract is inflamed and haemorrhagic; the lungs are congested; the lymph glands are swollen, congested and haemorrhagic.

Contaminated soil, food, water, flies, and dust are sources of infection. Scavengers such as buzzards may distribute infected material by feeding upon carcasses and vomiting the material at other locations.

Normally, transmission is made by the ingestion of contaminated food and water; through wounds or scratches; by inhalation; and mechanically by flies, ticks and birds.

Incubation is from 3 to 7 days.

Sheep are very susceptible, cattle somewhat less so, with the horse and mule intermediate; pigs have fair resistance (although serious outbreaks have occurred among them), while birds and fowl are practically immune.

Anthrax is prevalent wherever herbivorous domestic animals are raised (ie, world-wide). It is particularly prevalent in countries where no organised control of animal diseases exists.

Mortality is very high, probably from 40 to 80 per cent in affected animals.

Attenuated spore vaccines are effective immunising agents.

Penicillin and sulfathiazole may be used. Anti-anthrax serum used promptly in large doses is effective.

The disease has moderate communicability; once the disease is established, it often builds up to epizootic proportions. Control measures include strict quarantine of infected areas, observation and immunisation of exposed animals, prompt decontamination of discharges and areas inhabited by infected animals, and proper disposal of carcasses.

Brucella Group

The organisms of this group, Br. abortus, Br. melitensis, and Br. suis, have been described under anti-personnel agents. The following text is limited to those aspects which are concerned with pathogenicity for animals.

The three closely related species of Brucella produce the disease known as Brucellosis, which is sometimes given different names associated with each of the causative organisms.

(1)Brucella abortus is the most common cause of this disease in animals. The disease was formerly known as Bang's Disease and is also known as Contagious Abortion and Bovine Brucellosis. It is one of the most important diseases of livestock from an economic viewpoint. It is mainly a disease of bovine animals (cattle and oxen), but other animals, including horses, sheep, goats, swine, and dogs, are susceptible to a lesser extent. The organism can also cause Brucellosis in man. The disease may exist and persist in the genital system of the bull without evident symptoms. In pregnant cows it is characterised by a chronic inflammation of the uterus (especially of the mucous membrane), which is usually, but not invariably, followed by abortion between the fifth and eighth month of pregnancy. In cases of abortion, a placentitis is produced which is serious not only because it endangers the life of the foetus but also because it results in injury to the uterus that may affect future pregnancies and also render the cow susceptible to secondary infection. When non-pregnant cows become infected, there are usually few recognisable symptoms to indicate that they have the disease. Infection may occur by the mouth, through the vagina during service, after a bull has serviced an infected cow or when the bull is a carrier, or by other means (see transmission below). The disease presents a serious carrier problem, as a large percentage of infected cows may harbour the organism in their udders for as long as 7 years.

(2)Brucella melitensis produces a chronic disease which is also known as Malta Fever, Mediterranean Fever, and Contagious Abortion in animals and Undulant Fever in man. This organism is closely allied to Br. abortus and gains entrance into the body mainly through contaminated food. Goats are the natural or true hosts for this organism. Sheep are next in susceptibility, followed by cattle, horses and swine. This species is more pathogenic for man than the other two. Affected goats and sheep develop a moderate, remittent, long-continued infection which may escape notice. They may appear in poor condition, and a chronic inflammation of the udder may be present in severe cases, but it is not until late in the disease that the milk undergoes any noticeable change or milk secretion diminishes. A chronic cough is sometimes present, and 50 to 90 per cent of pregnant females may abort. Lameness may occur in both males and females. The disease is rarely fatal, but its importance lies in the possible transmission to man, where symptoms are much more serious. It does present a serious carrier problem, because large percentages of apparently healthy animals are carriers, with the organism being present in the udder and blood and frequently eliminated in the milk and urine.

(3)Brucella suis causes Brucellosis of swine and is also pathogenic for man and, occasionally, for horses, dogs, and cows. The disease in swine resembles that of both man and cattle. As with the other forms, it can result in contagious abortion, although this does not always result; that is, abortion is not the disease but rather a symptom that is sometimes observed. The organism is said to be more virulent for man than Br. abortus and less so than Br. melitensis. The tissues, blood, milk and urine of infected animals, especially goats, cattle and swine, are sources of infection.

Transmission is made by ingestion of milk, food and water contaminated with the organism; inhalation of infected dust from contaminated pens and forage; skin infection from direct contact; and by mating in cattle.

Incubation is variable, depending on the strain of organism, the kind of animal, and the route of infection.

Susceptibility of animal species to different strains is variable and has been considered above. Animals that recover from the disease usually are highly resistant to re-infection; however, some may be re-infected.

The disease is world-wide, particularly in areas having a high population of cattle, sheep, goats and swine.

Mortality is very low.

Vaccines are in general use for cattle. The measure most widely used is calfhood vaccination with a Br. abortus strain of reduced virulence, known as BAI Strain 19.

Antibiotics and sulfonamides appear promising, but there is no really effective treatment.

Epizootics are common in herds and flocks. Methods of control consist of testing, segregation, and slaughter of infected animals, as the disease is readily transmitted from animal to animal.

Malleomyces Mallei

This organism and the disease it produces in man have been discussed under anti-personnel agents. The following discussion is limited to those aspects concerned with its pathogenicity for animals.

Glanders or Farcy (cutaneous) is an acute or chronic disease of equines (horses, mules, asses), communicable to dogs, cats, goats, sheep and man. It does not occur in cattle and swine. Definite symptoms of lobular pneumonia or broncho-pneumonia are often present in the early stages of the acute form of the disease; however, irrespective of the form of the disease, lesions of the spleen and liver often occur.

1. The acute form is limited to mucuous membranes of the nose and upper respiratory tract. It is characterised by fever and prostration, nasal discharge, ulceration of the nasal mucosa, and acute swelling of the lymph nodes, which may break down and form deep, pus-discharging ulcers and sinuses. The acute disease is usually fatal, death occurring within 4 to 6 weeks.

2. The chronic form has a more gradual onset than the acute form and is characterised by the formation of persistent abscesses, arthritis, ulceration of the nostrils, multiple skin swellings, ulcers and nodules, and by general enlargement of the regional lymph glands. Cutaneous Glanders (Farcy) is usually chronic.

Infected animals and contaminated food and water are sources of infection.

Transmission is by the ingestion of contaminated food and water, by direct contact with infected animals (through skin abrasions), and by inhalation of infected material.

Incubation is from 2 to 5 days, occasionally 2 to 3 weeks.

Equines are the most susceptible animals. Recovery from the disease does not confer immunity.

The disease is most prevalent in parts of Europe, eastern and southern Asia, and northern Africa. Introduction of the disease must always by guarded against whenever an animal is brought into the herd from an external source.

Acute Glanders is usually fatal. The chronic form has a mortality of 50 to 70 per cent.

Immunisation is very unsatisfactory.

Treatment is unsatisfactory at the present time. Sulfadiazine and streptomycin may be of some value against the chronic form. Radical surgery is usually preferred in chronic cases. (The law in most civilised countries does not permit any form of treatment being tried, because all diagnosed cases must be destroyed as soon as possible.)

Epizootics may occur where large numbers of animals are herded together; where sanitary requirements for stabling, feed, and watering are lax; and where diagnostic procedures are inadequate.

VIRUSES

Foot-and-Mouth Disease Virus

This is a very small virus which is 8 to 12 milli-microns in diameter. At least six types of the virus have been recognised.

Foot-and-Mouth Disease, also known as Aphthous Fever, is an acute, contagious, highly infectious, febrile disease of cloven-footed animals (cattle, sheep, swine). Man is only slightly susceptible and, if infected, shows only mild symptoms. The disease causes a marked and rapid weight loss, a rapid decrease in milk flow, and a severely lowered reproductive capacity. It is characterised by an acute fever and by vesicle formation on the feet and mucous surfaces of the mouth and cheeks and on the udder. In most cases, a long time elapses before the animals return to normal.

Infected animals and contaminated food, water, milk and pastures are the sources of infection.

Transmission is by the ingestion of food, water and milk contaminated with urine, saliva, vesicular fluid, faeces and by direct contact.

Incubation is usually from 24 hours to 7 days, occasionally 2 or 3 weeks.

Cloven-footed animals have a susceptibility of from 80 to 100 per cent.

The disease is enzootic in various parts of Europe, Asia, Africa, and South America; it has also occurred in Mexico and Canada.

While the mortality in mild cattle epizootics is said to range from 3 to 5 per cent, malignant forms have caused a mortality as high as 50 to 70 per cent. The mortality for sheep is around 5 per cent and for swine is from 50 to 70 per cent.

Vaccines are effective in developing an active immunity, but each vaccine is specific only against the strain from which it was made. Promising strides in the development and production of vaccines have been made in recent years.

Treatment is supportive only.

The disease has very high communicability. It tends to spread rapidly over a wide geographic area. The disease is controlled mainly by quarantine and vaccination of exposed animals and the slaughter of infected animals.

The organism is quite resistant under natural conditions. Contaminated hay and bran may remain infective for weeks, and the organism may persist in moist soil for months. It is resistant to low temperatures but is destroyed within 30 minutes at 78°C or above. It resists drying for several weeks. Two per cent sodium hydroxide solution is preferred for decontamination.

Rinderpest Virus

The Rinderpest virus is very small. Rinderpest is also known as Cattle Plague. It is an acute, febrile, highly contagious and highly fatal disease of bovine animals (cattle and oxen), sheep, goats, and water buffaloes. It is characterised by sudden onset, croupous inflammation of the digestive tract, inflammation and erosion of the mucous membranes of the mouth, and bloody diarrhoea. The virus is present in practically all the body tissues and fluids of infected animals.

Infected animals are the sources of infection.

Transmission is by ingestion of food and water contaminated with urine, faeces, saliva, and eye and nasal secretions and by direct contact.

Incubation is from 3 to 9 days.

Cattle, oxen, sheep, goats,and water buffaloes are almost completely susceptible.

The disease is world-wide, except for the Western Hemisphere.

Mortality ranges from 25 to 100 per cent.

Excellent specific vaccines have been produced which induce an active immunity. A good immunising serum can be prepared which confers a passive immunity of short duration. Recovery from the disease is followed by an immunity of indefinite duration.

Massive doses of anti-serum may be used to protect susceptible animals but are of little value after symptoms have appeared.

Epidemicity is very high in non-immunised animals, as the disease is highly infective, is easily transmitted from animal to animal, and tends to spread gradually over wide areas.

The virus will survive for nearly 3 months under controlled optimum conditions. Exposure to 78°C kills it within a few minutes. It will resist drying and sunlight for only a few days but when frozen and then dried will remain viable for months. Effective disinfectants are 2 to 5 per cent phenol and 0.1 per cent bichloride of mercury.

Rift Valley Fever Virus

The virus is 23 to 35 milli-microns in size and will pass through most filters.

Rift Valley Fever, or Enzootic Hepatitis, is a highly infective and fatal disease of sheep. Goats and cattle are sometimes affected, and the disease is easily transmitted to man, in whom it usually takes a mild form. In sheep, the infection is characterised by a rapid course, high fever, loss of appetite, and progressive weakness; it may produce abortion in pregnant females. The disease is more severe in lambs than in older animals.

Infected animals are the sources of infection.

The disease is transmitted by the bites of mosquitoes and possibly by the inhalation of infected food.

Incubation is from 1 to 4 days for adult animals; 12 to 24 hours for lambs.

Sheep are highly susceptible, while goats, cattle and man are less so. Mice, rats, cats and apes are reported to be susceptible, while horses, rabbits, guinea pigs, and birds seem to be immune.

The disease is found in southern and central Africa, particularly in Kenya.

Mortality is 90 to 95 per cent in lambs, 35 to 45 per cent in adult sheep, and 5 to 20 per cent in cattle.

A living virus vaccine is in use in South Africa.

Treatment is supportive only.

Epidemicity is dependent on the presence of diseased animals and the specific mosquito vector.

The virus is said to be destroyed by exposure to a temperature of 73°C for 40 minutes.

Vesicular Stomatitis Virus

The virus is 0.07 to 0.1 micron in size.

Vesicular Stomatitis, also known as Pseudo Foot-and-Mouth Disease or Mouth Thrush, is a contagious, weakening, febrile disease, primarily of horses and mules and occasionally of cattle and swine. It is characterised by vesicular eruptions of the mucous membranes, particularly of the mouth. Occasionally, lesions may be found on the feet and udders. This disease has great similarity to Foot-and-Mouth Disease; the main difference is that the horse is not susceptible to Foot-and-Mouth Disease.

Horses, mules and cattle infected with the disease are sources of infection.

Transmission is by ingestion of contaminated food and water and by direct contact.

Incubation is from 2 to 9 days.

The percentage of horses and mules estimated to be susceptible is 70 to 100.

The disease is probably world-wide.

Mortality is from 0 to 10 per cent.

Immunisation is in the experimental stage only.

Treatment is unsatisfactory. Treatment is only supportive at the present time.

The disease tends to spread gradually over a limited geographical area.

Contaminated food remains infective over long periods. The organism is resistant to low temperatures. The virus remains viable in 0.5 per cent phenol for 23 days. It is destroyed by 2 per cent lye in 1 minute and by 0.05 per cent crystal violet. In general, it has much the same resistance as that of Foot-and-Mouth virus.

Vesicular Exanthema Virus

The virus is 0.07 to 0.1 micron in size.

Vesicular Exanthema is a contagious, febrile, weakening disease of swine. It is characterised by the formation of vesicles or lesions in the mouth and nostrils, on the snout, feet and udders, and around the coronary band. The vesicles produced by this virus are very similar to those produced by Foot-and-Mouth Disease and Vesicular Stomatitis. Animals become lame and hoofs are sometimes shed. Sows may abort in advanced pregnancy. Cattle, sheep and guinea pigs are not susceptible to the disease.

Infected swine and uncooked garbage containing pork trimmings are the sources of infection.

Transmission is by direct contact, by the ingestion of contaminated food and water, and by persons going from farm to farm.

The incubation period is from 2 to 7 days.

The infectivity rate ranges from 50 to 100 per cent in swine.

The estimated herd mortality is 0 to 5 per cent for swine.

No immunisation has been developed to date. Immunity following an attack of the disease is comparatively short.

Treatment is supportive only.

The disease is controlled to some extent by quarantine and slaughter but probably will not become completely manageable until the practice of feeding raw garbage is stopped.

See Rift Valley Fever Virus for stability.

Hog Cholera Virus

The virus is 0.03 to 0.035 micron in size.

Hog Cholera, also known as Swine Fever, is a highly acute, contagious, febrile disease of swine. It is usually chronic in older swine. It is characterised by high fever, yellowish discharges from the eyes, diarrhoea, loss of appetite, severe haemorrhages, viremia and extreme weakness.

Infected swine are the sources of infection.

Transmission is by contaminated garbage (infected pork) and through food, water, hog wallows and pens that have been contaminated from eye and nasal secretions, urine, blood and faeces of infected swine. Aerosol transmission is possible. Hogs that have recovered from the disease are possible carriers.

Incubation is from 5 to 6 days.

The percentage of exposed swine that are susceptible is 70 to 100 per cent. Animals recovering from the disease have an immunity of uncertain duration. The disease is highly infectious for susceptible swine only.

Hog cholera has a world-wide distribution and is the greatest menace with which swine-raisers have had to contend. It is widespread in North America, France, Africa, Germany and the United Kingdom.

Fatalities are estimated to be 80 to 90 per cent in the United States and Africa. In European countries the death loss is often lower.

Several excellent active and passive immunisation methods have been developed and are available.

Administration of hyperimmune serum is the best treatment and is moderately effective.

This highly contagious disease spreads rapidly and easily among swine. It tends to spread gradually over a wide geographic area. Control of the disease is by rigid measure of isolation and disinfection whenever an outbreak occurs and by crystal violet vaccine inoculation.

The virus is stable for months at low temperatures. It is destroyed by use of 1 to 2 per cent caustic soda, 2 per cent cresol for 1 hour, and 5 per cent milk of lime and by exposure to 83°C for 1 hour. It is resistant to phenol and glycerin.

African Swine Fever Virus

The virus is small and very resistant to environmental changes.

African Swine Fever, also known as Wart Hog Disease and East African Swine Fever, is a highly contagious and excessively acute disease of domestic swine. It is characterised by fever, pronounced haemorrhages of the lymphatic glands, the kidneys, and the mucosa of the alimentary tract, and by marked cyanosis (a patchy or diffuse purplish colour) of areas on the skin. Clinically, African Swine Fever is similar to but more acute than American Hog Cholera; its immunology is distinct from that of Hog Cholera.

The bush pig and wart hog harbour the virus but usually do not manifest symptoms. Contact with these wild pigs apparently initiates the disease in domestic swine, and once established it spreads rapidly by contact.

The disease is transmitted by contact with urine and faeces from infected pigs, the feeding of raw garbage, or pigs actually eating carcasses of virus-carrying wild pigs; it is mechanically spread by caretakers and others who pass from infected premises without taking proper precautionary measures. It is thought that infection may be transmitted from wart hogs to pigs by an insect vector; however, the vector has not been established. In a few cases where hogs have survived the disease, there is evidence that they may act as carriers for as long as 10 months after recovery.

The incubation period is usually 4 to 7 days when exposure is made by contact with infected pigs.

In epidemics the disease may occur in 100 per cent of susceptible domestic swine. The disease is of such a highly contagious nature that usually, in a few days after the first case is noticed, the majority of the pigs in a herd become infected. However, it has been noted that the virus coming from the wart hog or wild pig takes some time to attain high virulence; the occurrence of an isolated death being followed a week or so later by one or two more deaths and then after a further interval, the appearance of many cases with rapid spread.

African Swine Fever is not known to exist outside the African continent. Most of the cases have been confined to East Africa and South Africa.

Fatalities range from 95 to 98 per cent and may be up to 100 per cent.

Attempts at immunising swine against this virus have been for the most part unsuccessful. Even pigs that have survived infection are not consistently immune to subsequent exposures. Hog Cholera antiserum affords no protection against this disease, although in a few cases it appeared to delay the reaction. Although it has not been proved, it is felt that hyperimmune serum prepared from the virus at any given time might be found to be of no effect in subsequent outbreaks, owing to frequent spontaneous changes in the antigenic character of the virus.

At present there is no effective treatment for the disease.

The disease is highly contagious. It is prevalent in areas where opportunities exist for cohabitation between wild and domestic hogs. Control of the disease has been successful in East and South Africa by immediate slaughter of all infected and exposed animals, disposal of all carcasses by burning or deep burial, disinfection of the infected premises, and quarantine of infected areas. Measures to prevent contact of domestic swine with wild pigs have resulted in a marked decrease in the incidence of the disease.

The virus is very stable at low temperatures; it has remained viable after storage in a cold dark room for 6 years. It is killed by sunlight but will remain viable in infected carcasses, food and water for some length of time.

Fowl Plague Virus

The virus is small, measuring 0.06 to 0.09 micron.

Fowl Plague, also known as Fowl Pest, is an acute, contagious, highly fatal disease of fowl (particularly chickens and turkeys), characterised by haemorrhages in various tissues of the body, swelling and blood poisoning.

Infected fowl are sources of infection.

Transmission is by the ingestion of food, water and soil contaminated by the blood, urine, faeces, and eye and nasal secretions from infected fowl or birds; by infection through wounds; and possibly by blood-sucking insects.

Incubation is from 2 to 7 days; it may be as short as 24 hours.

The disease has an infectivity rate of 50 to 80 per cent.

The disease is world-wide.

The mortality rate is very nearly 100 per cent.

No satisfactory method of immunisation has been established.

No treatment has been developed.

Fowl Plague is highly contagious. Once the disease is introduced into a flock, it tends to spread rapidly and kills all or nearly all of the flock in a short period of time. Control measures include rigid quarantine and slaughter and burning of infected fowls. All houses and equipment should be disinfected and the contaminated soil treated with caustic soda.

The virus is stable in dried blood or tissues for many months but is destroyed by exposure to 83°C for 5 minutes. It is also killed by sunlight in a short time. It survives for 18 days on feathers of birds dying of the disease; it will remain viable for months if kept in a cool, dark place free from air. It is destroyed by 0.1 per cent bichloride of mercury in 30 minutes or by 5 per cent phenol in 10 minutes.

Newcastle Disease Virus

The virus is 0.08 to 0.12 micron in size.

Newcastle Disease, Pseudo-pest of fowl or Pneumoencephalitis, is an acute, highly contagious, febrile disease of fowl (chickens, turkeys and pheasants). The course of the disease is of short duration. It is characterised by severe respiratory and nervous symptoms, including difficult breathing, depression and stupor, twitching of the head and neck, marked weakness, a twisted neck, and perhaps paralysis. Egg production is interrupted for weeks.

Infected fowl are sources of infection.

Transmission is by direct contact, by the ingestion of food and water contaminated with faeces, and by the inhalation of contaminated dust.

Incubation is usually from 4 to 8 days, although it may be up to 13 days.

The infectivity rate ranges from 50 to 90 per cent. Chickens of all ages and turkeys are highly susceptible; pigeons, geese, ducks and other barnyard fowl are less susceptible.

The disease is world-wide.

Fatalitites range from 10 to 100 per cent. The disease found outside the United States is nearly 100 per cent fatal; strains found in the United States and the United Kingdom are usually much less pathogenic.

Vaccines prepared from live virus of low virulence produce an immunity of a few months' duration; revaccination of layers should be at 12 to 16 weeks of age. Recovery from the disease is accompanied by some immunity.

No treatment has been developed.

The disease is highly communicable and tends to spread gradually over wide geographic areas. The greatest incidence of the disease is in the winter months. Control measures consist of vaccination and maintenance of a high standard of sanitation.

The virus remains stable for years when dried in the frozen state (lyophilized); it is readily destroyed by being heated at 40°C (pasteurisation). Highly effective disinfectants are 3 per cent phenol, 3 per cent cresol, and 2 per cent sodium hydroxide solutions.

Equine Encephalomyelitis Viruses

There are several viruses, each specific for the different forms of the disease, which include Eastern and Western Equine Encephalomyelitis and Venezuelan Equine Encephalomyelitis. These viruses are also pathogenic for man, with high mortality rates. The disease in man has been described in anti-personnel agents, and the information below relates only to the infection which occurs in animals, mainly equines. In appearance the organisms are similar, being, small, neurotropic viruses.

Equine Encephalomyelitis (Sleeping Sickness) is an acute infectious disease of the central nervous system of horses and mules (also of man). It is characterised by fever, drowsiness or restlessness, poor co-ordination, and occasionally convulsions;

in the severe cases motor and sensory paralysis cause prostration, and death results from respiratory and cardiac failure.

Birds, and probably many wild and domestic animals, may become infected and serve as temporary reservoirs for the virus. The organism probably has a wider host range than any other pathogenic agent.

The infection is transmitted by blood-sucking arthropods (usually mosquitoes) and perhaps by ticks, from the animal reservoirs mentioned above to horses, mules and man. Transmission may be by contact with nasal secretions of infected horses.

Incubation is from 1 to 3 weeks.

Laboratory investigations indicate that many animals other than the equines are susceptible to the disease. Susceptibility of a herd may be as high as 80 per cent, but this varies with the kind of virus and the type of animals. Recovered animals appear to have a substantial immunity to the disease.

The disease occurs in the summer and early autumn. Eastern and Western types are found in Canada, the United States, and Central and South America; the Venezuelan type is found in Venezuela, Columbia and Trinidad.

Fatalities range from 27 per cent in the Western type to about 90 per cent in the Eastern type.

A very effective vaccine is available for producing active protective immunisation in horses. Passive immunity can also be conferred by injection of hyper-immune serum.

Treatment with immune serum is of great value if given before cardinal symptoms are present or in the very early stages; however, it is of little value when administered after symptoms appear. Sulfonamides and penicillin are ineffective; good nursing care and supportive treatment are very important.

The disease tends to spread rapidly over local areas when insect vectors are present.

The virus is inactivated by being heated at 78°C for 30 minutes; it is readily destroyed by formalin, bichloride of mercury, chloroform and ether. It is quite resistant to glycerin and phenol and to low temperatures.

Psittacosis Virus

This organism, usually classed as a virus, has been classed by some as a rickettsia but more recently has been placed tentatively in a group intermediate between viruses and rickettsiae. It is small and coccoid, is 0.2 to 0.4 micron in length, and is gram-negative. Although the name Miyagawanella psittacii has been applied to this virus, it has not been generally accepted.

Psittacosis or Parrot Fever is also called Ornithosis. The disease is often confused with Influenza, Atypical Pneumonia, and Typhoid Fever. It is a severe febrile disease in man; the onset may be sudden or may be gradual and insidious. It is characterised by acute pulmonary infection, chills, fever, anorexia, sore throat, severe headache, backache, constipation, great weakness, and prostration and is sometimes accompanied by delirium.

Parrots, parakeets, budgerigars, canaries, pigeons, and other birds are sources of infection.

The respiratory tract is the principal entry path. Transmission is made by contact with infected birds or by breathing air contaminated by faeces, urine, nasal discharges or the soiled feathers of sick, dying, or latently infected birds. Contaminated air may travel a considerable distance before it loses its infectivity. Apparently well birds (carriers) can transmit the infection. Transmission from man to man has occurred, and in rare instances man has become a carrier.

Incubation is usually from 6 to 15 days with 10 days as the mean, but it may be as long as 30 days.

All ages and races are susceptible. Children are readily infected but rarely develop severe illnesses. The disease is more severe in adults than in children. Recovery from an attack is generally followed by immunity, although re-infections have been reported.

The disease is widely distributed throughout the world. It is most prevalent in areas where parrots, parakeets, and similar birds are found. It usually occurs as a sudden outbreak among persons exposed to infected birds. Mild cases may occur from slight exposure to infected pigeons or to birds showing no symptoms of the disease.

Mortality is variable, from 9 to 20 per cent in untreated cases, and usually occurs among persons over 30 years of age.

Sulfadiazine, penicillin, aureomycin, chloromycetin, and terramycin shorten the course of the disease and greatly decrease the mortality.

The disease tends to be self-limiting. Transfer from man to man may occur by close contact, such as from patient to attendant.

The virus will remain viable on surfaces for 20 to 30 days and perhaps longer in dry faecal material. It is inactivated in 24 to 36 hours by 0.1 per cent formalin and 0.5 per cent phenol. Decontamination may be accomplished by burning infected material or by the disinfection of infected faeces. Two per cent cresol is an effective disinfectant.

Encephalitis and

Encephalomyelitis Viruses

These viruses are small neurotropic organisms. They are distinguishable antigenically, but some are so closely related that they appear to be variants of a common ancestor.

There are several types of encephalitic diseases, each produced by a specific virus. Their clinical pictures are similar, varying mainly in severity and the rate of progress of symptoms. They are characterised by inflammation of the meninges of the brain, headache, fever, dizziness, drowsiness or stupor, tremors or convulsions, severe prostration, occasional paralysis, and poor muscular co-ordination. The diseases are usually acute, prostrating and of short duration. The following are some of the more common types:

(1) St.Louis Encephalitis is endemic and epidemic in the central and western United States and prevails in summer and autumn when the arthropod vectors are most numerous.

(2) Eastern Equine Encephalomyelitis and Western in Canada, the United States, and Central and South America. They are viral diseases of lower animals, especially of horses and mules, and are transmissible to man.

(3) Venezuelan Equine Encephalomyelitis occurs in Central and South America. It is a disease primarily of equine animals; when the disease is transmitted to man, it usually induces a mild case with varied symptoms.

(4) Japanese B-type Encephalitis occurs in Japan, Korea, China, and some Pacific islands.

(5) Russian Far East Encephalitis occurs in spring and early summer, mainly in the Far East provinces of the Soviet Union and less frequently in Europe and Siberia.

Wild and domestic birds are the principal sources of mosquito infection for the virus types occurring in the United States. The equines, or horses, are also hosts but are not as important sources of mosquito infection as are birds.

Mosquitoes transmit all types except the Russian spring-summer type, which is tick-borne.

The incubation period is variable, ranging from 2 to 15 days.

Susceptibility ranges from 90 to 100 per cent and is usually high in the very young and the aged. Recovery from an infection results in an excellent short-term immunity to the specific virus, but not to any other type.

The diseases are usually prevalent in summer and early autumn; they are usually limited to areas and years of sustained high temperature and large numbers of mosquitoes. The highest rates are in rural and suburban localities.

Mortality is unknown; it is probably 5 to 60 per cent with all types. Japanese B and Eastern equine types have the highest case fatality.

Some effective virus vaccines have been developed on a small scale, but their widespread use is not practicable at present.

Treatment is supportive only; chemotherapy and anti-biotic treatments have not been developed.

Transmission from man to man is not known to occur but may be possible by means of respiratory droplet infection. Transmission of the Venezuelan Equine Encephalomyelitis by inhalation has occurred accidentally in laboratories.

Stability varies among the different types. All may be preserved frozen at −52°C. St.Louis and Japanese B viruses are inactivated at 74°C in 30 minutes; Russian Far East virus is inactivated at 78°C in 10 minutes, but the Western and Eastern equine viruses withstand this treatment and also resist 0.2 per cent chloroform, 1 or 2 per cent phenol, and 0.05 per cent mercuric chloride. St.Louis and Russian viruses are inactivated by 1 per cent formalin within a day; the Russian virus is inactivated by 1 per cent phenol in 10 days, but the St.Louis virus resists this treatment for at least 25 days.

Influenza Virus

The virus is large, having strains which range between 0.07 and 0.1 micron in size. Types A and B have been identified, and there are indications that Type C may come to be recognised.

Influenza or 'la grippe' is an epidemic disease, occasionally pandemic, which is characterised by catarrhal inflammation of the respiratory tract; sudden onset; fever of 1 to 7 days duration, marked prostration; and generalised aches and pains in back, limbs, and muscles. Sore throat, bronchitis, and pneumonia are complications of secondary bacterial infections.

Soiled articles and discharges from the mouth and nose of infected persons are the main sources of infection.

Transmission is probably by direct contact, by droplet infection (respiratory), or by articles freshly contaminated with nose and throat discharges of infected individuals.

The incubation period for Type A is from 1 to 2 days; for Type B it is from 12 to 18 hours.

The virus is highly infective and contagious; susceptibility is general, there being relatively little natural resistance. The resistance of some individuals exposed to the disease during epidemics appears to be due to previous infection. Acquired immunity following recovery from an attack lasts only for a few months to a year, and is effective only against specific strains of the virus. Artificial immunisation methods produce a specific immunity of short duration.

Prevalence is variable; the disease occurs sporadically as epidemics and pandemics. Epidemics may affect up to 50 per cent of a given population within 4 to 6 weeks.

The disease has a mortality range from 0 to 1 per cent, but it is often followed by complicating respiratory infections which can result in high mortality.

Vaccination with specific strains of the virus leads to an effective immunity of several months' duration against the same or closely related strains. The extreme antigenic variability of the influenza viruses makes it difficult to protect against all strains.

Treatment is supportive only, with control of secondary infections. There is no specific effective treatment.

Epidemics are widely prevalent under favourable conditions. Pandemics occur irregularly.

The virus is killed by being heated at 73°C for 20 to 30 minutes; it is resistant to freezing for several weeks. It can also be easily destroyed by steam, boiling, and ordinary antiseptics.

Variola Virus

The virus will pass through most filters and ranges from 0.15 to 0.2 micron in size.

Smallpox, or Variola, a highly contagious and often fatal disease, is characterised by severe fever and small blisters of the skin. The blisters later contain pus and form crusts which fall off in 10 to 40 days after the first lesions have appeared, leaving pink scars which gradually fade. Complications of the disease are secondary bacterial infections.

Lesions of the mucous membranes and skin of infected persons are sources of infection.

Transmission is made through contact with patients having the disease or with articles or persons freshly contaminated by discharges from lesions and skin of infected individuals.

Incubation is from 7 to 21 days, commonly 12 days.

The disease is highly contagious and infectious, with universal susceptibility. Recovery is usually followed by permanent acquired immunity.

According to the World Health Organisation, Smallpox has now been eradicated throughout the world but the chance is always present that it might recur. The frequency is greatest in winter and least in summer.

Fatalities range from 1 per cent with the mild type of disease to 30 per cent with the more severe types.

Artificial immunity by vaccination may be completely effective for 2 to 20 years, but re-vaccination every 3 years is advisable to maintain minimal immunity. A high degree of immunity is required against severe strains of the virus.

No specific treatment is available. Antibiotics and sulphur drugs are used to treat secondary infections.

Epidemicity is high, depending on the immunity status of population and exposure to the disease.

The virus is viable in water for several years at 22 to 27°C and is resistant in dry or wet form to very low temperatures. In the dry state it is more resistant than in the wet state. Decontamination can be effected by exposure to alcohol and acetone for 1 hour at room temperature, but the virus is resistant to some other disinfectants. Moist heat above 78°C and dry heat above 118°C are effective in 10 minutes.

Yellow Fever Virus

The virus particles are estimated to range from 0.017 to 0.028 micron in size and pass through most filters.

Yellow Fever (classic urban) is characterised by sudden onset, chills and fever, prostration, headache, backache, muscular pain, congestion of mucous membranes, severe gastro-intestinal symptoms, and jaundice (yellowing of skin) from liver damage. Haemorrhage from the stomach and gums often occurs. The disease is of short duration, with either death or complete recovery occurring within two weeks of onset. A forest or sylvatic form in South America and Africa is known as Jungle Yellow Fever and appears to be clinically identical to the classic type.

The blood from people and monkeys infected with Yellow Fever is the source of infection.

Transmission is usually by the bite of the female Aedes aegypti mosquito; in the forests of South America and Africa, it is usually by some other types of mosquito.

Incubation is from 3 to 6 days, rarely longer.

Susceptibility is general. Recovery from an attack is followed by lasting immunity. Infants born to immune mothers are immune for 3 to 4 months.

The disease is endemic in man, monkeys, and some other mammals in Western and Central Africa and among certain apes and monkeys in the tropical forests of Central and South America. Occasional human cases and epidemics occur in these areas and neighbouring urban or rural localities. The disease is unknown in the Pacific basin.

The overall case fatality rate, taking into account un-diagnosed mild infections (shown by the later presence in the blood of specific antibodies), is probably about 5 per cent. However, it is much higher in cases in which jaundice appears.

Innoculation with a modified living virus vaccine confers an active immunity that may last for at least 4 years and probably longer.

There is no specific treatment. Supportive treatment (bed rest and fluids) is essential for even the mildest cases.

Occasional epidemics still occur in Africa and South and Central America. These epidemics can be lessened by mosquito control and active immunisation.

The virus is resistant to freezing and drying but is destroyed by being heated at 78°C or above for 10 minutes. It is rather easily inactivated by common antiseptics. Mosquito control measures, including use of insecticides, are the preferred methods of control.

Dengue Fever Virus

This is a very small virus, about 17 to 25 milli-microns in size.

Dengue Fever, an acute, extremely disabling disease usually of sudden onset, is characterised by fever, chilliness, intense headache, backache, pain behind the eyes, joint and muscle pains, weakness and prostration, and an irregular rash. The fever, rarely exceeding 58°C, lasts for 5 or 6 days, and usually terminates abruptly after reaching a peak. Loss of appetite and constipation are common during the entire illness, and abdominal discomfort with colicky pains and tenderness may be manifested. The acute phase lasts only about a week, but convalescence may take several weeks. Dengue Fever is said to be temporarily the most incapacitating although the least fatal of epidemic diseases.

Sources of infection are the blood of infected persons one day before and up to 5 days following onset; infected mosquitoes; and, in some regions, possibly the blood of infected monkeys.

The disease is transmitted by the bite of the Aedes aegypti mosquito and certain other species of Aedes which have become infected by biting a patient.

Incubation is from 3 to 15 days, most often 5 to 8 days.

All persons except natives of endemic areas appear to be fully susceptible. Immunity to the invading strain is apparently permanent, but re-infection with a different strain is possible. However, if this occurs within a few months of the primary infection, the disease is much milder because of cross-immunity.

The disease is found mainly in the tropics and sub-tropics, extending to the Gulf Coast of the United States, but it may occur wherever the vector mosquitoes exist.

The fatality rate is very low.

A mouse-adapted virus vaccine has been found to be stable, safe, and effective in studies on human volunteers and will probably be useful in control of epidemics and for personnel travelling from non-Dengue to Dengue Fever areas.

There is no specific therapy, but supportive treatment is essential.

The virus does not spread directly from person to person. Epidemics occur in areas where vector mosquitoes are present in large numbers. The infectivity rate is extraordinarily high, being 75 to 100 per cent of the inhabitants of a locality being attacked. Any spread of the disease can be prevented by diligent mosquito control.

Blood from a patient remains infectious after storage in a refrigerator for several weeks. The virus may be preserved in a frozen and dried state at 23°C for at least 5 years. Mosquitoes do not transmit the infection through their eggs to the young, but the infected insects probably remain so for life. The virus is inactivated by ultra-violet light and by 0.5 per cent formalin.

Hepatitis Viruses

There are two strains of Hepatitis virus, virus A or Infectious Hepatitis and virus B or Serum Hepatitis (Homologous Serum Jaundice).

Infectious and Serum Hepatitis are characterised first by fever, loss of appetite, nausea, fatigue, headache, and abdominal discomfort. After a few days the fever subsides; then, because of liver damage, bile may be present in the urine, and jaundice (yellowing of the skin) appears.

Sources of infection of virus A are discharges from the nose, mouth, and gastro-intestinal tract of infected persons; sources of infection of virus B are blood, serum, or plasma from infected persons.

The usual mode of transmission for virus A is unknown but may be by contaminated water, food, or milk, or by direct personal contact. Virus B is transmitted by transfusions of infected blood, serum, or plasma, or in immunising serums prepared from infected human serum. The sharing of infected medical instruments, such as by drug addicts, is a common cause of transmission.

Incubation is long and variable – 15 to 40 days for virus A, 40 to 150 days for virus B.

Susceptibility is general. Immunity following an attack is undetermined, but second attacks are infrequent. No other animal than man is known to be susceptible.

The disease is common and is apparently worldwide in distribution. It is most prevalent in the autumn and winter months.

Mortality is less than 0.5 per cent.

There is no vaccine available at the present time. Gamma globulin provides protection against virus A for several weeks.

Treatment is supportive and includes diet therapy involving vitamins K and B.

Epidemics are most common in institutions and rural areas. The general incidence is about the same in rural and urban areas and is most common in children and young adults. Epidemics occcur in early winter, sometimes reaching a peak in late winter and extending into spring. Epidemics are controlled by the practice of good sanitation and personal hygiene, particularly in the sanitary disposal of respiratory discharges and faeces; by the elimination of flies; and by the use of disease-free blood or blood products in transfusions and in the preparation of vaccine.

The virus withstands a temperature of 74°C for at least 30 minutes. It is destroyed by boiling or autoclaving for 15 minutes. Virus A is unaffected by the normal concentration of chlorine added to drinking water. Virus B survives in the frozen state for several years and resists desiccation at room temperatures or exposure to 0.25 per cent phenol for at least a year. Virus B has been inactivated in serum by exposure to ultra-violet light for 45 minutes.

FUNGI

Coccidioides Immitis

In man and animals, this fungus occurs as thick-walled endospore-filled spherules, 20 to 80 microns in diameter; in artificial culture, it appears as a fluffy white cottony mould.

Coccidioidomycosis is a highly infectious disease. The usual primary form (known as Valley or San Joaquin Fever) is an acute, disabling, self-limiting respiratory infection resembling influenza with usually a low grade fever of 55 to 56°C, and a slight cough. The secondary, progressive form (known as Coccidioidal Granuloma) is a chronic, malignant, disseminated infection which involves any and all organs of the body, including the skin and bones, and produces numerous abcesses. The progressive form has a high fatality rate. A primary localised form of infection may occur on the exposed surfaces of the skin; it sometimes develops into the progressive, disseminated form of infection.

Dust, soil, and vegetation contaminated with spores of this fungus are sources of infection.

Transmission is made by inhalation of spores in dust from soils and dry vegetation and possibly through skin scratches or wounds.

The incubation period for the primary pulmonary form is 10 to 21 days, the average being about 12 days. The clinically active cases in about 1 per cent of Caucasions and a much higher percentage of Negroes develop into the progressive form of the disease, usually within a period of weeks or months after onset of the primary infection. The progressive form is not necessarily preceded by symptoms of primary infection.

Susceptibility to the primary infection is general. With the progressive form of the disease, Negroes and Orientals appear to be much more susceptible than Caucasions, and females are less apt to develop granulomatous lesions than are males. Probably up to 60 per cent of infected persons have no symptoms but become immune.

Recognised endemic areas are the arid south-western United States, northern Mexico, Central America, and the Chaco region of South America, involving Bolivia, Paraguay, and Argentina. Incidence is at its highest in hot, dry weather.

Fatalities are about 50 per cent in the secondary, progressive form.

Vaccine therapy is only in the experimental stage of development, and prospects for a good vaccine seem poor.

Supportive treatment, with complete rest, is the standard procedure. Progressive coccidioidomycosis is very resistant to specific chemotherapy, but some cases have shown improvement under treatment with prodigiosin, stilbamidine, or ethyl vanillate. Localised lesions may be treated by X-ray or surgery. Prognosis is excellent in the primary pulmonary infection, good in the primary cutaneous type, and very poor in the disseminated, progressive disease, especially in the Negro.

The disease is non-contagious. Small epidemics may occur in hot, dry seasons when large numbers of individuals, such as military units, are stationed in endemic areas or engaged in manoeuvres in these areas. Dust control, such as paving roads and runways, oiling sports fields, and planting lawns in endemic areas is effective in reducing the number of infections.

Spores of the fungus are highly resistant to drying and will live for months or years in culture or in the soil (dust). They are destroyed by autoclaving for 15 minutes or by exposure to formaldehyde fumes for 48 hours. Decontamination of large areas, even if possible, would not appear to be practicable

Histoplasma Capsulatum

The fungus appears as small (1 to 5 microns), oval, yeast-like intracellular bodies in the tissues of man and animals, and in cultures on sealed, blood agar slants at 53 to 54°C. In culture at room temperature, it forms a white, cottony, filamentous mould-like growth.

Histoplasmosis is a chronic, local or systemic, infectious disease of man and other animals. It is characterised by low grade granulomatous lesions of the skin or mucous membranes and/or Tuberculosis-like lesions of the lungs and by involvement of internal organs, especially the spleen and liver. Less than 1 per cent of naturally occurring primary infections develop into the progressive and usually fatal form of the disease.

Dust contaminated with spores of this fungus is a source of infection. The agent has been recovered from man, dogs, cats, rodents, skunks, and opossums and from the soil and air.

Transmission is usually by inhalation of spores in dust from soils and dried organic matter; it may also be transmitted by ingestion or through skin scratches.

In the few reported epidemics, symptoms appeared within 5 to 18 days following exposure.

Susceptibility is general. The occurrence and severity of clinical symptoms depend upon the dosage of infectious agent. The extreme infectiousness of the disease is shown by a positive histoplasmin skin test in 80 per cent or more of the adult population in local endemic areas. The effect of dosage is illustrated by essentially 100 per cent epidemic involvement among groups exposed to highly contaminated dust in enclosed places.

Histoplasmosis is endemic in the United States in regions including the central Mississippi and Ohio River valleys and in other scattered areas including the St. Lawrence River valley, eastern North Carolina, and parts of Mexico and Panama. However, cases have been reported also from many other parts of the world, including South America, Europe, Africa, Australia, and the Pacific Islands.

The acute pulmonary form of the disease has a very low fatality rate unless it disseminates. The progressive form is usually fatal.

Immunisation has not been developed.

Supportive treatment is used. Ethyl vanillate has been effective in many cases. Among other promising drugs being developed is hydroxystilbamidine. Surgery is curative in localised lesions.

The disease is non-contagious.

Spores are viable for several months and probably years in dry soil and for a period of months in tap water at temperatures ranging from near freezing to 54°C. They are destroyed by heat at 73°C for 15 minutes and by 1 to 2 per cent formalin in 24 hours.

Nocardia Asteroides

This aerobic fungus has characteristics of both moulds and bacteria and has been classified in an intermediate position. It usually forms orange colonies consisting of gram-positive, weakly acid-fast, branching filaments, 1 micron or less in diameter. In both culture and tissues it tends to fragment readily into bacillary and coccoid forms. The acid-fast rod-like fragments may be mistaken for tubercle bacilli.

Nocardiosis, a severe pulmonary infection, is similar in many respects to Tuberculosis but tends to form numerous abscesses instead of tubercules and is characterised by chronic pneumonia. It tends to spread to other organs of the body, especially the brain, where abscesses are formed. The disease often takes other forms, such as localised subcutaneous abscesses and tumours. Pulmonary infection is characterised by a general malaise, fever, a productive cough, night sweats, loss of appetite, and loss of weight. Brain involvement (tumour or abscess) presents symptoms of headache, nausea, and vomiting.

Soil, dust or vegetation contaminated with the organism are the main sources of infection.

The disease is transmitted by contaminated dust and possibly by droplet infection or through pus and other discharges from infected individuals. Skin infections usually result from contamination of wounds or scratches.

The incubation period in man is unknown; experimental infection in guinea pigs is usually fatal within a week.

Susceptibility to pulmonary infection is probably general and is dependent to a great extent upon dosage. Special predisposing factors may be required for production of progressive infection, since the organism is apparently widely distributed, but relatively few cases are reported. It is not known whether recovery provides any immunity.

The disease is world-wide.

The death rate is very high in untreated or advanced cases of generalised infection, perhaps close to 100 per cent. Early diagnosis and vigorous treatment with sulfonamides result in the cure of most cases.

Active immunisation in man has not been tried; in animals, experimental immunisation has produced an active protective immunity.

The sulfonamides, particularly sulfadiazine alone or combined with sulfamerazine, offer effective treatment. Supportive treatment is essential.

The disease is not known to spread from man to man or from animal to man, but infection by droplet spray or through the infection of wounds by contaminated dressings and other material is possible.

As the fungus exists normally in soil, it is presumed to be quite resistant to adverse environmental conditions.

TOXINS

Botulinum Toxin

This is the protein-like exotoxin formed by the botulinum bacillus. Through repeated purification procedures, it has been obtained in a crystalline form and is the most powerful poison known. The crude material or 'mud' is a brownish, amorphous mass. There are at least five distinct types, A, B, C, D, and E, of which types A, B, and E are known to be toxic to man; C and D are toxic for animals and probably for man.

Botulism is a highly fatal, acute poisoning. It is characterised by vomiting, constipation, thirst, general weakness, headache, fever, dizziness, double vision, dilation of the pupils, paralysis of the muscles involved in swallowing, and difficulty of speech. Respiratory paralysis is the usual cause of death.

Sources of the toxin are bacteria Clostridium botulinum and Cl.parabotulinum, which are rod-shaped, slightly motile, sporulating, gram-positive, anaerobic bacilli. The principal reservoir of the bacteria is the soil. The bacteria grow and form their toxin under anaerobic conditions, usually in improperly canned, non-acid foods such as meats, sausages, and some vegetables, including corn, string beans, spinach, and olives.

Transmission is through eating of food contaminated with botulinum toxin. The bacteria do not grow or reproduce in the human body, and poisoning is due entirely to the toxin already formed in the ingested material. Fresh foods are not involved; freshly well-cooked foods are not involved, as heating destroys the toxin. Possibly the toxin could be introduced through breaks in the skin or by inhalation.

Symptoms of poisoning usually do not appear until between 12 and 72 hours after food containing the pre-formed toxin has been consumed, the length of time depending upon the amount of toxin contained in the consumed food.

All persons are susceptible to poisoning. The few who recover from the disease have an active immunity of uncertain duration and degree.

The disease has world-wide distribution; it is prevalent wherever improperly canned food products are consumed.

Mortality is directly related to the amount of toxin consumed.

Passsive immunisation with anti-toxin appears to be encouraging as a protective measure for humans but is of little therapeutic value. Active immunisation with botulinum toxoid is of proved protective value.

Treatment is mainly supportive. Anti-toxin therapy is of doubtful value, particularly where large doses of the poison have been consumed.

The disease is not contagious. Epidemics occur only where widespread distribution and consumption of a contaminated food product have occurred.

The toxin is stable for a week in non-moving water where it is not aerated. It persists for a long time in food when it is not exposed to air. The toxin is destroyed when boiled for 15 minutes, but botulinum spores resist boiling for 6 hours. Pressure cooking will destroy the spores. Botulinum toxin differs from other bacterial toxins in that it is not destroyed by gastro-intestinal secretions.

Staphylococcus Toxin

This toxin is produced in food by certain strains of staphylococci. It is an entero-toxin, since it has a specific action on the cells of the intestinal mucosa. Unlike most bacterial exotoxins, it is stable at boiling temperature and antigenically is irregular in eliciting the formation of immune bodies.

Food poisoning (not infection) is produced following the ingestion of food in which various strains of staphlococci have been growing. It is usually characterised by sudden, sometimes violent onset, with severe nausea, vomiting, stomach cramps, severe diarrhoea, and prostration. Patients usually feel normal 24 hours after the attack begins.

The source of contamination is not known in most cases but is probably of human origin. Food implicated as sources of food poisoning are chiefly pastries (creamy), milk (raw), milk products, and meat. Food handlers who are nasal or skin carriers of pathogenic staphylococci or who have an open staphylococcal lesion on their hands, arms, or face have been traced as sources of poisoning. The implicated foods are usually allowed to remain at a warm temperature before consumption, thus providing an incubation period for formation of the toxin.

The consumption of contaminated custard-filled pastry, processed meats (particularly ham), and perhaps milk from cows with infected udders are the usual modes of transmission. Improper food handling is responsible for many outbreaks.

Incubation is relatively short. One half-hour to 4 hours, usually 2 to 4 hours, elapse between the taking of food and the appearance of symptoms.

Most persons are susceptible, but individual reactions are variable.

The toxin is world-wide and is probably the principal cause of acute 'food poisoning'.

Fatalities are rare but can occur.

There is no immunisation.

Treatment is supportive.

Food poisoning is non-contagious. Most outbreaks are small and are confined to persons who have eaten the same contaminated food.

The toxin is resistant to freezing, to boiling for 30 minutes, and to potable quantities of chlorine. The organism remains viable after 67 days of refrigeration.

Chemical Warfare Agents

Chemical agents are inorganic substances used in warfare to attack the organs of the human body in such a way that they prevent those organs from functioning normally. The intention is either to prevent the human body from functioning at all or to somehow prevent some aspect or another of its normal processes from proceeding normally. The results are usually disabling to a varying degree, or fatal.

Throughout the history of warfare attempts have been made to somehow use chemical agents as weapons of war. Most attempts were unsuccessful for it was not until the growth of the chemical industry, and dye manufacturing in particular, during the latter half of the 19th Century that the wherewithal to produce really effective military chemical agents came to hand. Until then vague notions of using 'sulphurous vapours' or some other addition to battlefield smoke, were from time to time put forward for consideration but none of these proposals appears to have come to anything like fruition.

By the outbreak of the First World War in 1914 the first military chemical agents were already in the arsenals of the major powers. The French were the first to use chemical agents in the form of tear gas grenades that were hurled at German infantry during the Battle of the Frontiers in August 1914. Their effect was minimal, mainly due to a complete lack of understanding of how to utilise such weapons but their use was duly noted by the Germans and artillery shells filled with lacrymatory chemicals were duly fired at British troops in the Neuve-Chapelle area in October 1914. Once again the results were negligible and were not even reported upon by the British recipients at the time. Mention can be found of the use of similar gas shells against Russian troops near Bolimov in January 1915 but the low temperatures prevalent at the time rendered the agent innocuous and ineffective.

It was not until the 22nd April 1915 that modern chemical warfare as we now know it made its proper debut. The place was the salient in the trench lines near Ypres and the recipients were French and British troops. The agent involved was Chlorine which was released from gas cylinders and allowed to float towards the Allied trenches borne upon the prevailing breeze. The result upon the unsuspecting and unprepared Allied troops was devastating and a new myth of German 'horror' weapons was born.

Once the immediate surprise and cries of disgust had died down the first protective respirators appeared and the age-old military contest of defence versus attack began all over again. As the use of more and more chemical agents became more and more widespread, the tactical situation settled down to a state where the employment of chemical agents became a 'normal' part of Western Front life for both sides. Despite all the discomfort and inconvenience involved, chemical agents ceased to be the killers they had initially been and became merely weapons that reduced the efficiency of fighting men and restricted their movements. Even the arrival of new and highly efficient chemical weapons such as the various mustard agents did little to affect the course of battle once protective capes and other defensive measures had been adopted.

After the Armistice of 1918 had taken effect a tally of the casualty totals for the Western Front revealed that chemical weapons had been responsible for around 91,000 deaths, about 5 per cent of all the fatal casualties inflicted on both sides. The total of chemically-wounded was higher (300,000) but still far less than the casualties inflicted by the more conventional weapons such as artillery and small arms. Only in one theatre did chemical-inflicted casualties form a more significant proportion of total losses. That was in Russia. The hapless Tsarist armies, who were never able to produce or field any effective defensive measure or other counter against chemical weapons, lost well over half a million men to chemical weapons and they sustained some 62 per cent of all casualties caused by chemical weapons during the First World War. Those terrible Russian casualty lists cast their shadow to this day.

After 1918 chemical agents continued to be used virtually wherever warfare was in progress. The Allies used chemical bombs during their short intervention into the Russian Civil War, the French and Spanish used various chemical weapons against dissident tribesmen in their North African colonies and as every schoolboy knows, the Italians used chemical weapons in Ethiopia in 1935. Less well known was the use of chemical weapons by the Japanese during their forays into China during the 1930s.

The Geneva Protocol of June 1925 was a well-intentioned attempt to outlaw chemical weapons from war but like so many other humanitarian gestures it has been effective only in part. It is true that no use was made of chemical warfare during the Second World War but virtually all the major participants produced military chemical agents and weapons and were equipped to conduct chemical warfare should the need arise. Every major theatre had chemical weapon dumps ready to supply chemically-filled shells and bombs if they were ever used by the other side.

Exactly why the Germans did not resort to the use of chemical weapons during the latter stages of the Second World War is still one of the major mysteries surrounding that conflict. The remorseless and determined conduct of all other aspects of that conflict involving the German nation was such that the introduction of chemical warfare would have been entirely within the bounds of understanding, but no such resort was ever made. When it is considered that the Germans had at their disposal the first nerve agents the mystery becomes even deeper. The military rewards of the surprise introduction of such agents at some points of the conflict could have been considerable. The first days of the Normandy landings in June 1944 and the last desperate holding battles before Berlin in May 1945 were typical examples of occasions where the use of nerve agents could have been responsible for turning the tide against the Allies at critical times. Whether the final outcome of the war would have been any different is another matter, but the use of chemical weapons, and the nerve agents in particular, could have prolonged the conflict considerably and weakened Allied resolve.

The years since 1945 have seen relatively little use of chemical weapons until recently. Chemical agents were used as defoliants to remove jungle growth and prevent their use as cover for guerilla activities in Malaya and Vietnam but there were few reports of anything more lethal. That is until reports began to appear of the use of unspecified chemical agents in the Far East and Afghanistan. These agents were used to kill the occupants of remote areas and villages and prevent their providing aid or harbour to insurgents who resented the presence of authorities allied to the Soviet sphere of influence. Terms such as 'Yellow Rain' became commonplace to the Western media but no really definite confirmation of the use of such agents was forthcoming.

It was left to the Iran-Iraq conflict to spawn yet another round of the use of chemical weapons in war. Definite evidence of the employment of nerve agents and mustard agents has been forthcoming from the Persian Gulf on many occasions. Chemical warfare is with us still.

Modern chemical weapons can be delivered by several means. The original delivery system, the gas cylinder, is still a possible delivery system but overlooks the fact that many modern chemical agents are not gases but liquids or solids in some form or another that give off vapour at most ambient temperatures, or are dispensed in aerosol or droplet form. Therefore chemical agents may now be delivered from aircraft underwing tanks or containers, aircraft-delivered bombs, from artillery shells or by rockets, both guided and unguided. In fact most modern delivery methods are basic and differ little from delivery systems used for other less lethal payloads such as screening smoke.

However chemical weapons require careful handling and storage if they are not to be a hazard to their users as well as to an enemy. Many military chemical agents are corrosive and are thus inclined to leak from unprepared containers. It would appear that even moderate handling damage that would pass unnoticed with orthodox munitions can lead to leaks as well. The term 'leak' is relative. Even a pin-prick that will allow minute contents of a chemical agent container to escape constitutes a hazard to handlers and storemen. Chemical weapons such as shells and bombs therefore require special handling containers, storage and handling methods.

This handling and storage hazard was one reason for the development of the so-called binary chemical munitions. These binary weapons contain the relatively innocuous ingredients of nerve agents (usually VX) held in separate compartments during handling and storage. The ingredients are automatically mixed only after the binary weapon has been released from its delivery system. Thus the nerve agent is formed only after a chemical shell has been fired or a chemical bomb has been dropped. The United States has been in the forefront of such developments but it would be surprising if the Soviet Union was not far behind. The display of Soviet chemical weapons held at the Soviet Shikhany 'facility' during October 1987 revealed only what appeared to be a range of munitions dating from the 1950s or thereabouts – more advanced Soviet chemical weapons than the rather basic examples displayed at Shikhany can be anticipated.

Chemical warfare is no longer a method of inflicting casualties upon an enemy. The degree to which nearly all military forces are now able to call upon in-the-field defensive measures ranging from protective clothing to hard shelters means that a chemical attack will have few opportunities of inflicting significant numbers of casualties once the initial surprise has been lost. Instead the introduction of chemical warfare can act as a disabling measure that reduces the fighting efficiency of an enemy.

Even the most efficient form of protective clothing is uncomfortable to wear after a short period and the build-up of body heat inside the clothing reduces the body's ability to function properly even under normal conditions – under battle conditions that reduction in normal body functioning is accelerated. Vision is another factor that is soon degraded. Every respirator produced to date reduces normal vision to some degree or other and some more than others. Added to these combat efficiency degradation factors are difficulties in communication, the inability to recognise comrades, officers and NCOs, difficulty in serving and aiming weapons, and difficulty in even basic self-maintenance operations such as eating, drinking and body waste disposal. It is not long before even the best-trained and motivated soldier becomes a helpless and useless burden to his military organisation.

Then there is the added logistical burden of decontamination after a chemical attack. If uninvolved personnel are not to be affected by chemical agents that have contaminated vehicles, weapons and equipment they have to be cleansed and rendered safe to handle. This takes time, manpower, effort and materials. Decontamination can absorb time and manpower to such an extent that even the most prepared of military organisations have to lose some measure of efficiency while the cleaning-up process proceeds. Thus even a small and relatively harmless chemical agent attack can produce results out of all proportion to the efforts involved from the attacker.

Then there is the question of persistency. Some chemical agents are harmful for short periods of time while others remain potentially harmful for much longer. Some tactical scenarios require agents that inflict casualties and require defensive measures to be taken by the defender and yet persist for only a short period – 15 minutes is usually taken as a short period. Other tactical situations might well require agents to remain harmful for days or longer. Ground contaminated by persistent agents (such as the various mustard agents) can remain potentially lethal or harmful to any troops attempting to cross it for days, so such tracts of ground can remain 'no-go' areas that remain closed to manoeuvring forces for prolonged periods. Thus an attacking force could screen its flanks with persistent agents and use the bulk of their own forces at the main point of attack. In contrast a defender could cover large areas with persistent agents to hamper an advance.

There is no perfect military chemical agent. Each agent outlined in this section has advantages and disadvantages for the military user. Commanders have to trade off these advantages and disadvantages to suit their particular tactical situation.

The following list of military chemical agents should not be taken as complete. It contains most of the known military chemical agents but new horrors are constantly being devised in laboratories all around the world and are kept under security wraps until the time comes when they might be used. Some of the agents in the list are unlikely to be used for military purposes but the possibility that they might be resorted to is always present.

CHOKING AGENTS

Choking agents injure unprotected man chiefly in the respiratory tract – ie, in the nose, throat, and particularly the lungs. In extreme cases, membranes swell, lungs become filled with liquid, and death results from lack of oxygen; thus these agents 'choke' an unprotected man. Fatalities of this type have been referred to as 'dry-land drownings'.

The choking agent originally employed as a war gas, Chlorine gas, is now rarely (if ever) used as a military chemical agent.

Phosgene (CG)

CG, normally a chemical agent with a short duration of effectiveness, was used extensively in the First World War. In fact, more than 80 per cent of the First World War chemical agent fatalities were caused by CG.

Chemical name. Carbonyl chloride.

Not readily hydrolised under usual field conditions; however, rain destroys its effectiveness, and heavy vegetation, jungle, and forests cause considerable loss of CG by hydrolysis on leaf surfaces.

Odour. New mown hay or grass; green corn.

Rate of action. Delayed. Although immediate symptoms may follow exposure to a high concentration of CG, a delay of 3 hours or more may elapse before exposure to a low concentration causes any ill effects.

No skin or eye toxicity.

CG exerts its effect solely on the lungs, and results in damage to the capillaries. It causes seepage of watery fluid into the air sacs. When a lethal amount of CG is received, the air sacs become so flooded that air is excluded and the victim dies of anoxia (oxygen deficiency). If the amount of CG is less than lethal, and proper care is provided, the watery fluid is reabsorbed, the air cell walls heal, and the patient recovers. However respiratory problems of one form or another may remain chronic for years. The severity of poisoning cannot be estimated from the immediate symptoms, since the full effect is not usually apparent until 3 or 4 hours after exposure. Most deaths occur within 24 hours.

Duration of effectiveness is short. However, vapour may persist for some time in low places under calm or light winds and stable atmospheric conditions (inversion).

Delayed action casualty agent.

Diphosgene

As a shell filling, DP has the advantage of a high boiling point which permits filling in the field, whereas CG, with a low boiling point, must be kept refrigerated during filling operations under summer conditions. However, DP does have certain disadvantages. Since it is slightly lacrymatory, troops are not as easily surprised as with CG. Furthermore. its lower volatility adds to the difficulty of setting up an effective surprise concentration. (DP is converted to CG in the body and exerts its effect after this conversion.)

Chemical name. Trichloromethyl chloroformate.

Rate of hydrolysis. Slow at ordinary temperatures.

Odour. New mown hay or grass; green corn.

Skin and eye toxicity. No effect on skin; slight lacrymatory effect.

Delayed rate of action. Although symptoms may follow exposure to a high concentration of DP, a delay of 3 hours or more may elapse before exposure to a low concentration causes any ill effects.

Since DP is converted to CG in the body, the physiological action is the same for both agents. The casualty effect is exerted solely on the lungs and results in damage to capillaries.

Delayed or immediate-action casualty agent, depending upon dosage rate.

Chlorine

Chemical name. Chlorine.

Rate of hydrolysis. Slow.

Odour. Pungent and unmistakable, like bleaching powder.

Skin and eye toxicity. Irritates eyes.

Powerful irritant, first on upper and then on lower respiratory tract.

Duration of effectiveness. Short.

Chlorine is mentioned here for historical purposes as it is now little used as a military chemical agent and is rarely encountered in a military context.

NERVE AGENTS

While the nerve agents differ in molecular structure, they have the same physiological action on man in that they upset the balance between the sympathetic (adrenergic) and parasympathetic (cholinergic) nervous systems which together are the autonomic nervous system. Normally in the body there is controlled discharge within the nervous system due to destruction by cholinesterase of acetylcholine, a product of nerve cell metabolism. The nerve agents were found by German scientists to react with the cholinesterase in an irreversible reaction in tissue fluid, permitting accummulation of acetylcholine and continual stimulation of the nervous system. Rapid use of so-called autonomic blocking agents which act directly on the effector nerve cell will nullify the effect of acetylcholine. No apparent chemical reaction seems to occur between these autonomic blocking agents and acetylcholine. Atropine salts are the most commonly used autonomic blocking agents. Because the major action of these chemical agents is on the parasympathetic nerve system, they are often popularly called nerve agents.

Tabun (GA)

First discovered in 1936, Tabun is a colourless to brownish liquid giving a colourless vapour.

Chemical name. Dimethylaminoeth-oxy-cyanophosphinne oxide.

Reacts slowly with water but fairly rapidly with strong acids or alkalies.

Odour. Faintly fruity; none when pure.

Skin and eye toxicity; (a)Eye effect. Very high toxicity; much greater through eye than through skin. Vapour causes pupil of eye to contract, resulting in difficulty in seeing in dim light, or generally distorted vision.

(b)Skin effect. Liquid decontamination of the smallest drop is essential. Vapour penetrates skin readily.

Rate of action. Very rapid.

Individuals poisoned by GA display approximately the same sequence of symptoms regardless of the route by which the agent enters the body (whether by inhalation, absorption or ingestion). These symptoms, in normal order of appearance, are: running nose; tightness of chest; dimness of vision and pinpointing of the eye pupils; difficulty in breathing; drooling and excessive sweating; nausea; vomiting; cramps. and involuntary defaecation and urination; twitching, jerking and staggering; and headache, confusion, drowsiness, coma and convulsion. These symptoms are followed by cessation of breathing and death. Symptoms appear much more slowly from skin dosage than from respiratory dosage. Although skin absorption great enough to cause death may occur in 1 to 2 minutes, death may be delayed for 1 to 2 hours. Respiratory lethal dosages kill in 1 to 10 minutes, and liquid in the eye kills nearly as rapidly. The number and severity of symptoms which appear are dependent on the quantity and rate of entry of the nerve agent which is introduced into the body. (Very small skin dosages sometimes cause local sweating and tremors with little other effect.)

Duration of effectiveness depends upon munitions used and the weather. Heavily splashed liquid persists 1 to 2 days under average weather conditions.

Sarin (GB)

A colourless liquid; its vapour is also colourless.

Chemical name. Methylisopropoxyfluoro-phosphine oxide.

Very rapidly hydrolysed in alkaline solutions.

Almost no odour in pure state.

Skin and eye toxicity. (a)Eye effect. Very high toxicity; much greater through eye than through skin. Vapour causes pupil of eye to contract; vision difficult in dim light.

(b)Skin effect. Liquid does not injure skin but penetrates it rapidly. Immediate decontamination of the smallest drop is essential. Vapour also penetrates the skin.

Rate of action. Very rapid - death usually occurs within 15 minutes after fatal dosage absorbed.

Individuals poisoned by GB display approximately the same sequence of symptoms regardless of the route by which the poison enters the body (whether by inhalation, absorption or ingestion). These symptoms, in normal order of appearance, are: running nose; tightness of chest; dimness of vision and pinpointing of the eye pupils; difficulty in breathing; drooling and excessive sweating; nausea, vomiting, cramps and involuntary defaecation and urination; twitching, jerking and staggering; and headache, confusion, drowsiness, coma and convulsion. These symptoms are followed by cessation of breathing and death. Symptoms appear much more slowly from skin dosage than from respiratory dosages. Although skin absorption great enough to cause death may occur in 1 to 2 minutes, death may be delayed for 1 to 2 hours. Respiratory lethal dosages kill in 1 to 10 minutes, and liquid in the eye kills nearly as rapidly. The number and severity of symptoms which appear are dependent on the quantity and rate of entry of the nerve agent which is introduced into the body. (Very small skin dosages sometimes cause local sweating and tremors with little other effect.)

Evaporates at approximately the same rate as water. Depends upon the munitions used and the weather.

Soman (GD)

GD is a colourless liquid which gives off a colourless vapour.

Chemical name. Methylpinacolyloxy-fluorophosphine oxide.

Fruity odour; with impurities, odour of camphor.

Skin and eye toxicity.

(a)Eye effect. Very high toxicity; vapour causes pupil of eye to contract. resulting in difficulty in seeing in dim light and generally distorted vision. Toxicity much greater through eye than through skin.

(b)Skin effect. Extremely toxic by skin absorption. Liquid does not injure the skin but penetrates it rapidly. Immediate decontamination of the smallest drop is essential.

Very rapid rate of action. Death usually occurs within 15 minutes after fatal dosage is absorbed.

Individuals poisoned by GD display approximately the same sequence of symptoms regardless of the route by which the poison enters the body (whether by inhalation, absorption or ingestion). These symptoms, in normal order of appearance, are: running nose; tightness of chest; dimness of vision and pinpointing of the eye pupils; difficulty in breathing; drooling and excessive sweating; nausea, vomiting, cramps and involuntary defaecation and urination; twitching, jerking and staggering; and headache, confusion, drowsiness, coma and convulsion. These symptoms are followed by cessation of breathing and death.Symptoms appear much more slowly from skin dosage than from respiratory dosages. Although skin absorption great enough to cause death may occur in 1 to 2 minutes, death may be delayed for 1 to 2 hours. Respiratory lethal dosages kill in 1 to 10 minutes, and liquid in the eye kills nearly as rapidly. The number and severity of symptoms which appear are dependent on the quantity and rate of entry of the nerve agent which is introduced into the body. Very small skin dosages sometimes cause local sweating and tremors with little other effect.

Duration of effectiveness depends upon the munitions used and the weather. Heavily splashed liquid persists 1 to 2 days under average weather conditions.

V-Agents

Included in the category of nerve agents are the V-agents – the standard V-agent is VX. V-agents are generally colourless and odourless liquids which do not evaporate rapidly or freeze at normal temperatures. Because of their low volatility, their vapour effect is limited and the duration of their effectiveness is increased. V-agents are absorbed by vegetation.

In liquid or aerosol form, V-agents affect the body in a manner similar to that of the G-agents. V-agents are usually disseminated as liquid droplets which produce casualties when absorbed through the skin. Since liquid G-agents evaporate quickly from the skin, the dosage required to produce casualties by that route is high and the time to appearance of casualties is correspondingly short as compared with the much less volatile V-agents. If evaporation is excluded, the time to appearance of casualties would be roughly similar at the same dose level with both V-agents and G-agents. Aerosol V-agents produce casualties by inhalation and by absorption of agent droplets through the skin.

The chemical name for VX is ethyl S-2 diisopropylaminoethyl-methylphosphonothioate.

BLOOD AGENTS

Blood agents are absorbed into the body primarily by breathing. They affect bodily functions through action on the enzyme cytochrome-oxidase, thus preventing the normal transfer of oxygen from the blood to body tissue.

Hydrogen Cyanide (AC, also known as HCN)

Chemical name. Hydrogen cyanide or hydrocyanic acid.

Rate of hydrolysis. Low under field conditions.

Odour. Similar to peach kernels.

Skin and eye toxicity. Moderate.

Very rapid rate of action. Death occurs within 15 minutes after a lethal dosage has been received.

AC interferes with utilisation of oxygen by the body tissues by inhibition of the enzyme cytochrome-oxidase. AC causes a marked stimulation of the breathing rate. A variation of AC in powdered form, known as Zyklon B, was used in the German extermination camps prior to 1945.

Duration of effectiveness. Short; the agent is highly volatile, and in the gaseous state it is lighter than air.

AC can be used to generally clog respirator and other filters.

Cyanogen Chloride (CK)

Chemical name. Cyanogen chloride.

Rate of hydrolysis. Very low. Odour. Its irritating and lacrymatory properties are so great that the odour can go unnoticed.

Skin and eye toxicity. Too low to be of military importance; highly irritating to eyes and mucous membranes.

Rate of action. It is assumed that the effect of CK arises from its conversion to AC in the body. In general, CK may be considered a rapid-acting chemical agent.

The general action of CK is similar to that of AC – it interferes with utilisation of oxygen by the body tissues. However, it differs from AC in that it has a choking effect, a strong irritating effect, and causes a slow breathing rate.

Duration of effectiveness. Short. Vapour may persist in jungle and forest for some time under suitable weather conditions.

CK can be used for filter element penetration.

Arsine (SA)

Chemical name. Arsenic trihydride, arsine.

Rate of hydrolysis. Rapid, but an equilibrium condition is reached quickly. (Under certain conditions, SA forms a solid product with water which decomposes at 30°C).

Odour. Mild garlic-like odour.

No skin or eye toxicity.

SA interferes with functioning of the blood, and damages the liver and kidneys. Slight exposure causes headache and uneasiness. Increased exposure causes chills, nausea and vomiting. Severe exposure damages blood, causing anaemia.

Duration of effectiveness. Short.

BLISTER AGENTS

The blister agents are used for casualty effect; the use of ground may be restricted, movements slowed, and use of material or installation hampered. These agents affect the eyes and lungs and blister the skin. During the First World War, Mustard (H) was the only blister agent in major use. It was recognised by a distinctive odour and had a fairly high duration of effectiveness under normal conditions. Since then, blister agents have been developed which are odourless and vary in duration of effectiveness. Most blister agents are insidious in action; there is little or no pain at the time of exposure except with Lewisite (L) and phosgene oxime (CX) which causes immediate pain on contact. The development of casualties is somewhat delayed. CX produces a wheal (similar to a bee sting) rather than a water blister which the other blister agents produce. Protection from blister agents is extremely difficult since they attack any part of the body which comes in contact with the liquid or vapour.

Arsenicals

The arsenicals are a group of related compounds in which Arsenic is the central atom. The arsenical chemical agents discussed under blister agents include Lewisite (L), Mustard-Lewisite mixture (HL), phenyldichloroarsine (PD), ethyldichloroarsine (ED), and methyldichloroarsine (MD). Disadvantages of these agents are that they hydrolise rapidly and are not as toxic as other blister agents.

Levinstein Mustard (H)

This is Mustard made by the Levinstein process. It contains about 30 per cent sulphur impurities, which give it a pronounced odour. The properties of H are essentially the same as those listed for HD below.

Distilled Mustard (HD)

HD is H which has been purified by washing and vacuum distillation. HD however, has less odour and a slightly greater blistering power (negligible in the field) than H, and is more stable in storage.

Chemical name. 2,2'-dichloro-diethyl sulfide.

Rate of hydrolysis. Very slow at ordinary temperature.

Odour. Garlic-like.

Skin and eye toxicity. Eyes are very susceptible to low concentrations; higher concentrations are required to produce incapacitating effects by skin absorption rather than by eye injury.

Delayed rate of action - usually 4 to 6 hours until first symptoms appear. Latent periods have been observed, however, up to 24 hours and, in extreme cases, up to 12 days.

Mustard acts first as a cell irritant and finally as a cell poison on all tissue surfaces contacted. The first symptoms of HD poisoning usually appear in from 4 to 6 hours; the higher the concentration, the shorter the interval of time between the exposure to the agent and the first symptoms. The physiological action of HD may be classified as local and general. The local action results in conjunctivitis or inflammation of the eyes; erythema (redness of the skin) which may be followed by blistering or ulceration; and inflammation of the nose, throat, trachea, bronchi, and lung tissue. Susceptibility also varies with individuals. Injuries produced by HD heal much more slowly and are more liable to infection than burns of similar intensity produced by physical means or by other chemicals. This is due to the action of HD in making the blood vessels incapable of carrying out their functions of repair, and by the fact that necrotic (dead or dying) tissue acts as a good medium for bacterial growth.

Duration of effectiveness. Depends upon the munitions used and the weather. Heavily splashed liquid persists 1 to 2 days under average weather conditions, and a week or more under very cold conditions.

Nitrogen Mustards

The Nitrogen Mustards are a group of related compounds which may be considered as derivatives of ammonia because the hydrogen atoms are replaced by various organic radicals. In each of these chemical agents, nitrogen is the central atom. Three members of the group are described individually below.

Nitrogen Mustard (HN-1)

Chemical name. 2,2'dichloro-triethylamine.

Rate of hydrolysis. Quite slow.

Odour. Faintly fishy or musty.

Skin and eye toxicity. Eyes are very susceptible to low concentration; higher concentrations are required to produce incapacitating effects by skin absorption rather than by eye injury.

Delayed rate of action – 12 hours or longer.

Irritates the eyes in dosages which do not significantly damage the skin or respiratory tract, insofar as single exposures are concerned. This irritation appears in a shorter time than that from HD. After mild vapour exposure, there may be no skin lesions. After severe vapour exposures, or after exposure to liquid HN-1, erythema may appear earlier than in HD contamination. There may be irritation and itching as with HD. Later, blisters may appear in the erythematous areas. The skin lesions are similar to those caused by HD. Effects on the respiratory tract include irritation of the nose and throat, hoarseness progressing to loss of voice, and a persistent cough. Fever, laboured respiration, and moist rales (an abnormal crackling sound from the lungs caused by the accumulation of fluids) may develop. Broncho-pneumonia may appear after the first 24 hours. Following ingestion or systemic absorption, the HN-1's cause inhibition of cell mitosis resulting in depression of the blood-forming mechanism and injury to other tissues. Severe diarrhoea, which may be haemorrhagic, occurs. Lesions are most marked in the small intestine and consist of degenerative changes and necrosis in the mucous membranes. Ingestion of 2 to 6 milligrammes causes nausea and vomiting.

Duration of effectiveness. Depends on the munitions used and the weather. Somewhat shorter than HD where heavily splashed liquid persists 1 to 2 days under average weather conditions, and a week or more under very cold conditions.

Nitrogen Mustard (HN-2)

HN-2 is highly unstable and is no longer seriously considered as a chemical agent. It is rated as somewhat more toxic than HN-1.

Rate of hydrolysis. In winter, hydrolysis is fairly rapid until 50 per cent complete. Alkalies induce hydrolysis.

Skin and eye toxicity. HN-2 has the greatest blistering power of the nitrogen mustards in vapour form but is intermediate as a liquid blistering agent. Toxic eye effects are produced more rapidly than by HD.

Rate of action. Skin effects delayed 12 hours or longer.

Irritates the eyes in dosages which do not significantly damage the skin or respiratory tract, insofar as single exposures are concerned. This irritation appears in a shorter time than that from HD. After a mild vapour exposure, or after exposure to liquid HN-2, erythema may appear earlier than in HD contamination. There may be irritation and itching as with HD. Later, blisters may appear in the erythematous areas. The skin lesions are similar to those caused by HD. Effects on the respiratory tract include irritation of the nose and throat, hoarseness progressing to loss of voice, and a persistent cough. Fever, laboured respiration and moist rales may develop. Broncho-pneumonia may appear after the first 24 hours. Following ingestion or systemic absorption, the HN-2's cause inhibition of cell mitosis resulting in depression of the blood-forming mechanism and injury to other tissues. Severe diarrhoea, which may be haemorrhagic, occurs. Lesions are most marked in the small intestine and consist of degenerative changes and necrosis in the mucous membranes. Ingestion of 2 to 6 milligrammes causes nausea and vomiting.

Duration of effectiveness. Depends upon the munitions used and the weather. Somewhat shorter than HD whose heavily splashed liquid persists 1 to 2 days under average weather conditions, and a week or more under very cold conditions.

Nitrogen Mustard (HN-3)

HN-3 is the most stable in storage of the three Nitrogen Mustards and would seem to be admirably suited for use in artillery shells.

Chemical name. 2,2',2—trichloro-triethylamine.

Rate of hydrolysis. Very slow.

Skin and eye toxicity. Eyes are very susceptible to low concentrations; higher concentrations are required to produce incapacitating effects by skin absorption.

Rate of action. Most symptoms delayed 4 to 6 hours as after exposure to HD but in some cases lacrimation, eye irritation, and photophobia develop immediately.

HN-3 irritates the eyes in dosages which do not significantly damage the skin or respiratory tract, insofar as single exposures are concerned. This irritation appears in a shorter time than that from Mustard. After mild vapour exposure there may be no skin lesions. After severe vapour exposure, or after exposure to liquid HN-3, erythema may appear earlier than in HD contamination. There may be irritation, and itching as with HD. Later, blisters may appear in the erythematous areas. The skin lesions are similar to those caused by HD. Effects on the respiratory tract include irritation of the nose and throat, hoarseness progressing to loss of voice, and a persistent cough. Fever, laboured respiration, and moist rales may develop. Broncho-pneumonia may appear after the first 24 hours. Following ingestion or systemic absorption, the HN-3's cause inhibition of cell mitosis resulting in depression of the blood-forming mechanism and injury to other tissues. Severe diarrhoea, which may be haemorrhagic, occurs. Lesions are most marked in the small intestine and consist of degenerative changes and necrosis in the mucous membranes. Ingestion of 2 to 6 milligrammes causes nausea and vomiting.

Duration of effectiveness. Considerably longer than HD (see HD).

Mustard-T Mixture (HT)

HT is a mixture of 60 per cent HD and 40 per cent T. T, a sulphur and chlorine compound similar in structure to HD, is a clear yellowish liquid with an odour similar to HD. HT has a strong blistering effect, has a longer duration of effectiveness, is more stable, and has a lower freezing point than HD. Its low volatility makes effective vapour concentrations in the field difficult to obtain. Properties are essentially the same as those of HD.

HT causes blisters, irritates the eyes, and is toxic when inhaled.

Duration of effectiveness. Depends on munitions used and the weather. Somewhat longer than HD whose heavily splashed liquid persists 1 to 2 days under average weather conditions, and a week or more under very cold conditions.

Phosgene Oxime (CX). Dichloroformoxime.

CX may appear as a colourless, low-melting point (crystalline) solid or as a liquid. It has a high vapour pressure, slowly decomposes at normal temperatures (depending on the temperature and humidity), boils at 53 to 54°C, melts at 39 to 40°C, and is readily soluble in water. It has a disagreeable penetrating odour.

CX is a powerful irritant which produces immediate pain varying from a mild prickling sensation to a feeling resembling a severe bee-sting. It causes violent irritation to the mucous membrane of the eyes and nose. When it comes in contact with the skin, the area becomes blanched in 30 seconds and is surrounded by a red ring. A wheal forms in about 30 minutes and the blanched area turns brown in about 24 hours, with a scab forming in about a week. The scab generally falls off in about 3 weeks. Itching may be present throughout healing, which in some cases may be delayed beyond 2 months.

Lewisite (L)

Chemical name. Dichloro (2-chlorovinyl) arsine.

Rapidly hydrolysed in liquid or vapour state.

Odour. Geranium-like; very little odour when pure.

Skin and eye toxicity. L has approximately the same blistering action to the skin as HD, even though the lethal dosage for L is much higher.

Rapid rate of action.

L produces effects similar to HD but, in addition, acts as a systemic poison, causing pulmonary oedema, diarrhoea, restlessness, weakness, subnormal temperature and low blood pressure. In order of appearance and severity of symptoms it is: a blister agent; a toxic lung irritant; and, when absorbed in the tissues, a systemic poison. Liquid L causes an immediate searing sensation in the eye and permanent loss of sight if not decontaminated within one minute. L produces an immediate and strong stinging sensation to the skin; reddening of the skin starts within 30 minutes. Blistering does not appear until after about 13 hours. Like HD, it is a cell poison. Skin burns are much deeper than with HD. When inhaled in high concentrations it may be fatal in as short a time as 10 minutes.

Duration of effectiveness. Somewhat shorter than HD. Very short duration under humid conditions.

Mustard-Lewisite Mixture (HL)

Mustard-lewisite mixture is a variable mixture of HD and L which provides a low-freezing mixture for use in cold weather operations or as a high-altitude spray. Properties are listed for the eutectic mixture (the mixture having the lowest possible freezing point) which is 63 per cent L and 37 per cent HD by weight. Other mixtures, such as 50:50, may be prepared to meet pre-determined weather conditions and have advantages over the eutectic mixture because of the increased HD content. Mixtures of H and L are not satisfactory due to the poor storage characteristics.

L is rapidly hydrolysed in the liquid or vapour state; HD hydrolyses slowly at ordinary temperatures.

Garlic-like odour.

Skin and eye toxicity. Very high.

Rate of action. Produces immediate stinging of skin and redness within 30 minutes; blistering delayed about 13 hours.

HL liquid causes severe damage to the eyes. Contamination of the skin is followed after a short time by reddening, then by blistering which tends to cover the entire area of the reddened skin. The respiratory lesions are similar to those produced by Mustard, except that in the most severe cases pulmonary oedema may be accompanied by pleural effusion. Liquid on the skin, as well as inhaled vapour, is absorbed and may cause systemic poisoning. This change is manifested in capillary permeability which permits the loss of sufficient fluid from the blood stream to cause blood thickening, shock and death.

Duration of effectiveness. Depends on the munitions used and the weather. Somewhat shorter than HD where heavily splashed liquid persists 1 to 2 days under average weather conditions and a week or more under very cold conditions.

Phenyldichloroarsine (PD)

Although phenyldichloroarsine is classed here as a blister agent, it also acts as a vomiting agent.

Rate of hydrolysis. Rapid.

Odour. None.

Skin and eye toxicity. About 30 per cent as toxic as HD to the eyes. On bare skin PD is about 90 per cent as blistering as HD but it is decomposed immediately by wet clothing.

Rate of action. Immediate effect on eyes; effects on skin delayed from 30 minutes to 1 hour.

Similar to the vomiting agents but with added blistering action. Its limited use during the First World War did not indicate any marked superiority of PD over the other vomiting agents used.

Duration of effectiveness. Depends on the munitions used and the weather. Somewhat shorter than HD under dry conditions; short duration when wet.

Ethyldichloroarsine (ED)

Ethyldichloroarsine was introduced by the Germans in March 1918 in an effort to produce a volatile agent with a short duration of effectiveness that would be quicker acting than DP or HD and would be more lasting in its effects than PD.

Rate of hydrolysis. Rapid.

Odour. Fruity, but biting and irritating.

Skin and eye toxicity. Vapour is irritating but not harmful to eyes and skin except on prolonged exposure. Liquid ED has approximately one-twentieth the blistering action of liquid L.

As with other chemical agents containing arsenic, ED is irritating to the respiratory tract and will produce lung injury upon sufficient exposure. The vapour is irritating to the eyes and the liquid may produce severe eye injury. The absorption of either vapour or liquid through the skin in sufficient amounts may lead to systemic poisoning or death. Blistering of the skin is produced by prolonged contact with either liquid or vapour.

Duration of effectiveness. Evaporates approximately as rapidly as water. Short duration under wet conditions.

Methyldichloroarsine (MD)

Methyldichloroarsine is similar to ethyldichloroarsine.

Rate of hydrolysis. Very rapid.

Odour. None.

Skin and eye toxicity. Blistering action slightly less than that of HD. Has effect on eyes similar to that of L (produces corneal damage) but less severe. Concentration required for blistering effect is too high to attain in the field.

Rate of action. Immediate irritation of eyes and nose. Blistering effect is delayed several hours.

As with L and the other similar arsenicals, MD is irritating to the respiratory tract and produces lung injury upon sufficient exposure. The vapour is irritating to the eyes and the liquid may produce severe eye injury. The absorption of either vapour or liquid through the skin in sufficient amounts may lead to systemic poisoning or death. Blistering of the skin is produced by prolonged contact with either liquid or vapour.

Duration of effectiveness. Relatively short. Evaporates much faster than water.

Simulated Mustard (MR)

After the First World War a number of substances having physical properties similar to HD, and generally containing a dye and an odouriferous constituent, were tested in an attempt to find a substitute for mustard in the testing of dispersion apparatus and munitions, and for training purposes. Almost all of the substances had some disadvantageous feature, such as instability, corrosiveness to metals, staining to fabrics, relatively high cost or, in the case of those containing aniline and nitrobenzene, some toxicity. The use of molasses residuum solution (MR) as a simulant for HD was first suggested in

March 1937. Since that time MR has been used successfully in tests of aircraft smoke tanks, thin-case bombs, and chemical land mines. MR is well adapted to training needs.

MR is a mixture consisting of a 25 per cent solution by volume of molasses residuum in water. Molasses residuum is obtained in the manufacture of ethyl alcohol from molasses. As obtained from industry, the concentrated molasses residuum is a dark brown viscous liquid with a characteristic molasses odour. It is soluble in water to the extent of 90 per cent. The undiluted material is more viscous than HD, but when one volume of molasses residuum is diluted with three volumes of water it forms a dark brown liquid of thin, syrupy consistency which, although it has a lower specific gravity than either pure or crude mustard, has a viscosity and surface tension sufficiently close to those of HD to ensure comparable flow characteristics.

MR has a low freezing point and can be used in moderately cold weather without danger of freezing.

The solution has a distinctive molasses odour. When the solution is sprayed from an aircraft the molasses odour can be detected on the contaminated area for about 45 minutes to one hour. The solution has sufficient colour to produce easily defined patterns. The patterns obtained by dispersion from aircraft smoke tanks, chemical land mines, and thin-case bombs are similar to those produced by HD.

The solution is harmless by contact to man and animals but, because of the cresol used as a stabilising agent, should not be swallowed.

VOMITING AGENTS

The three vomiting agents listed below are normally solids which, when heated, vapourise and then condense to form toxic aerosols. Under field conditions, vomiting agents cause great discomfort to their victims; when released indoors, they may cause serious illness or death. The vomiting agents are primarily used for mob and riot control. The three principal vomiting agents are discussed below.

Diphenylchloroarsine (DA)

Rate of hydrolysis. Slow in mass but rapid when finely divided.

No pronounced odour.

Skin and eye toxicity. Irritating; not toxic.

Rate of action. Very rapid; within 2 or 3 minutes after 1 minute of exposure.

In progressive order; irritation of the eyes and mucous membranes, viscous discharge from the nose similar to that caused by a cold, sneezing and coughing, severe headache, acute pain and tightness in the chest, and nausea and vomiting. For moderate concentrations the effects last about 30 minutes after an individual leaves the contaminated atmosphere. At higher concentrations, the effects may last up to several hours.

Duration of effectiveness. Short because agent is disseminated as an aerosol.

Adamsite (DM)

Chemical name. Diphenylaminochloroarsine (also phenarsazine chloride).

Rate of hydrolysis. Quite rapid when in aerosol form.

No pronounced odour.

Skin and eye toxicity. Irritating; relatively non-toxic.

Rate of action. Very high. Only about 1 minute is required for temporary incapacitation.

In progressive order; irritation of the eyes and mucous membranes, viscous discharge from the nose similar to that caused by a cold, sneezing and coughing, severe headache, acute pain and tightness in the chest, and nausea and vomiting. The effects develop more slowly than with DA and, for moderate concentrations, last about 30 minutes after an individual leaves the contaminated atmosphere. At higher concentrations, the effects may last up to 3 hours.

Duration of effectiveness. Short, because agent is disseminated as an aerosol.

Diphenylcyanoarsine (DC)

Rate of hydrolysis. Very slow.

Odour. Similar to a mixture of garlic and bitter almonds.

Skin and eye toxicity. Irritating; not toxic.

Rate of action. Very rapid. Higher concentrations are intolerable in about 30 seconds.

In progressive order; irritation of the eyes and mucous membranes. viscous discharge from the nose similar to that caused by a cold, sneezing and coughing, severe headache, acute pain and tightness in the chest, and nausea and vomiting. DC is more toxic than DA. For moderate concentrations the effects last about 30 minutes after an individual leaves the contaminated atmosphere. At a higher concentration, the effects may last up to several hours.

Duration of effectiveness. Short, because agent is disseminated as an aerosol.

TEAR AGENTS

The tear agents cause flow of tears and irritation of the skin. Since there is small likelihood of tear agents producing casualties, they are rarely used except for training and riot control. They have little more than nuisance value in war in view of the effectiveness of the modern protective respirator. The principal tear agents are discussed below.

Chloroacetophenone (CN)

Not readily hydrolised.

Odour. Fragrant; similar to apple blossoms.

Skin and eye toxicity. Irritating; not toxic in concentrations likely to be encountered in the field.

Rate of action. Practically instantaneous.

In addition to powerful lacrymatory effects, CN is an irritant to the upper respiratory passages. In higher concentrations, it is irritating to the skin and causes a burning and itching sensation, especially on moist parts of the body. High concentrations can cause blisters. The effects are similar to those of sunburn, are entirely harmless, and disappear in a few hours. Certain individuals may experience nausea following exposure to CN.

Duration of effectiveness. Short, because the agent is disseminated as an aerosol.

CNC

No chemical name; solution of chloroacetophenone in chloroform.

Odour. Similar to chloroform.

Skin and eye toxicity. Irritating; not toxic.

Rate of action. Instantaneous.

Causes flow of tears, irritates respiratory system, and causes stinging of skin.

Duration of effectiveness. Short, because the agent is disseminated as an aerosol.

CNS

No chemical name; mixture of chloroacetophenone, chloropicrin and chloroform.

Odour. Like flypaper.

Skin and eye toxicity. Irritating; not toxic.

Rate of action. Instantaneous.

In addition to having effects described under CN, CNS also has the effects of chloropicrin (PS) which acts as a vomiting agent, a choking agent, and a tear agent. CNS may cause lung effects similar to those of CG and may also cause nausea, vomiting, colic and diarrhoea which may persist for weeks. The lacrymatory effects of PS are less marked than those of CN and are relatively unimportant in CNS, as shown by the fact that these effects are no greater than with CNC which contains no PS.

Rate of detoxification. The effects of chlorpicrin are long-lasting and cumulative, and may cause the effects of CNS to be prolonged for weeks. Such a prolonged effect may be highly undesirable in training and riot control so CNS is now rarely used.

Duration of effectiveness. Short.

CNB

CNB was adopted in 1920 and remained in use until it was replaced by CNS. The advantage claimed for CNB was that its lower chloroacetophenone content made it more satisfactory than CNC for training purposes. Actually, the same result can be obtained with CNC merely by using a lower concentration.

No chemical name. Solution of chloroacetophenone in benzene and carbon tetrachloride.

Odour. Like benzene.

Skin and eye toxicity. Irritating; not toxic.

Powerfully lacrymatory.

Duration of effectiveness. Short.

BBC. Bromobenzylcyanide.

Rate of hydrolysis. Very slow.

Odour. Like soured fruit but not unpleasant.

Skin and eye toxicity. Irritating; not toxic.

Rate of action. Instantaneous.

BBC produces a burning sensation of the mucous membranes and severe irritation and lacrimation of the eyes with acute pain in the forehead. BBC is considerably less toxic than CG.

Duration of effectiveness. Depends upon the weather and the munitions used. Heavily splashed liquid persists 1 to 2 days under average weather conditions.

CS. O-chlorobenzalmalononitrile.

CS is a white crystalline powder. It has a minimum purity of 96 per cent, is insoluble in water and ethanol, but is soluble in methylene chloride.

Rate of hydrolysis. Unknown.

Odour. Pepper-like.

Skin and eye toxicity. Highly irritating but not toxic.

Rate of action. Very rapid.

CS produces immediate effects even in low concentrations. The onset of incapacitation is 20 to 60 seconds and the duration of effects is 5 to 10 minutes after the affected individual is removed to fresh air. During this time affected individuals are incapable of effective concerted action. The physiological effects include extreme burning of the eyes accompanied by copious flow of tears; coughing, difficulty in breathing, and chest tightness; involuntary closing of the eyes; stinging sensation of moist skin; running nose; dizziness or swimming of the head. Heavy concentrations will cause nausea and vomiting in addition to the above effects.

Duration of effectiveness. Short.

CS has now largely replaced CN and other tear agents as a riot control agent.

Incapacitating Chemical Agents

Incapacitating chemical agents are capable of producing physiological or mental effects that prevent exposed personnel from performing their primary military duties for a significant period of time; there is complete recovery from these effects. The incapacitating agents fall into two general groups: (a)Those which produce temporary physical disability such as paralysis, blindness or deafness. (b)Those which produce temporary mental aberrations. Unlike the lethal agents, these incapacitants can produce purely temporary effects without permanent damage. In this respect they resemble the riot control agents.

The incapacitating agents suggest employment where military necessity requires control of a situation but where there is good reason for not harming the surrounding population or even troops. They also suggest covert uses either to confuse defence or retaliatory forces, or to affect the rationality of an important leadership group at some particularly crucial point. Despite the military failure of agent BZ it seems very likely that other agents have been, or will be, produced to meet similar specifications.

Incapacitating agent BZ

The American Chemical Incapacitating Weapons System included those agents and related munitions and disseminating devices designed to produce an incapacitating effect upon man to meet military requirements. Agent BZ was type-classified Standard B to provide a limited capability to meet these requirements. However, trials revealed that BZ, and similar agents, were extremely unreliable and unpredictable in effect and BZ has been withdrawn from use. Existing American stocks of BZ are being destroyed.

BZ (Benactyzine) is a crystalline solid at normal temperatures and is sufficiently stable to be disseminated as a smoke from a pyrotechnic device. Other members of this glycolic acid ester series may be liquid at usual temperatures and may be absorbed through the skin. Once in the body BZ acts by blocking the activity of cholinergic synapses in a manner similar to that of atropine but unlike atropine BZ produces predominantly central rather than peripheral effects.

General symptoms from agent BZ are: interference with ordinary activity; dry, flushed skin; tachycardia (ie abnormally rapid beating of the heart); urinary retention; constipation; slowing of mental and physical activity; headache; giddiness; disorientation; delusions; hallucinations; drowsiness, sometimes maniacal behaviour; and an increase in body temperature. For some individuals these symptoms may persist for several days. The symptoms may also vary widely from individual to individual.

NBC protection equipment

PROTECTIVE MASKS AND CLOTHING

BELGIUM

Seyntex NBC Protective Clothing

DESCRIPTION
Seyntex produces a wide range of NBC protective clothing, most using conventional charcoal cloth as the basic protective layer. Included in the range are a two-piece smock and trousers suit and a similar two-piece suit with a front closure. The latter may be produced in camouflage colours and the outer material of both is flame-retardent. Charcoal cloth is also used for a facelet mask, socks and full-length or half-length casualty bags. Butyl is used for NBC protective boots and for semi-permeable or impermeable NBC gloves.

Seyntex also produces disposable NBC clothing including a suit with hood, a cape and gloves. All are produced using three-layered laminates with welded seams. Various NBC protective garments and suits for civil defence use are also produced.

All Seyntex NBC garments are vacuum-packed to small dimensions and packed into cardboard boxes.

STATUS
In production.

Seyntex NBC protective overall

Seyntex NBC protective two-piece suit with front closure

MANUFACTURER
Seyntex nv/sa, Seyntexlaan 1, Industriepark Zuid, B-8880 Tielt, Belgium.

BULGARIA

Protective Shoulder Hood

DESCRIPTION
Although this protective hood originated in Bulgaria it is also used by some other Warsaw Pact forces. It is intended to be worn over the integral hoods of protective suits or over protective masks. As such its use is mainly confined to heavy-duty decontamination teams or personnel handling toxic agents, but it has also been issued to missile fuel handlers. The hood is a one-piece garment made from heavy rubber. It covers the head, neck and shoulders and provides protection against toxic agents, corrosive chemicals and fuels in either liquid, aerosol or vapour form.

STATUS
Probably still in limited production. In service with Bulgarian and some other Warsaw Pact armed forces.

MANUFACTURER
State factories.

CANADA

NBC C4 Protective Mask

DESCRIPTION
In co-operation with the Canadian Defence Research Establishment, Canadian Arsenals Limited produced an advanced NBC protective mask meeting all Canadian Armed Forces, NATO and QSTAG 695 requirements, and known as the NBC C4. The NBC C4 will replace the C3 unit now in service with Canadian and other armed forces (see following entry).

The NBC C4 is lightweight, easy to don, is compatible with optical and telecommunications equipment, and is easy to decontaminate and service in the field. The mask uses the C2 face mask canister, has a bromobutyl rubber facepiece, a reflex face seal and agent resistant engineering thermoplastics for maximum protection against field concentrations of all known NBC agents, screening smoke and particulates. It also offers resistance to thermal radiation. The filter canister is replaceable without removing the mask from the face and offers low breathing resistance.

The mask is provided with shatterproof polycarbonate lenses with an anti-scratch coating. They are shaped and positioned for optimum forward and side vision and are compatible with combat spectacles and optical equipment. The mask design allows air circulation to virtually eliminate lens fogging. Soft transparent silicone is used for the nosecup to provide comfort and easy identification of the wearer.

The NBC C4 has two speech transmitters. A primary speech transmitter has been designed and positioned to produce maximum intelligibility. A side speed transmitter is compatible with telecommunications equipment.

The mask is held in place on the head by what is described as a unique harness design. The nosecup is manufactured from soft silicon to aid fast donning and for comfortable use. A drinking device compatible with the NATO system is provided.

The NBC C4 is produced in four sizes from extra small to large. Canisters may be worn either on the right- or left-hand side and the non-magnetic canister thread is NATO standard according to STANAG 4155. The mask outline is streamlined overall and is easy to decontaminate. Refurbishing can be carried out in the field although a leak tester is required.

The NBC C4 weighs 0.705 kg, including the filter canister.

STATUS
In production for the Canadian Armed Forces.

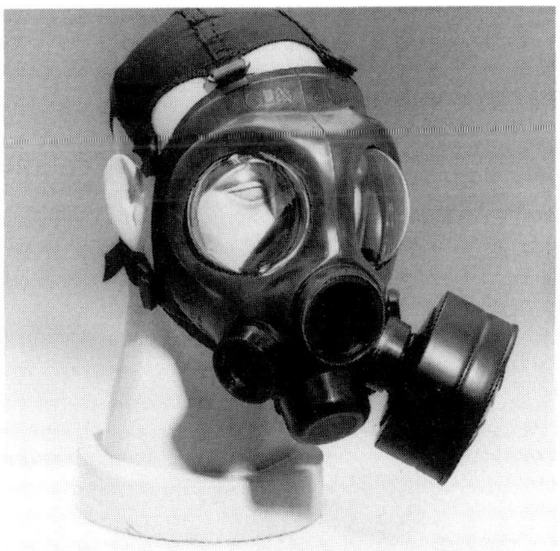

NBC C4 protective mask

MANUFACTURER
(Mask) Canadian Arsenals Limited, 2 Place Félix Martin, Montreal, Canada H2Z 1Z2.
(C2 canister) Racal Filter Technologies Limited, 1380 California Avenue, Brockville, Ontario, Canada K6V 5Y6.

Gas Mask CML-BIO C3

DESCRIPTION
Developed for use by the Canadian Armed Forces, this mask provides protection against all known NBC agents. It evolved from a series of masks that was first introduced during the Second World War and has as its main component a mustard-resistant injection moulded bromobutyl rubber face-piece. The face-piece is produced in three sizes, large, medium and small, and is treated to remain non-tacky at temperatures as high as +53°C. The twin eye-pieces are flat shatter-proof glass lens each 63.5 mm in diameter and set close to the eye so that optical equipment such as gun sights, optical instruments and binoculars can be used. An internal nose cup is designed to allow incoming air to pass over the interior of the eye-pieces, but expelled air is passed directly through an outlet valve to maintain a fog free condition inside the mask. The outlet valve assembly also incorporates a speech transmitter assembly incorporating a double layer of polyester film for clear transmission of speech. The C2 or C5 filter canister is normally sealed by a screw-type cap when not in use mounted on the left-hand side of the mask. The filter contains a pleated paper particulate filter and an activated impregnated charcoal bed. The mask is held in place on the head by a webbing harness, and when not in use is carried in a stout fabric bag which has pockets for accessories such as the special spectacles required by spectacle users for combat.

A special adaptor to accommodate standard NATO filters is available. Supplied with all masks is a drinking device that can be supplied to up-grade C3 masks already in service.

Weight of the mask is 0.715 kg, including the filter canister.

STATUS
In production. In service with the Canadian Armed Forces and some NATO and other armed forces.

MANUFACTURER
(Mask) Canadian Arsenals Limited, 2 Place Félix Martin, Montreal, Canada H2Z 1Z2.
(C2 canister) Racal Filter Technologies Limited, 1380 California Avenue, Brockville, Ontario, Canada K6V 5Y6.

Gas Mask CML-BIO C3

Protective Coverall

DESCRIPTION
This protective coverall is a one-piece garment intended to be worn in place of normal combat clothing in conditions where NBC protection might be required. It has an attached hood and is opened up the front by a zip fastener and a Velcro front closure. Velcro is also used at the wrist and ankle openings and elastic webbing stirrups are provided on each leg to ensure that a satisfactory overlap of the trouser leg and the overboot is always maintained. The hood is designed to fit closely around the Canadian C3 mask but can be produced for other masks. Pockets are provided and the garment is produced in a number of sizes, each with a built-in suspender adjustment for height variations.

The coverall has a double layer protection. The outer shell of the garment is of lightweight 170 g/m^2 50/50 nylon/cotton combat cloth treated with a fluoro-chemical oil and water repellent finish. This finish ensures that liquid agents stay on the surface where they can be removed by field decontamination processes. The fabric is permeable to air and moisture vapour to maintain normal body heat balance. The second layer of the garment is an underlayer of open cell polyurethane foam 2.4 mm thick and containing re-activated charcoal bonded with latex. The foam layer is laminated to a knitted nylon substrate for added strength and to provide a smooth lining.

STATUS
In production. In service with the Canadian Armed Forces.

MANUFACTURER
Anachemia Canada Inc, Case postale 147, Lachine (Québec), Canada H8S 4A7.

Canadian Protective Coverall

NBC Protective Gloves

DESCRIPTION
Specially produced for the Canadian Armed Forces, these gloves consist of duo-stretch knitted cotton coated with high molecular density butyl rubber. The gloves are olive green in colour and have gauntlet-type cuffs designed to be worn under the sleeves of protective coveralls. They are produced in a range of sizes from small to extra-large. The fingers and palm of the gloves are of curved configuration.

Extra thin unlined gloves made of dipped butyl are available for use when highly tactile operations have to be carried out under CW conditions.

STATUS
In production. In service with the Canadian Armed Forces.

MANUFACTURER
The Acton Rubber Company Limited, Case postale 300, 881 Landry, Acton Vale (Québec), Canada J0H 1A0.

NBC Overboots

DESCRIPTION
The main protective material used in these overboots is high molecular density butyl rubber and they are intended to be worn over the normal combat boots. The individual overboot is 280 mm high and is produced, single width, in a wide range of sizes. The upper is of butyl coated duo-stretch nylon fabric while the outsole and upper reinforcing components are made of neoprene rubber. There is also an inner liner, insole cover and back strip to assist in quick donning and doffing. A cord lace ensures a close fit around the ankle with the upper intended to be worn inside protective outer garments. The sole is ribbed for traction on smooth surfaces.

STATUS
In production. In service with the Canadian Armed Forces.

MANUFACTURER
The Acton Rubber Company Limited, Case postale 300, 881 Landry, Acton Vale (Québec), Canada J0H 1A0.

Canadian NBC overboots

NBC Protective Material Kit

DESCRIPTION
The NBC protective material kit consists of a roll of flexible, laminated, camouflage-coloured, coated plastic, rolls of pressure-sensitive tape, a knife and a set of instructions. It is intended for use in the field to produce rapid protective covering for stores and equipment in the event of NBC attack. In an emergency it could also be used to produce NBC protective covering for personnel. The rolls of plastic can be produced in various widths and are simply unrolled and cut to suit the requirement. When the item involved has been covered the protective material can be secured in place using the pressure sensitive tape.

STATUS
In production. In service with the Canadian Armed Forces.

MANUFACTURER
Nor Baker Industries Limited, 175 Deerfield Road, Newmarket, Ontario, Canada L3Y 2L8.

CHINA, PEOPLE'S REPUBLIC

Type 69 Protective Mask

DESCRIPTION
The Type 69 protective mask is stated to provide the wearer protection against toxic gases and vapours, smoke, radioactive dusts and bacteria. It consists of a moulded rubber facepiece with wide vision eyepieces that are probably flexible. A charcoal-based filter canister is located on the side of a large voice transmission device that projects prominently to the front. The voice transmission device is stated to provide 90 per cent articulation at a range of 50 metres.

Weight of the mask is 650 grammes and the filter canister 150 grammes.

STATUS
In production. In service with the Chinese armed forces and offered for export.

MANUFACTURER
China North Industries Corporation, 7A Yue Tan Nan Jie, PO Box 2137 Beijing, Beijing, China.

The driver and crew of an armoured personnel carrier wearing Type 69 protective masks

Protective Mask M-65

DESCRIPTION
The M-65 protective mask features a prominent filter housing on the left-hand cheek of the face-piece and a prominent voicemitter at the front. Large lens eyepieces are provided and the mask is held in position by a six-strap head harness. The filter has an efficiency of 99.995 per cent and can withstand aerosol droplets down to 0.3 micron. The filter capacity for inhalation and exhalation is 3 litres a minute. Air flow resistance for inhalation is less than 176 Pa, and less than 98 Pa for exhalation at 30 litres per minute.

Weight of the M-65 is 0.6 kg.

STATUS
In production. In service with the Chinese armed forces.

DEVELOPMENT AGENCY
Research Institute for Chemical Defence, PO Box 1044, Beijing, China.

MANUFACTURER
Xinhua Chemical Factory, Taiyuan, Shanxi, China.

Protective mask M-65

Protective Mask M-85

DESCRIPTION
The M-85 protective mask appears to have a natural rubber face-piece held in place by a four strap head harness. A cylindrical filter canister is fitted to the left-hand side of the face-piece and large eye lenses are provided. A prominent voicemitter device is fitted to the front of the face-piece. The filter has an efficiency of 99.995 per cent and can withstand aerosol droplets down to 0.3 microns. Air flow resistance for the filter canister is less than 147 Pa at 30 litres per minute.

Weight of the M-85 is 0.7 kg.

This mask is marketed in the United States as the MF-11 (see page 49).

STATUS
In production.

DEVELOPMENT AGENCY
Research Institute for Chemical Defence, PO Box 1044, Beijing, China.

DEVELOPMENT AGENCY AND MANUFACTURER
Xinhua Chemical Factory, Taiyuan, Shanxi, China.

Protective mask M-85

Protective Suit, Permeable, M-82

DESCRIPTION
The material used for the Protective Suit, Permeable, M-82, is made of vinylon-cotton blend fibre (outer layer) and cotton flannel (inner layer). One surface of the inner layer is finished with an active charcoal-acrylate mixture and the other surface is finished with an oil-repellent agent. The basic suit uses two garments, a fly-fronted jacket with an integral hood, and over-trousers. The suit is completed by butyl rubber gloves and overboots, the latter held in position by laces. The suit is supplied in a camouflaged and waterproof finish.

The suit weighs 1.3 kg and has a life of between 20 and 50 hours with mustard or G-type nerve agents in vapour form, and up to six hours against mustard and nerve agents (both G and V agents) in liquid vapour form (2.5 grammes per square metre, 0.1 mg droplets at 20°C, relative humidity 80 per cent).

STATUS
In production. In service with the Chinese armed forces.

DEVELOPMENT AGENCY
Research Institute for Chemical Defence, PO Box 1044, Beijing, China.

MANUFACTURER
Huajiang Machinery Plant, PO Box 508, Yichang, Hubei, China.

Protective Suit, Permeable, M-82

Protective Suit, Butyl, M-66

DESCRIPTION
Intended for use when carrying out decontamination procedures, the M-66 protective suit is manufactured using a butyl-based material. It is a one-piece suit with an integral hood and boots. Butyl rubber gloves are also worn. Elasticated closures are provided at the cuffs and around the integral hood, while laces are provided at the ankles and waist.

The M-66 suit weighs 2.3 kg and has a life of up to 130 minutes against 20 mg droplets of HD at 36°C.

STATUS
In production. In service with the Chinese armed forces.

DEVELOPMENT AGENCY
Research Institute for Chemical Defence, PO Box 1044, Beijing, China.

MANUFACTURER
Guilin Rubber Products Factory, Guilin, Guangxi, China.

Protective Suit, Butyl, M-66

CZECHOSLOVAKIA

Protective Mask, Model M-10

DESCRIPTION
First issued during 1970, the Model M-10 mask has several advanced features, most of which are immediately noticeable. The large eye-pieces provide good visibility which is maintained by the use of an inner mask to prevent fogging. The lower part of the mask is very prominent and houses inlet valves on either side of the internally-fitted filter system. A single outlet valve at the front below the 'voicemitter' enables the wearer to communicate with relative ease. A head harness holds the mask in place.

The M-10 appears to be a close copy of the American M17 series of masks.

STATUS
In Czechoslovak Army service and has been exported.

MANUFACTURER
State factories.

Protective Mask, Model Chema-S

DESCRIPTION
Normally issued for civilan use, the Model Chema-S uses a moulded black rubber face mask held in place by adjustable head straps. The circular eye-pieces are held in place by crimped metal rings and the prominent metal outlet valve is held in place by a non-adjustable neck strap. From the inlet valve a corrugated rubber hose connects the mask to a Chema-S-1-37(KL) filter canister. The entire equipment is normally carried in a fabric satchel.

STATUS
Held in reserve for Czechoslovak armed forces but issued for civilian use.

MANUFACTURER
State factories.

Protective Mask, Model Chema-S6-37

DESCRIPTION
Intended primarily for civilian use, the Model Chema-S6-37 has a face mask moulded from grey or grey-green rubber. The mask is held in place by five adjustable elastic straps which join at a point at the back of the head, and a carrying strap is fitted near the chin. A circular metal valve housing also holds the connector to the filter canister which may be fitted directly or via a hose, but in both cases the usual canister is the Chema-S1-37 which is attached via a screw fitting.

STATUS
Civilian use only.

MANUFACTURER
State factories.

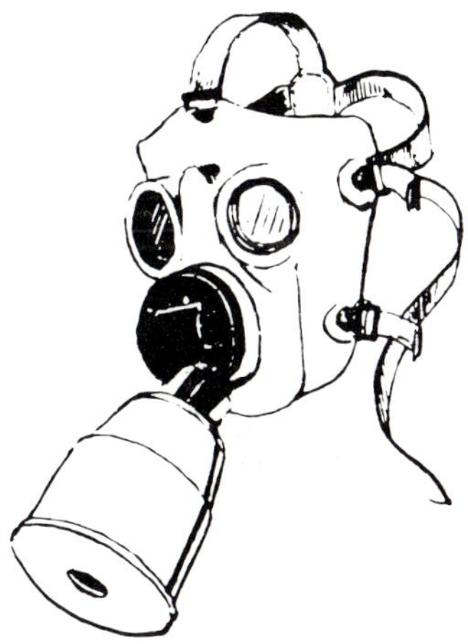

Protective mask, model Chema-S6-37

Protective Mask, Model FATRA-38

DESCRIPTION
The Model FATRA-38 is the Czechoslovak version of the Second World War German GM-38 from which it differs mainly in having six attachment points for the head harness. It is issued for civilian use. The filter used is the Eckhardt Model EF-2. This cylindrical metal canister is 117 mm high and 98 mm in diameter.

STATUS
Production complete. Civilian use only.

MANUFACTURER
State factories.

Multi-purpose Protective Cape

DESCRIPTION
The Czechoslovak protective cape is a progressive development of the Soviet Model OP-1 and in many ways resembles the East German Model SBU-61. However, it does have some unique features, including three-fingered integral gloves, and an inflatable waist-band which can be used when crossing water obstacles. It does not prevent the wearer from using his personal firearms. The cape is made from a lightweight synthetic material which is waterproof, airtight and heat resistant. As with other similar capes, the Czechoslovak design can be used either as a cape, as a coverall, or combined with one or more other capes to form small tents or shelters. It can also be used as a stretcher, and two or more capes can be combined to form a small floating raft for carrying light weapons or supplies. Used in conjunction with overboots and a protective mask, this cape is stated to provide excellent protection against liquid agents and will partially protect against nuclear fallout particles.

STATUS
In production and service with the Czechoslovak armed forces.

MANUFACTURER
State factories.

EGYPT

BSS and CM3 Protective Masks

DESCRIPTION
These two protective masks are Egyptian-produced versions of the Soviet ShM helmet-type protective head mask (see separate entry in this section). Both are identical to the Soviet originals in virtually every respect and use the same type of filter canister system and are carried in canvas shoulder bags. The BSS is the standard type of ShM mask covering the entire head while the CM3 has a diaphragm voicemitter and a head harness, making it similar to the Soviet Protective Mask, Communication (see separate entry in this section).

STATUS
In production. In service with the Egyptian armed forces and possibly some other Middle Eastern armed forces.

MANUFACTURER
Heliopolis Company for Chemical Industries, Haikstep, Heliopolis, Egypt.

FINLAND

KEMIRA NBC Respirator M76

DESCRIPTION
The KEMIRA NBC Respirator M76 has a natural rubber facepiece held in place on the head by a six-point adjustable head harness. The eyepieces are flexible polycarbonate lenses held in position by aluminium frames. The respirator is supplied with or without a stainless steel speech diaphragm and it is available in three sizes; small, medium and large.

The side-mounted aluminium canister casing has a 60 mm diameter screw fitting with a right-hand thread and contains activated charcoal and a folded glass/cellulose fibre dust particle filter. Accessories include a water-repellent fabric haversack or a HD-polyethelene carrying case, a spare NBC filter canister and a spare valve set.

STATUS
Production complete.

MANUFACTURER
KEMIRA Oy, Safety Equipment, SF-65230 Vaasa, Finland.

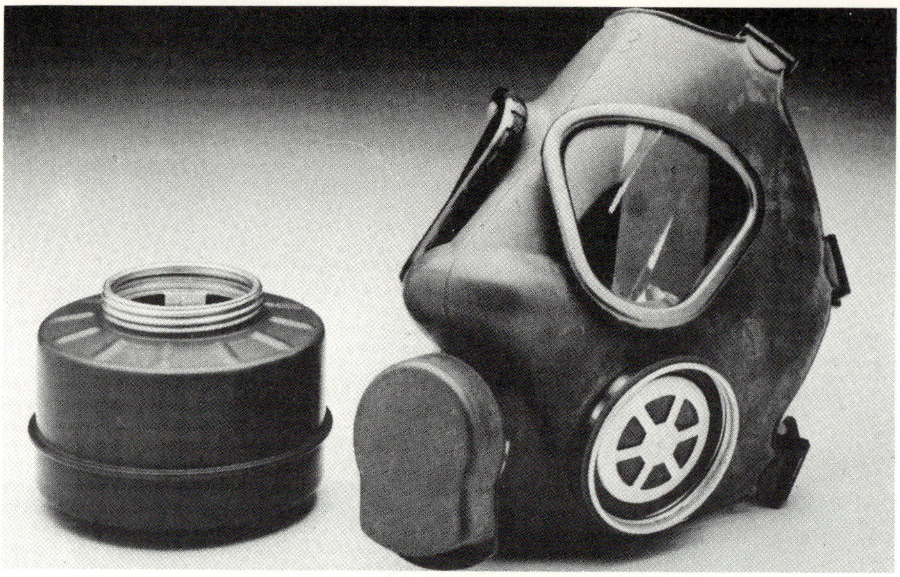

KEMIRA NBC Respirator M76 without speech diaphragm

FRANCE

GIAT Model ARF NBC Respirator

DESCRIPTION
During 1987 GIAT produced the first prototypes of a new NBC respirator for military use. An initial batch of 200 was delivered for French Army troop trials and full production is not expected before 1989-90.

The respirator has a rubber facepiece fitted with a one-piece flexible vision lens. The filter canister is mounted on the chin with a voicemitter located directly above the canister fitting. There is provision for a hose to be coupled to a remote filtering system or canister and a drinking device is incorporated. The facepiece is held in position on the head by a six-piece strap harness.

STATUS
Ready for production for French Air Force.

MANUFACTURER
Groupement Industriel des Armements Terrestres (GIAT), 10 place Georges Clémenceau, 92211 Saint-Cloud, France.

New GIAT NBC Respirator (T J Gander)

Protective Mask, Model ANP 51 M 53 and ME 82

DESCRIPTION
This mask is made from thick black rubber and covers the face. It is fitted with two large circular eye-pieces and a voicemitter in front of the mouth. There are several accessories that can be fitted to the mask, including special lenses for use with weapon sights or other similar optical devices. The filter canister can be either fitted directly onto the face mask or connected to it by a corrugated rubber pipe. The normal operational canister is the CF 63/67 which is circular and weighs 0.25 kg. Other similar canisters can be fitted which have less capability or are for training only, including versions for use with riot control agents and carbon monoxide. If required the mask can be connected to a vehicle collective protection air system. The mask and its basic equipment are normally carried and stored in a Model 63 carrying bag; the mask alone weighs 0.4 kg. A variation of the basic mask is the Model ANP (ANP – appareil normal de protection) 51 M 53b which has an integral microphone.

The ME 82 is an export model with the option of mounting the filter canister on the normal chin position or at the end of a hose. The facepiece uses an improved periphery seal, has a drinking device

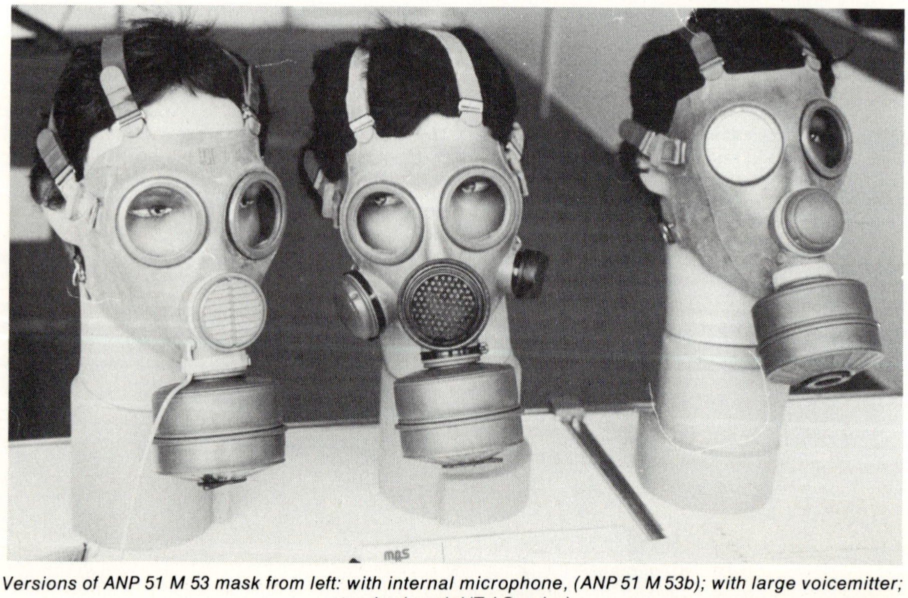

Versions of ANP 51 M 53 mask from left: with internal microphone, (ANP 51 M 53b); with large voicemitter; standard model (T J Gander)

and an extra side-mounted exhaust valve for increased air flow. It is possible to fit this mask with lenses to suit weapon sights or vision devices and a built-in microphone for use with tele-communications equipment.

STATUS
In production. In service with the French Army and other French armed forces and para-military units. Some have been exported.

MANUFACTURER
Groupement Industriel des Armements Terrestres (GIAT), 10 place Georges Clémenceau, 92211 Saint-Cloud, France.

ANP 51 M53 mask in use by French infantryman

Protective Mask Type 'Gendarmerie' (G1)

DESCRIPTION
This protective mask is intended for police and similar use and is proof against anti-riot and tear gas agents as well as radioactive dusts and aerosols. It consists of a one-piece clear plastic face-piece without any eye-pieces. The translucent face-piece is held in place by an easy-to-fit head harness and uses an internal fitting that entirely covers the nose and mouth to prevent internal misting. Incoming air is drawn through two CFL 82 filter canisters, one each side of the face-piece, and expelled through a front-mounted outlet valve that also acts as a speech transmission device. The edges of the face-piece are flexible to ensure a tight fit over the face.

The mask alone weighs 500 grammes. Each of the two CFL 82 filter canisters weighs 130 grammes so the combined weight is 760 grammes.

The 'Securité Civile' SC83 mask is a similar mask intended for use by civilian personnel.

STATUS
In production for French National Police.

MANUFACTURER
Groupement Industriel des Armements Terrestres (GIAT), 10 place Georges Clémenceau, 92211 Saint-Cloud, France.

Protective mask type 'Gendarmerie' (G1)

Paul Boyé NBC Protective Uniform

DESCRIPTION
With the NBC protective uniform produced by Paul Boyé, the concept of personal NBC protection for the soldier has been taken one step further from the idea of special clothing to meet the protection requirement to the adoption of a special uniform. With the Paul Boyé uniform the soldier is equipped with combat clothing that can provide a high degree of protection against the effects of chemical and biological agents and is easily decontaminated after being affected by nuclear fall-out.

The uniform consists of two outer garments, two layers of gloves and special socks. The outer garments are a jacket and trousers made of three layers of material. The outer layer is made of high density (75 g/m^2) polyamide 75 taffeta that is treated with a water repellant and dyed standard NATO green with an infra-red reduction element. The centre layer is also liquid repellant to trap any aerosols or liquid that may penetrate the outer layer. It is formed from an unwoven 95 g/m^2 cellulose material. The inner layer is bonded to a knitted support and consists of polyurethane foam impregnated with activated charcoal.

The jacket weighs approximately 0.9 kg and the trousers 0.8 kg. An integral hood that fits round an NBC mask is provided as is a zip fastener with a velcro-fastened over-flap closure. Various reinforcement panels are provided. The trousers are held by adjustable braces and have velcro-fastened adjustment flaps at the bottom.

The gloves are in two separate layers, an outer glove and an inner glove. The outer glove is made of goatskin 0.6 to 0.8 mm thick. The inner glove has three layers of material. The outer layer is polyamide mesh with a density of 20 g/m^2. The middle layer is compressed polyurethane foam impregnated with activated charcoal, and the inner layer is cotton jersey. The NBC socks have much the same make-up as the inner gloves but the foam layer is thinner to allow the socks to be worn under normal combat boots.

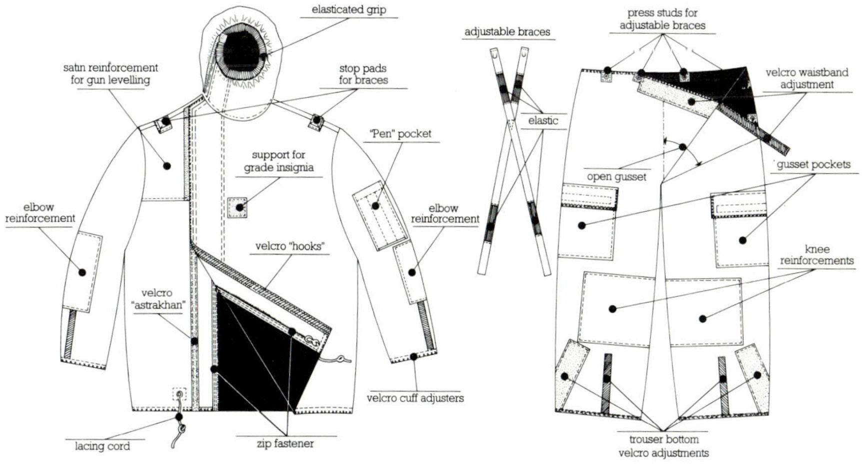

Main components of Paul Boyé NBC uniform outer garments

All the items in the NBC suit are issued in sealed envelopes.

While the NBC uniform is effective against most chemical and biological agents, it has only limited effectiveness against nuclear flash and fire. To overcome this a version of the Paul Boyé NBC uniform has been produced that has the outer garments made from a layer of 50/50 Kermel (a fire retarding fabric) and IF80. IF80 is a non-flammable fabric that retains its fire-retardent properties. Using this mixture of flame-resistant fabric the NBCF (F – fire) uniform has complete protection against all NBC hazards. With this fire-resistant uniform it is suggested that the outer layer should be worn separate from the inner layers which would then be supported on a removable lining held in place by velcro strips. A one-piece undergarment made of foam rubber impregnated with active carbon can be worn under these fire-proof suits. Two versions of a NBCF suit known as the T3P are in production for the French Air Force. One is for use for aircrew and the other (with a hood) for use by ground maintenance personnel.

STATUS
In production. The NBC uniform is in service with the French Army and Switzerland (45 000 suits purchased in 1986; known as the *habit de protection C 86*). The fire-resistant suit is in production for the French Air Force.

MANUFACTURER
Manufactures de Vêtements Paul Boyé, 53 quai de Bosc, 34202 Segte, France.

Bachmann-TMB Model 63 Disposable NBC Suit

DESCRIPTION
The Model 63 disposable NBC suit is meant for short-term emergency use only and allows the wearers to escape from NBC contaminated areas. After use the suit must be disposed of by burning, burial or other means. The suit is worn in combination with a standard NBC respirator.

The Model 63 is formed from a flexible layer of transparent material involving no fabric threads. It consists of a full length outer garment, gloves, a head cowl to accommodate the ANP 51 M 53 respirator, overboots, and a spare pair of gloves. The material is lightweight and provides only short term protection against NBC agents.

The suits are issued individually wrapped in packages weighing 0.5 kg each.

STATUS
In production. In service with the French armed forces.

MANUFACTURERS
Bachmann SA, 69 avenue Danielle-Casanova BP 15, 94200 Ivry-sur-Seine, France.
Techniques Michel Brochier SA, ZI –Route de Chartinieres, Dagneux 01120 Montluel, France.

French troops wearing the Model 63 disposable NBC suit

Bachmann-TMB Heavy Duty NBC Decontamination Suit

DESCRIPTION
This NBC decontamination suit is made from heavy duty butyl-based fabric welded at all seams. It may be repeatedly decontaminated using sprays or heated showers.

STATUS
In production.

MANUFACTURERS
Bachmann SA, 69 avenue Danielle-Casanova BP 15, 94200 Ivry-sur-Seine, France.
Techniques Michel Brochier SA, ZI –Route de Chartinieres, Dagneux 01120 Montluel, France.

Heavy duty decontamination suit

GERMANY, EAST

Protective Mask, Model GM-38

DESCRIPTION
The GM-38 was first issued in 1938 and is still in use in East Germany. The face mask is made from black synthetic rubber and is held in place by an adjustable five-strap harness which is judged to be relatively comfortable to wear. At mouth level there is a cast metal housing containing the outlet and inlet valves, onto the front of which the FE-42 filter canister is screwed. The FE-42 is 108 mm in diameter and 89 mm long, and weighs about 0.45 kg. The GM-38 was originally issued in a corrugated metal cylinder but simple fabric carriers are also used.

An essentially similar model is used in Czechoslovakia as the Model FATRA-38 (qv).

STATUS
Production complete. In service in East Germany and elsewhere.

MANUFACTURER
State factories.

Multi-purpose Protective Cape, Model SBU-61

DESCRIPTION
This protective cape is similar to the Soviet cape-coverall OP-1 but differs in some fittings. It is normally carried in a roll on top of the standard field pack and is so arranged that it can be quickly released down the wearer's back ready for fitting. Once released it can be worn as a cape, overgarment or as a coverall, in a manner similar to the OP-1. It also has the same general versatility as the OP-1. Two can be combined to form a two-man tent about 1.8 metres long, 1.8 metres high and 1.25 metres wide. The securing items are nearly all toggle-type buttons. The cape is made from rubberised fabric and is available in grey, olive drab or dark blue and weighs about 2.5 kg. By itself it provides reasonable protection but as it is not completely airtight, even when worn as a coverall, extra protective under-clothing must be worn for full protection.

STATUS
Probably no longer in production. In service with the East German forces.

MANUFACTURER
State factories.

East German motor cycle troops wearing SBU-61 capes

Protective Overboots, Model 61

DESCRIPTION
The Model 61 protective overboots are worn in conjunction with the multi-purpose cape Model SBU-61 as part of the East German protective equipment. As such they are worn over the normal field boots and are secured above the knee by straps. Further straps fasten to the wearer's belt and then tie-straps secure the overboot to the ankle. The Model 61 overboots are made from olive green rubberised cotton and are stated to provide good protection against most NBC agents.

STATUS
Probably no longer in production. In service with the East German forces.

MANUFACTURER
State factories.

Protective Overboots, Heavy

DESCRIPTION
These boots are made from heavier material than the norm and have heavy black soles. They are manufactured from rubberised fabric and can be removed without touching the possibly-contaminated sole. Normally the boots are secured in place by elastic bands but extra cords secure them to the wearer's belt. A second cord passes through loops at the ankle for a tight fit.

STATUS
Probably still in production. In service with the East German forces.

MANUFACTURER
State factories.

GERMANY, WEST

Helsa-Werke NBC Protective Clothing

DESCRIPTION
Helmut Sandler GmbH and Co KG have been producing NBC protective material since 1977 and are currently producing complete NBC protective outfits for the Norwegian Army. The order was placed during 1984 and is worth Nkr100 million. Deliveries will continue until 1989.

The NBC outfits include a smock overall with an integral hood, overtrousers, overboots and gloves. The smock and overtrousers are generously cut and are provided with pockets. They are made from two layers of material. The outer layer is a material known as Koblenz art. 41 01 26 while the inner filter layer is known as Koblenz 24 60 74 which is impregnated with activated charcoal. The outer layer is water and oil repellent and prevents the ingress of moisture but allows body transpiration.

STATUS
In production for the Norwegian Army

MANUFACTURER
Helsa-Werke, Helmut Sandler GmbH & Co KG, Postfach 60, D-8586 Gefrees, Federal Republic of Germany.

NBC protective suit produced using Helsa-Werke NBC protective material

Kärcher NBC Protective Suit

DESCRIPTION
The Kärcher NBC Protective Suit is a one-piece coverall garment with an integral hood, intended for use by NBC decontamination personnel or by special NBC units carrying out rescue or other missions. Special attention has been given in its design for full NBC protection and wearer comfort.

The coverall is made from a contamination-proof material that meets West German Mil. Spec. 8305-0177. At the front of the coverall there is a gas-tight zip fastener which can be closed from top left to bottom right. This arrangement allows the wearer to self-check that the fastener is properly closed. The zip fastener is completely covered by a flap, made of the same material as the coverall. When removing the coverall the flap is folded back to ensure that the wearer does not come into contact with contaminated material. The complete suit can be donned and doffed without external assistance.

An NBC protective mask (with an integral drinking device) is firmly attached over the coverall's hood by adjustable straps on a head harness. NBC boots are fixed to the suit and the cuffs are also fixed to the coverall and serve as the seat for NBC protection gloves. A double seal effect at the cuffs is achieved by clamps with a second cuff over the clamps. This arrangement allows gloves to be changed easily.

The complete suit weighs approximately 5 kg. It is available in two sizes, Size 1 for persons up to 1.75 metres in height and Size 2 for persons over 1.75 metres.

STATUS
In production. In service with various countries in North Africa and the Middle East.

MANUFACTURER
Alfred Kärcher GmbH & Co, Alfred-Kärcher-Strasse 28-40, D-7057 Winnenden, Federal Republic of Germany.

Kärcher NBC protection suit

GREECE

COBRA NBC Respirator

DESCRIPTION
The COBRA NBC respirator was developed from a line of earlier respirator designs in answer to a Hellenic Army General Staff requirement. The COBRA is the latest in a line of respirator models that began with the B-3 in 1955 and developed through the B-31 (1965, for police force use), the B-32 (1968, also for police force use) and the B-33 (for industrial use).

The COBRA is a twin visor mask developed to meet NATO requirements as well as Greek and West German specifications. The face mask and head harness are made using a special quality of halogen butyl rubber mixture with a matt black finish. Air-tight sealing of the mask on the face is achieved through an inner lip sealing frame on the mask body. In addition a rib on the outer rim of the mask ensures a good fit of the mask inside an NBC clothing hood. The mask is supported on the face by a five-piece head harness which provides a fast fitting time. Special spectacles can be provided if required.

The curved eye-pieces provide a wide field of vision and are kept free of misting by directing incoming air over their internal surfaces. Exhaled air cannot come into contact with the eyepieces due to an inner mask that covers the nose and mouth.

The filter canister is fitted under the high efficiency voicemitter assembly which also contains the exhalation valve. The filter canister used is the SUPER-COMBI NBC with a claimed efficiency of 99.997 per cent.

The COBRA is provided with a drinking device for water and liquid foods. The device is mounted on the right-hand side of the facepiece and consists of an inner mouthpiece, an intake valve, an external connecting tube and a special canteen top.

The COBRA respirator can be carried in the Type 87 B carrying case. This case has space for the respirator, two filter canisters, spares, an instruction book and NBC first aid equipment (eg an atropine self-injector).

STATUS
Ready for production.

MANUFACTURER
BIANA SA, Personal Protective Equipments, 3rd km Koropi – Vari Ave., PO Box 5, 194 00 Koropi, Greece.

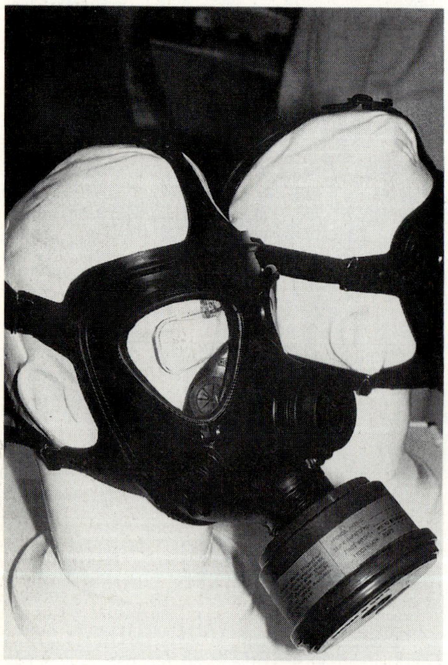

COBRA NBC respirator

Two views of the COBRA NBC respirator, with and without a filter canister fitted

ISRAEL

Shalon-Chemical Industries Protective Masks

DESCRIPTION
The Israeli firm of Shalon-Chemical Industries Limited produces a range of individual NBC protective masks to suit a variety of roles. The most widely used version is the No 15A1 which is intended for general military and police use. The No 15-S-80 is a special version for armoured vehicle crews with the filter canister separate from the face mask. Other mask versions are the No 10A1 intended for children in the 4- to 12-year age range and the No 4A1 for adults from 12 years' upwards.

All these masks are issued with the No 80 filter-canister. On the No 15A1 the filter is mounted below the prominent 'voicemitter', and on the No 15-S-80 the filter is on the end of a flexible hose plugged directly into the armoured vehicle air system. All the masks are equipped with a leakage-proof drinking system; the wearer uses a water-bottle plugged into the mask from which the liquid is directed to the mouth by an internal tube. Also internal is an additional mouth and nose seal for extra protection. On the children's models the filter and outlet valve positions are reversed.

When not in use the masks are carried slung on the chest by a fabric strap. A special training version of the filter canister is supplied with each mask.

STATUS
In production. In service with the Israeli Defence Forces and with several foreign countries.

MANUFACTURER
Shalon-Chemical Industries Limited, 25 Nahmani Street, 65794 Tel Aviv, Israel.

NBC mask No 15-S-80 for tank crewmen

NBC mask No 15A1

Civil Defence Hood

DESCRIPTION
The civil defence hood produced by Eagle Military Gear Overseas Limited at their Ashkelon factory has a minimum protection factor of 10,000. It consists of a hood covering the head, neck and part of the shoulders, and is provided with a transparent face panel producing minimal vision distortion. The hood has its own forced ventilation system produced by a small unit worn on a belt and drawing air through a filter canister. The ventilation unit is powered by a small electrical battery. A drinking device may be fitted to the hood.

The hood has been designed for rapid donning and doffing and can be used with a minimum of pre-use training. A special version is available for children.

STATUS
In production.

MANUFACTURER
Eagle Military Gear Overseas Limited, IBM House, 31 Shaul Hamelech Boulevard, PO Box 33737, Tel Aviv 64927, Israel.

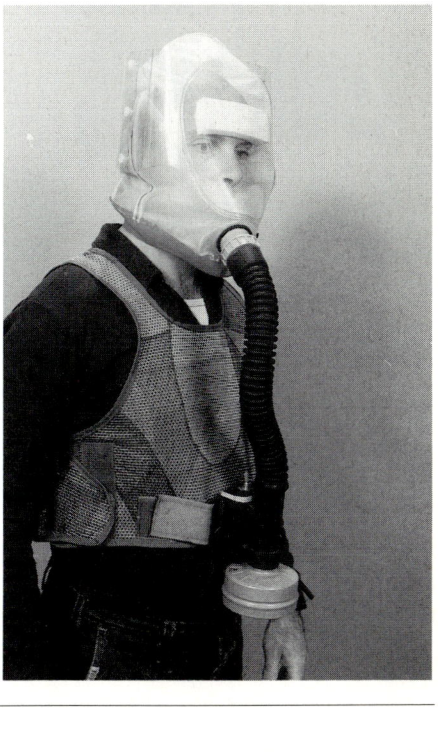

Civil defence hood fitted with drinking device

NBC Disposable Protective Clothing

DESCRIPTION
Unlike most NBC protective clothing, Chemoplast Limited clothing is manufactured using a clear three-layer laminate using polyethelene, polyamide and polyethelene. Continuous heat welding is used throughout.

A disposable NBC suit consists of overtrousers sealed at the ankle and waist by elasticated bands and an overjacket with an integral contoured hood. Gloves are provided as an extra. The suit material is rigorously tested for gas permeability, mechanical strength, thickness tolerances, heat seal strength and any incidental defects. The suits are issued in sealed pouches.

Also produced is a disposable NBC personal cape manufactured using a polyethelene/polyamide laminate. The cape is intended for emergency use only and acts as a coverall for the entire head and body with the user in a crouching position.

STATUS
In production.

MANUFACTURER
Chemoplast Limited, POB 2110, Afula 18301, Israel.

GUMSUR NBC Protective Clothing

DESCRIPTION
GUMSUR NBC protective clothing is intended for emergency use by both military and civilian personnel and is available in a range of sizes to suit adults and children. The clothing consists of a blouse, overtrousers, gloves and footwear covers, all intended to be worn over normal clothing or uniforms. All items are available for issue as a complete outfit contained in a sealed pouch. The material used for all the garments is a transparent two-layer film laminate with a total thickness of 0.07 mm. The inner layer is low density polyethelene with the outer layer being polyamide. All seals are high frequency welded.

The blouse has a cut-out contoured hood which fits around any NBC respirator or mask. A quick-fastener cord is provided for easy and secure closing of the hood. Once fitted the blouse permits freedom of movement. The protective gloves are intended to be disposable while the footwear covers have thickened soles.

GUMSUR also produce a protective personal cape that can be used as an emergency individual shelter with the user crouching under the cape and holding down or securing the cape periphery.

STATUS
In production.

MANUFACTURER
GUMSUR Limited, POB 1713 Holon 58117, Israel.

ITALY

Sekur NBC Protective Suit

DESCRIPTION
The Sekur NBC protective suit, described by the manufacturer as a sweat-free outfit, was developed in close co-operation with the Italian Ministry of Defence and associated laboratories.

The suit is worn with a respirator and consists of a jacket with integral hood, overtrousers, gloves (high tenacity butyl latex approximately 0.8 mm thick) and overboots. The suit can be worn over normal combat clothing and the hood will fit over the M 59, M 73 and M 90 facemasks (qv).

The suit is available in three sizes and allows free movement, being easy to don or remove. The material used is water-repellent and impervious to fresh and salt water, petrol and motor oils. It is stated to be easy to clean and decontaminate and it has a long shelf life. For storage it is packed in a vacuum-sealed plastic bag contained in a light waterproof carrying case. Weight of a suit is approximately 3 kg and it is coloured NATO infra-red reflecting olive green.

The materials used for the suit are a protective outer layer of polyester cotton Fabric 5122 with a protective inner layer of tear-resistant gas-absorbing charcoal impregnated Foam 5230, heat welded to a knitted nylon fabric to provide protection from friction caused by underlying garments. The materials provide protection against chemical agents, including nerve agents, for up to 24 hours of continuous use in a contaminated environment. Air flow resistance for transpiration is lower than 1 to 1.5 mBar (0.25 metres per second air flow speed).

Various material variations are possible, including the use of flame retardent materials and varying thicknesses of charcoal foam.

STATUS
In production. In service with the Italian Army, Navy and Air Force.

MANUFACTURER
Sekur SpA – Pirelli Group, Via di Torrespaccata 140, 00169 Rome, Italy.

Overboots worn with Sekur NBC protective suit

Sekur NBC protective suit

Sekur NBC suit being worn with M 59 facemask

Sekur-Pirelli M 59 and M 73 NBC Facemasks

DESCRIPTION
The Sekur-Pirelli M 59 NBC facemask uses a moulded rubber face-piece with two large flexible triplex glass eye-pieces. It has a side-mounted outlet valve and uses the Pirelli M 58 NBC filter. The M 58 includes a particulate filter and activated charcoal to produce an efficiency of more than 99.997 per cent. Apart from the NBC version there are also available filters for use against carbon monoxide and inorganic gases and particulates. The mask has a front-mounted speed transmitting device and the mask is available in three sizes. The connecting screw thread for the filter canister is a standard NATO item. Weight is 0.49 kg.

The M 73 facemask is intended for use by tank crew, artillery personnel and the users of optical and other instruments. The eyepieces of the M 59 are replaced by circular optical inter-changeable lenses. The M 73 also has a flexible tube leading to a separate filter canister or filter system. It is available in three sizes.

Both types of mask are normally carried in a M 66 case. This is a tough canvas case that carries not only the mask and filter canister but also a full range of NBC accessories and kits. Weight is approximately 0.5 kg.

A range of accessories is available including a synthetic resin carrying case, a drinking device for water or nourishing liquids and a microphone that is clipped to the front of the voice transmitter device. A version using non-magnetic materials is also available.

Sekur-Pirelli M 59 NBC facemask

M 59 and M 73 in production and in service with the Italian armed forces and some civil defence organisations.

MANUFACTURER
Sekur SpA – Pirelli Group, Via di Torrespaccata 140, 00169 Rome, Italy.

Sekur-Pirelli M 73 NBC facemask

Sekur-Pirelli M 59 NBC facemask with microphone attachment

Sekur-Pirelli M 59 NBC facemask with drinking device in place

Sekur-Pirelli M 90 NBC Protective Mask

DESCRIPTION
Development of the Sekur-Pirelli M 90 NBC protective mask began during the mid-1980s and was directly contracted and partially funded by the Italian Ministry of Defence. It has undergone extensive operational testing with the Italian armed forces and was due to enter production during 1989.

The M 90 respirator consists of the M 90 facepiece, the M 91 filter canister, a carrying bag and several accessories including a canteen connector, containers for nourishing liquids and other accessories. The entire equipment was designed to provide protection against all known NBC agents over prolonged periods of continuous wear, including when the wearer is asleep. The level of user comfort is claimed to be such that the design will allow wearing the mask for periods of over five days.

The M 90 facepiece has an external surface made of bromobutyl rubber and there is a reverse seal made from natural rubber compounds. Both these components are produced during a single manufacturing process and have a smooth external surface for ease of decontamination, using boiling water if necessary. The facepiece is held on the head by a five-strap chloroprene rubber harness and complete donning time is claimed to be nine seconds. A small bead around the periphery of the facepiece ensures an effective seal between the mask and the hood of a protective NBC suit.

The M 91 filter canister is located in a downwards-facing central position on the lower front of the facepiece to ensure that the canister does not interfere with the field of vision and to allow the ground area close to the wearer's feet to be easily observed. The central location also assists in lens demisting and allows corrugated hoses for centralised filter systems to be fitted without problems arising.

The M 90 combines in a single unit the filter canister connector, the respiratory valves and the single speech diagphragm. It is claimed that combining these units adds stability to the mask and helps preventing it from swinging and twisting when sudden head movements are made. The unit is constructed using a low density glass reinforced polymer with a high impact strength and good resistance to chemical agents. The unit can incorporate a microphone in a protected position by inserting it under the front screen of the unit.

Sekur-Pirelli M 90 NBC protective mask

In use external air is purified by the M 91 filter and enters the facepiece interior on both sides between the facepiece and the nose cup. Tissot channels draw the air towards the eyepieces and it is then directed down to reach the wearer's respiratory tract through nose cup check valves. A protrusion on the sides of the nose cup separates the Tissot channels from the check valves area. Exhaled air flows from the mask through the exhalation valve to the front screen of the respiratory unit, cleaning particles from the speech diagphragm as it passes through. For nuclear flash protection the speech diagphragm is protected by a set of three overlapping screens.

The two eyepieces are made from allyl-diglycol carbonate with the eyepiece profile square to the normal visual axis to allow the use of optical instruments. Purpose-designed corrective spectacles for use within the mask are available.

The M 91 filter canister efficiency is claimed to exceed all NATO requirements (efficiency against 0.3 micron aerosols is above 99.997 per cent), having a protection factor above 10 000, and was designed to counter threat levels three to four times greater than those covered by NATO specifications. Weight of the M 91 filter canister is approximately 0.3 kg and it contains approximately 290 cubic centimetres of treated activated charcoal. A patented device included in the facepiece connector allows the safe replacement of a filter canister in NBC contaminated atmospheres. The M 90 can accommodate any filter canister with a QSTAG 496 NATO standard thread.

The carrying case is made of butyl rubber-coated fabric and is waterproof and impervious to NBC agents. It weighs 0.48 kg.

A drinking facility is located on the right-hand side of the facepiece. The drinking facility is used with a direct connector fitted to a standard canteen and the same facility can be used with sealed and disposable flexible containers for nourishing liquids. The canteen connector uses a self-cleaning needle valve.

The M 90 facepiece is available in three sizes; small, medium and large. Weight is approximately 0.54 kg. All parts can be dismantled and reassembled using standard tools such as a screwdriver and a pin wrench. All components can be replaced. Shelf life is up to 10 years.

STATUS
Due to enter production during 1989.

MANUFACTURER
Sekur SpA – Pirelli Group, Via di Torrespaccata 140, 00169 Rome, Italy.

Sekur-Pirelli Heavy Duty NBC Clothing

DESCRIPTION
Designed for use by personnel who have to work in a decontamination role or in hazardous conditions, the Pirelli heavy duty NBC clothing is impermeable to any NBC agents. The main garment is a one-piece coverall with an integral hood, made of nylon cloth coated on both sides with butyl rubber. Heavy rubber boots are glued to the ends of the trouser section of the over-garment to provide a complete seal and the butyl latex gloves are sealed to the garment sleeves using rubber rings.

Being impermeable to liquids, the outfit is also impermeable to body heat and perspiration so it cannot be worn for other than very short periods. To overcome this inconvenience the suit can be equipped with an internal ventilation and filtered air system powered by a portable turbo unit which employs two standard NBC filter canisters.

Protection against liquid agents is afforded for more than six hours. The external finish is a NATO infra-red reflecting green colour.

STATUS
In production.

MANUFACTURER
Sekur SpA - Pirelli Group, Via di Torrespaccata 140, 00169 Rome, Italy.

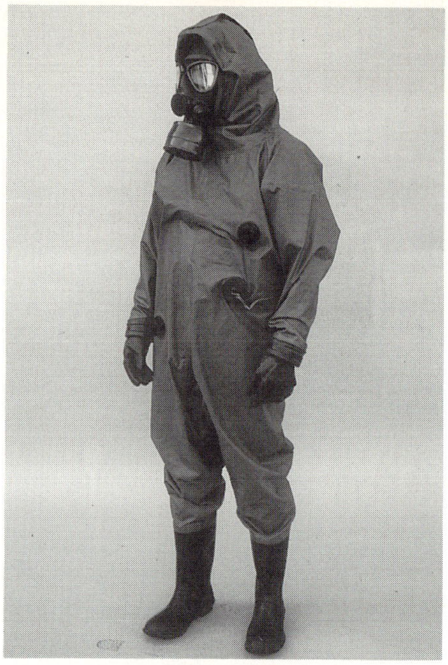

Sekur-Pirelli heavy duty NBC clothing

KOREA, SOUTH

Daewoo NBC Respirators

DESCRIPTION
The Daewoo Corporation produces and offers for export two models of NBC respirator. One is the KM9A1, a licence-produced version of the American M9A1 respirator (see separate entry under *United States* in this section). The KM9A1 is identical to the American original and weighs 1.2 kg complete with its carrier case.

The Daewoo K1 is a Korean design that uses a West German type of canister mounting adapter. The face piece has flat wide vision lenses, an efficient voicemitter system, a drinking system assembly and there is provision for wearing spectacles. The airflow resistance of the inhalation and exhalation system is stated to be low enough to allow normal breathing when the wearer is running.

Weight of the K1 is 1.29 kg ± 20 g.

STATUS
Both models are in production.

MANUFACTURER
Daewoo Corporation, 541, 5-Ga Namdaemoon-Ro, Jung-Gu, CPO Box 2810, Seoul, Republic of Korea.

NETHERLANDS

NBC Protective Gloves

DESCRIPTION
The NBC protective gloves produced by Vredestein Icopro BV are made of high quality butyl rubber and are manufactured using injection moulding. The gloves have been designed to be worn with cotton inner gloves and provide an optimum balance between tactile sensing and protection against chemical agents. The gloves have a penetration resistance of up to 8 hours against mustard (H) agents.

The gloves are available in five sizes and have an overall length of 290 mm. The degree of elongation at breaking point is up to 600 per cent.

Under development are NBC gloves with a length of 360 mm and NBC overboots.

STATUS
In production.

MANUFACTURER
Vredestein Icopro BV, PO Box 200, 6880 AE Velp, Netherlands.

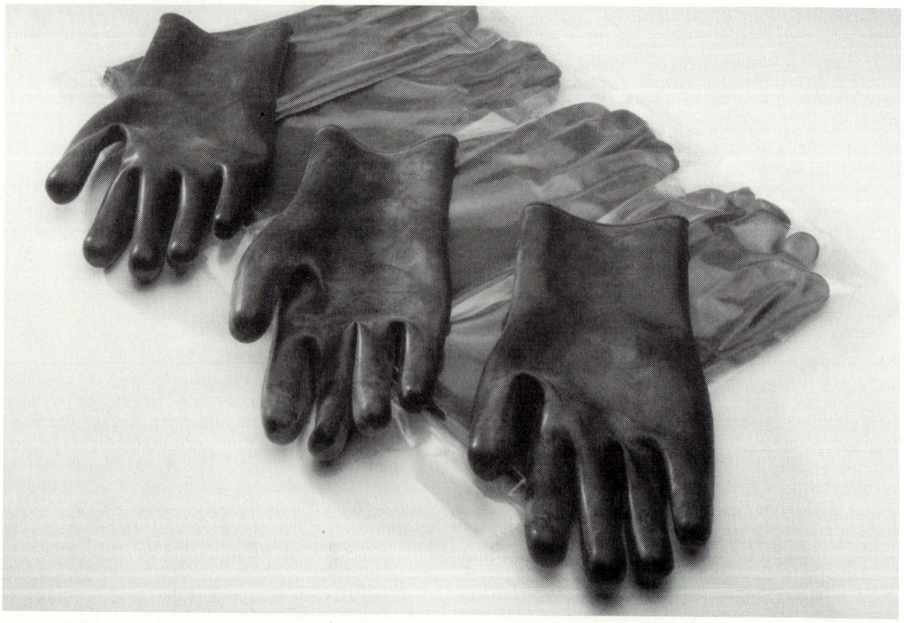

Vredestein NBC protective gloves

ROMANIA

Protective Mask, Diaphragm

DESCRIPTION
This protective mask is rather complex in that it is moulded to fit the wearer's face to the extent of using a rubber nose cup. The mask itself is made from natural rubber and is held in place by an intricate head harness consisting of seven straps. The diaphragm housing also contains the inlet and outlet valves and the connector for the corrugated rubber hose that joins to a filter canister. While the close moulding of the mask provides good protection against most agents, the eye-pieces have no anti-dim provision other than the usual kit and visual efficiency is low as a result. The mask is also stated to be uncomfortable to wear. Weight of the face mask alone is around 0.5 kg. The oval-shaped metal filter canister weighs about 0.9 kg and is 180 mm long and 110 mm across its widest point.

STATUS
In limited Romanian service, probably including civilian use.

MANUFACTURER
State factories.

TAIWAN

T62 Protective Mask

DESCRIPTION
The T62 protective mask is a locally-produced version of the American ABC-M17 mask and differs from it in few respects. It is produced with a special canvas carrying satchel which also carries a cleaning kit. The mask is stated to weigh 0.41 kg and the filter canister 0.265 kg.

STATUS
In production. Probably in service with Taiwan armed forces and available for export.

MANUFACTURER
Hsing Hua Company Limited, PO Box 8746 Taipei, Taiwan, Republic of China.

UNION OF SOVIET SOCIALIST REPUBLICS

New Soviet NBC Respirator

DESCRIPTION
During 1985 reports began to reach the West regarding a new type of Soviet NBC respirator. Few detailed illustrations of the respirator have been published and no indication of the designation involved has been forthcoming. From the few illustrations published the respirator appears to use a rubber facepiece similar to that used on the British S6 with relatively small flat vision lenses. Although it cannot be confirmed it appears that the respirator is held in place by a head harness instead of a hood as has been the usual Soviet practice. A prominent voice-emitter device is fitted to the front of the facepiece and a bulky filter canister is mounted on the cheek; it is assumed that right- and left-hand filter canister fittings can be used.

STATUS
In production. In service with the Soviet armed forces.

MANUFACTURER
State factories.

Soviet Army riflemen wearing the new Soviet NBC respirator

Helmet-type Protective Mask, Model ShM

DESCRIPTION
The Model ShM is the standard Soviet protective mask and is also used by some other Warsaw Pact forces. As the mask completely covers the head it provides good protection against most agents but is rather uncomfortable to wear for long periods and general visibility is limited when wearing it. The helmet itself, which can be worn under a steel helmet, is made from grey or beige rubber. Crimped metal rings secure the eye-pieces to the facemask and as there is only a limited anti-fogging system for the mask interior, the eye-pieces can be fitted with extra internal gelatine lenses; an anti-dim set is issued with the mask. Incoming air is filtered through an MO-2 or MO-4U canister filter element that can be fitted either directly onto the mask or onto a length of fabric-reinforced corrugated rubber hose. A GP-2 carbon monoxide filter canister can be fitted between the mask and the standard canister and training or carbon monoxide canisters can be fitted in place of the standard canister if required. The

Model ShM mask with outlet valve and hose connection to filter canister

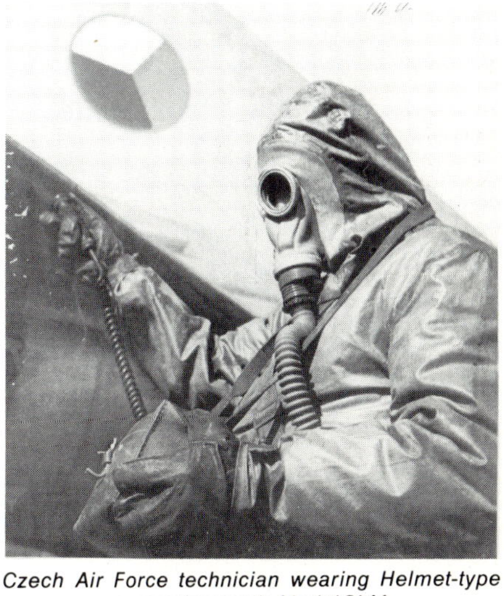

Czech Air Force technician wearing Helmet-type protective mask, Model ShM

carbon monoxide GP-2 filter canister may be either rectangular or cylindrical, and weighs about 0.635 kg. Incoming air is directed across the eye-pieces to reduce fogging to some extent.

Despite its general efficiency, the Model ShM is rather heavy and uncomfortable and the external filter canister and hose make it rather awkward to fit and wear. The mask is issued in a fabric carrier which also contains the filter canister, hose and the anti-dim set. The combined weight is about 2 kg.

The Model ShM is produced in Egypt as the BSS (qv).

SPECIFICATIONS
Chemical and biological filter canisters:

Model	**MO-2**	**MO-4U**
Length	195 mm	205 mm
Width	135 mm	135 mm
Depth	70 mm	68 mm
Weight (approx)	0.875 kg	0.85 kg

STATUS
In production and service with the Soviet and Warsaw Pact forces.

MANUFACTURER
Various state factories.

Protective Mask, Communication

DESCRIPTION
Although this mask uses the same external filter system and canister as the Model ShM it consists of a face mask only, held in place by an adjustable head harness. Thus the ears are exposed so headphones can be worn and a diaphragm 'voicemitter' in front of the mouth permits speech to be transmitted more clearly than is possible with the Model ShM. The visual efficiency of this mask is reported to be low.

A similar mask is produced in Egypt as the CM3 (qv).

STATUS
In production and service with the Soviet and Warsaw Pact forces.

MANUFACTURER
Various state factories.

Civilian Protective Mask, Model GP-7

DESCRIPTION
It is intended that the Model GP-7 protective mask will eventually replace existing GP-4U, GP-5 and GP-5M masks in use by civilians and some non-combabtant military personnel. The GP-7 operates along the same lines as the GP-5 but has some advantages in terms of use and ergonomics and the filter resistance has been reduced. An improved 'floating' facial seal creating less pressure on the face has been introduced along with an improved speech device. Special attention has been placed on making the mask, which is available in three sizes, a good fit for a wide percentage of the user population.

The GP-7K filter unit is mounted on the left-hand side of the mask and is covered with fabric. The face piece is held in position on the head by a rear panel and five straps; the two cheek straps have metal self-tightening buckles. Each mask is provided with a protective fabric cover and a hold-all. Also provided with each mask are two thermal inserts and six anti-perspiration layers to absorb facial sweat.

The GP-7 face mask, the MGP, weighs 600 grammes, and the filter unit 250 grammes.

STATUS
In production.

MANUFACTURER
State factories.

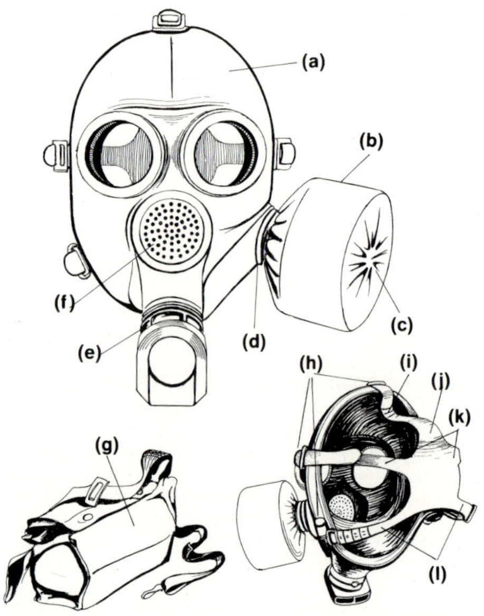

Parts of the GP-7 civilian protective mask: **(a)** *MGP face mask* **(b)** *GP-7K filter unit* **(c)** *filter unit fabric cover* **(d)** *inlet valve location* **(e)** *outlet valve* **(f)** *speech device* **(g)** *hold-all* **(h)** *buckles* **(i)** *forehead strap* **(j)** *strap assembly* **(k)** *temple straps* **(l)** *cheek straps*

Civilian Protective Mask, Models GP-5 and GP-5M

DESCRIPTION
The Model GP-5 may be regarded as the civilian counterpart of the military Model ShM as the rubber head-covering face-piece (designated ShM-62) is almost identical to the Model ShM. Like the Model ShM, the Model GP-5 uses air deflection across the eye-piece interiors and is issued in five sizes, but with the Model GP-5 the filter canister is fitted directly onto the inlet valve. The filter canister is cylindrical and is fitted via a female connector on the face-piece. A fabric carrier contains the face-piece and filter canister and a set of anti-dim discs for fitting inside the glass eye-pieces. Pockets on the carrier contain a small first-aid kit and an anti-gas kit.

The Model GP-5M is essentially similar to the Model GP-5 and there are few differences between the two.

STATUS
Production complete. Scheduled to be replaced by the Model GP-7.

MANUFACTURER
State factories.

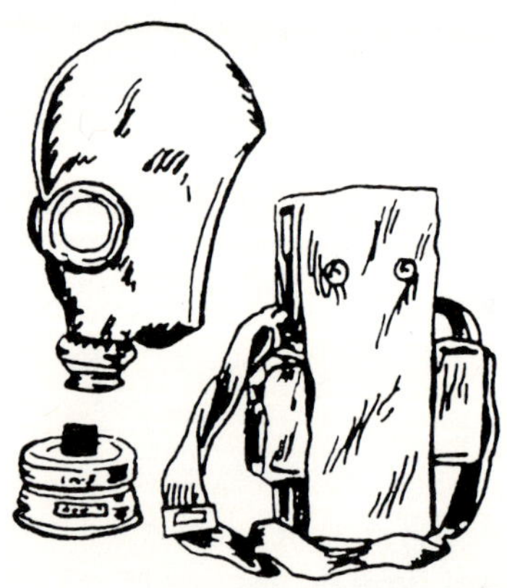

Model GP-5 civilian protective mask

Civilian Protective Mask, Model GP-4U

DESCRIPTION
Designed primarily for general civilian use, the Model GP-4U uses a moulded rubber face mask which is secured to the head by three straps fitted to six points on the mask. The round glass eye-pieces are held in place by crimped aluminium rings and as there is no other method provided or keeping the eye-pieces clear when worn, an anti-dim kit is issued with the mask. A single inlet and double outlet valves are fitted and air is drawn into the mask via a corrugated rubber hose which is secured to the face mask by wire and tape. A GP-4U canister is used for filtering. When not in use the Model GP-4U mask is carried in a fabric carrier fitted with a shoulder strap and a waist strap. The weight of the complete equipment in its carrier is around 2.5 kg.

The GP-4U filter canister used with this mask is cylindrical and weighs 0.56 kg. It is 150 mm high and 80 mm in diameter.

STATUS
Production complete. Scheduled to be replaced by the Model GP-7.

MANUFACTURER
State factories.

Special Protective Mask, Model ShMS

DESCRIPTION
The Model ShMS has provision for fitting optically corrective lenses, either spectacle lenses or special lenses to suit binoculars or weapon sights. Normally the Model ShMS is issued with optically flat lenses which are fitted into internal or external grooves, according to the circumstances, the corrective lenses fitting into the other groove. As a measure against dimming extra gelatine lenses can be fitted inside the internal lenses. The lenses are as small as 40 mm in diameter so the mask's visual efficiency is low but, as it retains the overall protection and filter system of the Model ShM, it is effective against nearly all agents. The mask is made from grey or beige rubber and is issued in five sizes. Weight of the fabric carrier, mask, tube filter canister, hose and anti-dim set is approximately 2 kg.

STATUS
In production and service with the Soviet and Warsaw Pact forces.

MANUFACTURER
Various state factories.

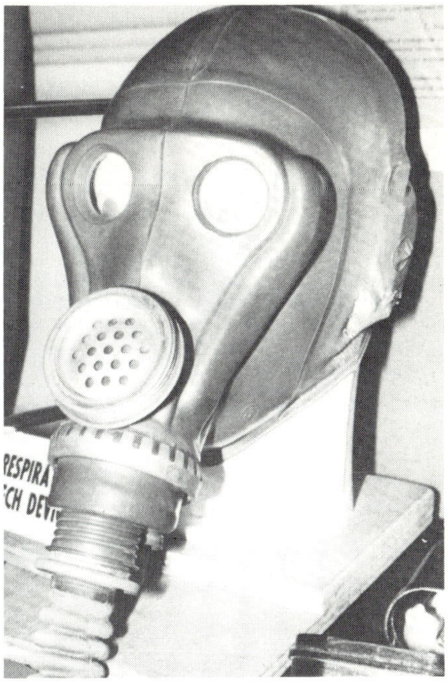

Special protective mask, Model ShMS, fitted with 'voicemitter' (T J Gander)

Protective Cape/Groundsheet

DESCRIPTION
This cape/groundsheet is normally carried in the same container/satchel as the protective mask, and it is thus issued to all Soviet troops. Two types of cape/groundsheet are in use, one of cloth-reinforced paper and the other of translucent green plastic. Both are essentially the same in shape and purpose but the cloth-paper version has reinforced hems into which canes can be inserted so the cape can be used as a shelter or small tent. The plastic cape has small black tie-wraps for the same purpose. The cloth-paper cape has a layer of cotton fabric bonded between two layers of brown paraffin-impregnated paper by water-resistant glue. When worn as a cape, both types are secured around the neck by tapes. Both types are considered expendable and weigh about 0.5 kg.

STATUS
In production. In service with the Soviet and Warsaw Pact forces.

MANUFACTURER
State factories.

Paper Protective Cape

DESCRIPTION
Although it has now been largely replaced by the later cloth-paper and plastic protective cape, the older paper protective cape is still in use. Normally it is carried in the same container/satchel as the wearer's protective mask, but the paper cape is intended to give only temporary protection against liquid agents and fallout particles, and is expendable. The cape is formed from two pieces of paper joined over the head and in a seam down the back. Cotton tie-wraps secure the cape in place once fitted. The cape is issued folded flat and can also be used as a groundsheet.

STATUS
No longer in production. Still used as reserve and training equipment by the Soviet and Warsaw Pact forces.

MANUFACTURER
State factories.

Protective Suit, Combined Arms

DESCRIPTION
This is the standard protective suit for all arms of the Soviet and Warsaw Pact forces and, although there are a few variations, the suit consists of a protective coat-overall (the OP-1), overboots and gloves.

The coat-overall OP-1 is a multi-purpose garment produced in five sizes which can be worn in several ways. With the sleeves inside the garment it can be worn as a cape or with the sleeves outside as a conventional overgarment. The lower portions may be wrapped around the legs and secured in place by straps. Apart from its protective use, the OP-1 can also be used as a raincoat, groundsheet, field shelter, flotation bag and as an emergency pannier for casualties.

The overboots are usually knee-length but hip-length versions are available. The knee-length versions are secured below the knee by tie-wraps and the hip length versions are usually secured to the wearer's uniform belt by straps.

The gloves come in two versions: a temperate climate version with five fingers, and a cold-weather version with two fingers and a thermal lining.

All items are manufactured from rubberised fabric and provide reasonable protection against most NBC agents. However, the suit is not completely airtight and extra protective undergarments must be worn for complete protection.

The coat-coverall OP-1 weighs approximately 1.6 kg, the overboots 1 kg and the gloves 0.4 kg.

STATUS
In production. In service with the Soviet and Warsaw Pact forces.

MANUFACTURER
State factories.

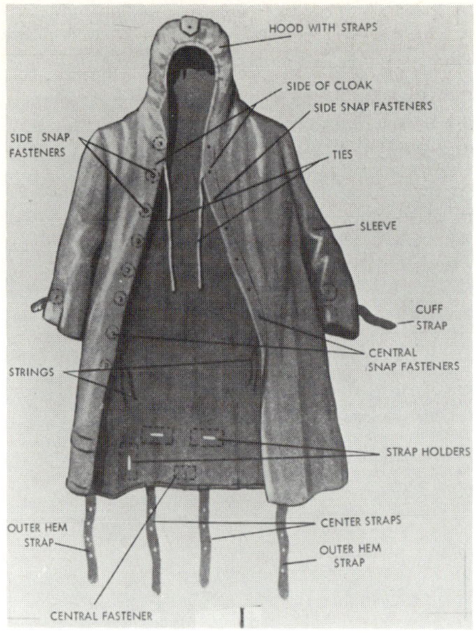

Main component parts of OP-1 protective suit

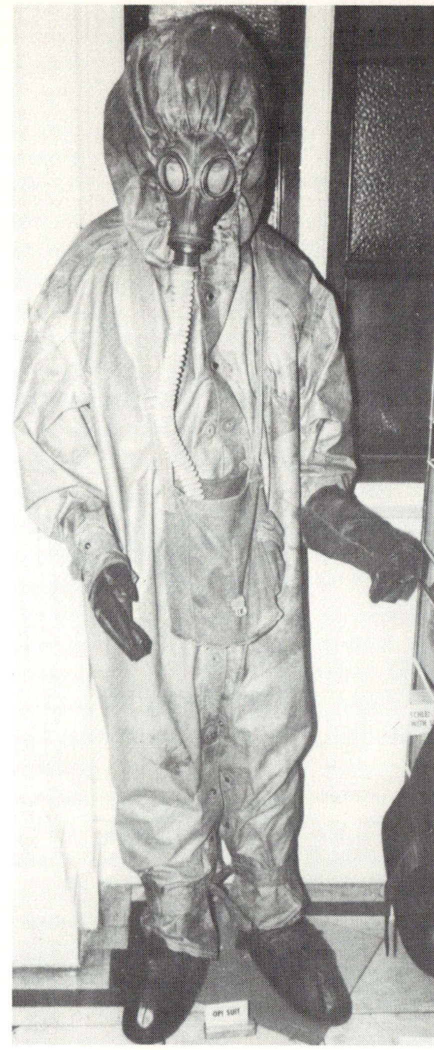

OP-1 protective suit, combined arms, shown with hood and Model ShM helmet mask (T J Gander)

Impregnated Coveralls, Model ZFO-58

DESCRIPTION
These coveralls are intended to protect against toxic chemical aerosols and vapours, but not against liquid agents. They can also provide partial protection against biological and radiation contamination. The coveralls are manufactured from chemically-impregnated porous cotton fabric, and are intended to be worn with underwear, socks and a hood liner impregnated with the same chemical. The suit is sealed by elasticated cuffs and tie-wraps and gas flaps are provided. The coveralls can be worn either by themselves or (more normally) under protective clothing, and they are issued in three sizes.

STATUS
In production. In service with the Soviet and Warsaw Pact forces.

MANUFACTURER
State factories.

Heavy Protective Suit

DESCRIPTION
Designed for use by decontamination personnel and personnel handling toxic agents and munitions, this suit is produced in two basic versions: a one-piece garment and a two-piece suit. As a further variation both can have integral boots. The suits are made of cloth coated by a thick layer of rubber. They have integral hoods, sewn-in bibs, wrist, ankle and neck straps, thumb loops and belts. To complete the suits there are cloth liners and heavy gloves and boots. The suits provide excellent protection against all agents but as with other types of similar impermeable suit they are most uncomfortable to wear. Body heat build-up limits the wearing time to a maximum of two hours and in warmer climates to as little as 15 minutes.

STATUS
In production. In service with the Soviet and Warsaw Pact forces.

MANUFACTURER
State factories.

Lightweight Protective Suit, Rubberised, L-1

DESCRIPTION
Normally worn by Soviet and Warsaw Pact reconnaissance personnel, the L-1 protective suit consists of a jacket with fitted hood, overtrousers with integral overboots, two pairs of two-fingered gloves and a carrying satchel. It is designed so that it can be donned rapidly. The trousers have a strap at the left and a bib joins to straps fitted with elasticised sections. The trouser legs can be fitted with additional straps just below the knee and there are overboot straps at the instep. The jacket, which can be worn over or under the trouser straps (over seems to be the more likely) has an elasticated waist and another strap provides a neck seal. The jacket cuffs and the gloves have elastic sealing bands. When worn with a face mask the L-1 suit provides complete protection against most NBC agents. The basic fabric is butyl rubber-covered cotton fabric, except the satchel, which is canvas. The L-1 protective suit weighs approximately 3 kg.

STATUS
In production. In service with the Soviet and Warsaw Pact forces.

MANUFACTURER
State factories.

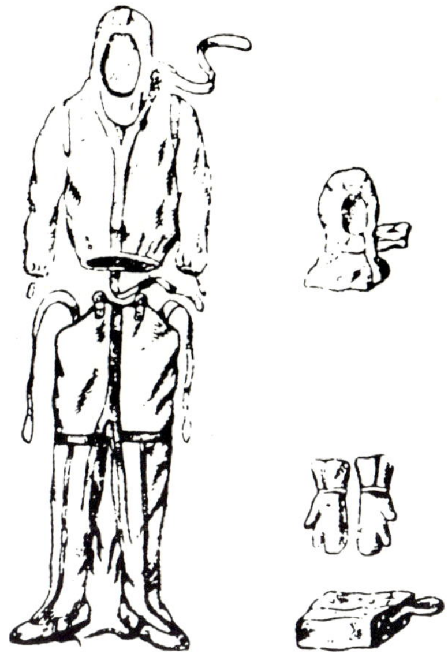

L-1 rubberised lightweight protective suit

Rubberised Protective Coverall

DESCRIPTION
This coverall is intended for use by personnel involved in NBC decontamination and toxic agent handlers. It consists of a single garment with an integral jacket, hood and trousers, and it also has a sewn-in bib, wrist, neck and ankle straps, thumb loops and a belt. It is normally worn with a hood liner, heavy rubber boots and gloves and a protective mask, which provide excellent protection against all known chemical agents, but as the suit is made from cotton fabric coated with green or black synthetic rubber, it is very uncomfortable. Wearing time is limited to as little as 15 minutes in a warm climate or up to two hours in colder surroundings, which can be increased by wearing cooling-type coveralls. The suit is produced in three sizes and weighs approximately 3.6 kg.

STATUS
In service with the Soviet and Warsaw Pact forces.

MANUFACTURER
State factories.

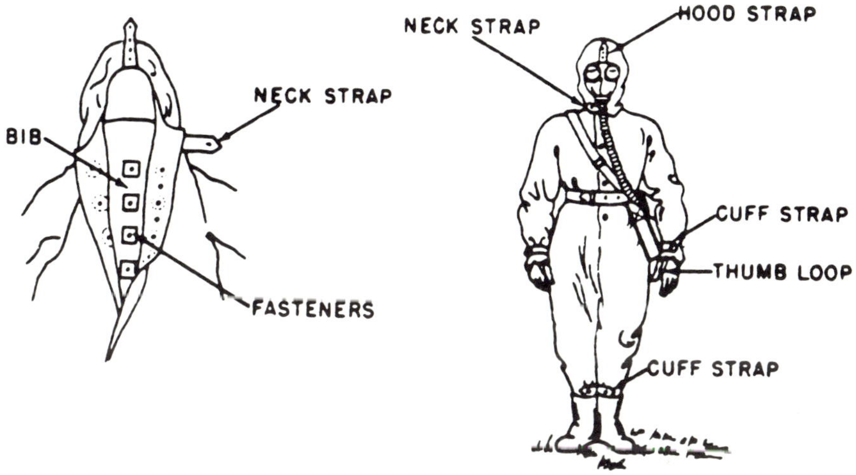

Rubberised protective coverall

Heavy Rubber Gloves

DESCRIPTION
These gloves are intended for use by decontamination teams or personnel handling toxic agents or fuels. Two types are produced, one with three fingers and one with five. Both are manufactured from heavy black synthetic or natural rubber on a fabric backing. The three-fingered glove is moulded flat and can fit either hand; a wrist strap forms a tight seal. The five-fingered gloves are issued 'handed' and lack the sealing strap. Both types of glove are intended for use with the heavy protective suit and both weigh about 0.34 kg a pair.

STATUS
In production. In service with the Soviet and Warsaw Pact forces.

MANUFACTURER
State factories.

Protective Apron, Rubberised

DESCRIPTION
Intended for use by personnel involved in decontamination of weapons and heavy equipment, this apron has replaced an earlier version made of either impregnated paper or polyvinyl chloride. The later version is manufactured from rubberised fabric and covers the body from the chest to the knees. It is 1 metre long and 0.85 metre wide, and weighs about 0.4 kg.

STATUS
In production. In service with the Soviet and Warsaw Pact forces.

MANUFACTURER
State factories.

Cooling-type Hooded Coverall

DESCRIPTION
The limits to the length of time rubberised protective garments can be worn can be partially off-set by special cooling overgarments which are designed to be worn over heavy protective clothing. The standard Soviet garment for this purpose, which is also issued to most Warsaw Pact forces, is made from cotton. It consists of a single-piece coverall with an integral hood, a five-button overlapping front opening, and securing straps or ties at the neck, wrist, ankle and mid-calf. There is a belt and a pocket for the protective mask filter canister. In use the suit is soaked in water, and the subsequent evaporation of the water cools the suit beneath. The efficiency of the cooling process depends on humidity and temperature, but the suit is not intended for use at temperatures below + 15°C.

STATUS
In production. In service with the Soviet and Warsaw Pact forces.

MANUFACTURER
State factories.

Cooling-type hooded coverall

UNITED KINGDOM

Respirator NBC S10

DESCRIPTION

The Respirator NBC S10 was designed and developed by Avon Industrial Polymers Limited for the British Ministry of Defence. The S10 will eventually replace the S6 respirator in service with the British Armed Forces. Its advantages include improved levels of protection, communication, vision, comfort, compatibility and maintainability. The S10 also has an integral drinking facility.

A production contract worth over £10 million was awarded to Avon Industrial Polymers in September 1987. The contract was placed by the United Kingdom Ministry of Defence following the initial development contract placed in January 1982. Initial production orders were for 200 000 respirators and the S10 was accepted into service by the Ministry of Defence in 1985.

For protection the S10 has a high performance integrally moulded reflex seal. The facepiece is available in four sizes and there are a minimum of moulding crevices which improves decontamination efficiency. The canister is supplied in a wrapped condition and it can be quickly replaced without removing the facepiece. The facepiece is held on the head by a fixed buckle harness arrangement that allows the respirator to be repeatedly donned and removed in seconds. Primary communication is via a horn-loaded voice transmitter but there is also a secondary speech transmitter for use with communications equipment. Vision is improved by the use of low profile coated polycarbonate lenses and a corrective lens system for spectacle wearers is available. A facepiece rib ensures a good interface with NBC protective clothing hoods. The drinking device has a double safety valve feature.

Respirator NBC S10

SF10 respirator for special forces fitted with tinted outset lenses to protect against flash

The filter canister can be worn on either the left or right of the facepiece and the position can be changed rapidly and easily – the S10 can also be used with remote filtration units. The filter canister S10 has the STAG 4155 thread.

The SF10 is a variant of the basic S10 developed for use by special forces. It differs from the basic version in having provision for an internal microphone (for use with a communications harness/radio transmitter and ear defenders) in place of the drinking facility and a cheek-mounted screw-in plug that provides access to a second filter canister mount that can accept an air-escape bottle for conditions of oxygen depletion – it also allows the use of twin filter canisters. The eyepieces can also accept tinted outset lenses to provide protection for the wearer against flash (such as that produced by stun grenades) and fragments; they can be quickly removed by hand. Total weight of the SF10 is approximately 0.8 kg.

The AR10 is a version produced for use by police forces. It has no drinking facility and uses flat polycarbonate lenses.

STATUS

In production. In service with the British Armed Forces and the Saudi Arabian National Guard.

MANUFACTURER

Protection Division, Avon Industrial Polymers Limited, Bradford-on-Avon, Wiltshire BA15 1AA, United Kingdom.

Respirator NBC S6

DESCRIPTION

Known to the British armed services as the Respirator NBC S6 No 1 Mark 1, the S6 has several advanced features not found on many other similar items of equipment, the most noticeable of which is pneumatic sealing for a close and comfortable fit. The shaped eye-pieces provide good vision and condensation is kept at a minimum by routeing incoming air over them. Their design is such that spectacles can be worn and no special fittings are needed to use weapon sights or binoculars. Weapon handling is taken into consideration by placing the filter canister on the left side of the mask (versions are produced to suit left-handed wearers). The face mask also mounts a speech transmission unit coupled to the outlet valve and all such fittings are made from anodised aluminium; the mouldings are natural rubber. The S6 is made in three sizes and the product has a shelf life of 25 years. It is stated that the S6 takes only 10 seconds to place into position. Normally it is carried in a chest, belt or shoulder-mounted haversack which weighs 0.57 kg

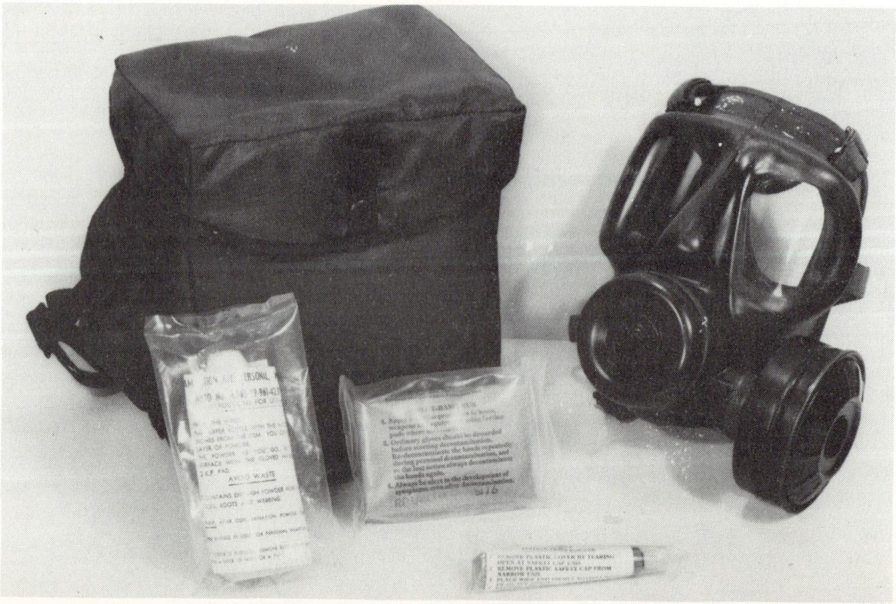

Respirator NBC S6 with satchel and cleaning/decontamination kit

together with a spare filter canister (see following entry). The respirator itself weighs 0.83 kg. A kit containing a microphone, rubber tube and connectors can convert the S6 for AFV use in about one minute.

The S6 is licence-produced in Turkey for the Turkish armed forces.

STATUS
In production. In service with the British Armed Forces and with Turkey.

MANUFACTURERS
The Leyland and Birmingham Rubber Company Limited, Golden Hill Lane, Leyland, Preston, Lancashire PR5 1UB, United Kingdom.
Makina ve Kimya Endustri Kurumu (MKE), Ankara, Turkey.

Respirator NBC S6 (T J Gander)

Haversack, Respirator, Anti-Gas, Mark II

DESCRIPTION
This shoulder-mounted haversack is made from waterproof polyurethane-coated nylon material and has a nylon webbing carrying strap. It was specifically designed to carry the S6 respirator and can also be used with the Respirator NBC S10. The haversack carries the respirator and all associated personal NBC defence decontamination kits.

Weight of the haversack (excluding equipment) is 0.4 kg. It is 260 mm high, 210 mm wide and 130 mm deep.

STATUS
In production. In service with the British Armed Forces and many overseas defence forces.

MANUFACTURER
Remploy Limited, 415 Edgware Road, Cricklewood, London NW2 6LR, United Kingdom.

Haversack, Respirator, Anti-Gas S6 Mark II in use with S10 respirator

Negretti Aviation Aircrew System AR5

DESCRIPTION
Designed and developed in conjunction with the British Ministry of Defence Chemical Defence Establishment (CDE) at Porton Down, Wiltshire, and the Institute of Aviation Medicine and Royal Aircraft Establishment, both at Farnborough, Hampshire, the Negretti Aviation NBC respiratory system AR5 provides full physiological protection in the air and on the ground against NBC agents in any known form. Protection is provided against agents entering the respiratory system or contacting the eyes and skin above the shoulders. The system uses an under-helmet hood concept, supplied with filtered air. This concept allows the existing range of personal equipment to be used.

The basis of the AR5 respirator is an oronasal mask mounted on a close-fitting polycarbonate faceplate. The edges of the faceplate are bonded to a flexible bromobutyl cowl worn over the head and beneath an aircrew protective helmet. Both the hood and mask are separately supplied with filtered air, or air/oxygen, by way of a chest mounted manifold. The respirator is matched to the supply system of a specific aircraft through an appropriate manifold which proportions and meters the airflow to the hood and mask.

The physiological requirements of aircrew operating at medium altitudes are maintained by adding oxygen to the filtered air supply to the mask. For high altitude application an alternative manifold is used which enables oxygen to be supplied to the crewman via a conventional pressure demand oxygen regulator. The positive air supply to the hood prevents the ingress of contaminants and reduces thermal stress in hot and humid conditions.

In the event of a failure of the ventilator supply the filtered air can be obtained by normal breathing demand. A quick disconnect in the mask supply line is provided for use in the event of ditching. An optional rip panel enables the complete faceplate to be rapidly removed during an emergency such as a parachute descent into water or a helicopter ditching. The hood is equipped with a neck seal and a bellows device is incorporated to permit free head movement.

Undistorted vision is achieved by the use of a high optical quality polycarbonate faceplate which is coated to prevent abrasion and chemical penetration. Misting or frosting during use at low temperatures is prevented by constantly purging the optical area with a supply of filtered air. If this supply fails part of the oxygen supply can be diverted to purge the faceplate by operating a valve on the manifold.

The oronasal mask incorporates a microphone and drinking facility. By utilising close-fitting spectacle frames corrective lenses can be worn.

Two versions of the AR5 system are available. The Mark 2 system is a non-oxygen version which uses a manifold which supplies filtered air to both the oronasal mask and hood. This version is used for applications where air only is required for breathing (eg helicopters) or where a filtered air/oxygen mixture from a pressure demand regulator is used

Negretti Aviation aircrew respiratory system AR5 in use with aviator carrying a portable ventilator

to supply the respirator (eg on transport aircraft). The Mark 3 system has a two-position change-over valve, permitting either filtered air to both the oronasal mask and hood (ground position) or filtered air to the hood and oxygen/air to the mask via a pressure demand regulator (flight position). The Mark 3 is primarily intended for fast jet use.

In order to supply filtered air to fast jet aircrew a built-in ventilator and filter pack is available. It incorporates a voltage regulator and ventilator failure warning device. Tactical and portable ventilators are also available.

A portable test set is available for the AR5 system.

The basic AR5 respirator assembly weighs 0.72 kg and the aircraft-mounted ventilator 0.78 kg. The portable ventilator weighs 5.9 kg (with battery) and the tactical ventilator 2.6 kg.

STATUS
In production. In service with the Royal Air Force, the Canadian Air Force and the US Marine Corps.

MANUFACTURERS
Negretti Aviation, Edinburgh Place, Harlow, Essex CM20 2DJ, United Kingdom.

CAM LOCK (UK) Limited, 70 Hawley Lane, Farnborough, Hampshire GU14 8EN, United Kingdom.

Protective Gloves NBC

DESCRIPTION
These protective gloves are in two parts, an inner light cotton liner and a black chloroprene rubber outer glove. The outer glove is impermeable to liquids and vapours, and can provide full protection for up to about six hours. The gloves are stored and issued in sealed plastic packs.

STATUS
In production. In service with the British Armed Forces.

MANUFACTURER
James North and Sons Limited, PO Box 3, Market Street, Hyde, Cheshire SK14 1RL, United Kingdom.

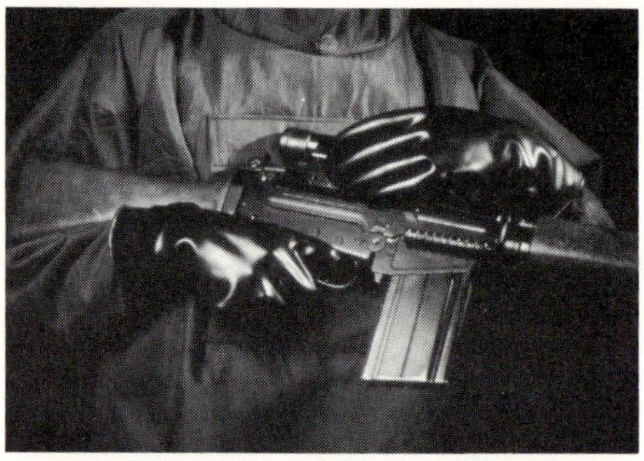

NBC protective gloves

Protective NBC Suit No 1

DESCRIPTION
There are three marks of the NBC Suit No 1, all made from the same charcoal-impregnated non-woven material inner lining which is impervious to liquids but permeable to air and water vapour, making garments less uncomfortable to wear than other comparable clothing. The material is also flameproofed to provide a measure of protection against nuclear flash and is shower-proof. The charcoal-impregnated non-woven material was developed at the Chemical Defence Establishment (CDE) at Porton Down in Wiltshire. The suits are pressure and vacuum packed to give extended shelf life, which in the case of the Mark 2 and 3, is four years, and for the Mark 4, six years. Both the Mark 2 and the Mark 3 are olive green and consist of overtrousers and a pull-over jacket. The Mark 4 uses a zip-up jacket in shades of disruptive camouflage. The No 1 Mark 2 suit is used by the Royal Navy on vessels with citadel systems. The Mark 3 suit was the standard ground forces issue but is gradually being replaced by the Mark 4. All three suits can be repaired with issue repair kits that contain six self-adhesive pvc patches.

The No 1 Mark 3 suit is supplied in four sizes to fit men from 1.5 to 1.9 metres in height plus special measures. The suit is a two-piece outfit which consists of a smock, with or without a hood and trousers. The smock uses two layers of material. The outer

NBC suit No 1 Mark 4

Donning the jacket of a NBC suit No 1 Mark 4

Suit No 1 Mark 3

NBC suit No 1 Mark 4 being worn by a British Army 81 mm mortar crew

layer is modacrylic and nylon while the inner layer is of charcoal-impregnated nonwoven cloth. The integral hood, when fitted, is lined and has a pull cord and cleat for adjustment and close fitting to the S6 and S10 respirators. The smock is generously cut and has a back gusset to allow freedom of movement. There is a large front flap pocket and a sleeve patch for locating detector paper. Adjustment is by Velcro with tape fastenings at the wrist and waist. The trousers are also made from two layers of material and have suspender tape fittings at the waist. There are Velcro tab fastenings at the base of the leg. Weight of the smock is 0.86 kg and trousers 0.7 kg. It is possible to supply the suit with the outer material finished in a disruptive camouflage pattern.

The Mark 4 NBC suit will eventually replace the earlier marks of NBC suit in service. It is a two-piece suit comprising jacket and trousers and is designed to be worn over combat clothing. The jacket has a front opening with a dual zip fastening. It has Velcro 'touch and close' fasteners to give effective closures at the wrist and for adjustment to size at the bottom edge of the jacket. Pouch pockets are provided each side of the front opening of the garment. The hood is a close fitting head piece, lined throughout with a plain weave cotton cloth. The hood is elasticated for close fitting with the S6 and S10 NBC respirators. The trousers have Velcro fasteners for waist adjustment and also to provide effective closures at the ankle and calf. Two tapes are stitched at the back of the trousers so that they can be used either as a belt or braces.

The Mark 4 NBC suit uses two layers of material. The outer layer is a flame-resistant modacrylic nylon twill fabric to reduce nuclear flash. The inner layer is the chief protective material constructed from a plyamide non-woven material impregnated with charcoal and fire retardent chemicals. Extensive testing has shown that when the suit is worn over normal combat clothing it provides a high degree of protection against chemical agents likely to be met in the field for a minimum of 24 hours.

STATUS
Mark 2: production complete. In service with the Royal Navy and some Commonwealth navies.
Mark 3: in production. In service with British Armed Forces and the armed forces of Australia, Egypt, Norway and Spain. Substantial numbers have been supplied to the American Department of Defense.
Mark 4: in production for the British Armed Forces.

MANUFACTURERS
J & S Franklin Limited, Franklin House, 151 Strand, London WC2R 1HL, United Kingdom.
Lantor (UK) Limited, St.Helens Road, Bolton BL3 3PR, United Kingdom.
Remploy Limited, 415 Edgware Road, Cricklewood, London NW2 6LR, United Kingdom.
James North and Sons Limited, PO Box No 3, Market Street, Hyde, Cheshire SK14 1RL, United Kingdom.

NBC Suit No 1 – Desert Pattern

The NBC Suits No 1 Mark 3 and Mark 4 – Desert Pattern are for use in desert environments. The Mark 3 suit comprises a smock and trousers, the smock having a hood with a draw cord and cleat to close around a respirator. The Mark 4 suit comprises a jacket with a full length front zip, and trousers. Both suits have an inner layer of charcoal impregnated non-woven fabric and an outer flame-retardent modacrylic/nylon fabric. This combination provides good chemical protection with a low physiological burden in hot or humid areas. The outer fabric is sand coloured or can be printed with a variety of disruptive pattern camouflage patterns to suit local terrains and requirements.

STATUS
In production.

MANUFACTURER
J & S Franklin Limited, Franklin House, 151 Strand, London WC2R 1HL, United Kingdom.

NBC Suit No 1 – Desert Pattern

NBC Clothing for Aircrew

DESCRIPTION
The complete set of NBC clothing for aircrew which was developed at the Chemical Defence Establishment (CDE) at Porton Down, Wiltshire, consists of an inner coverall, a hood and socks.

The NBC inner coverall permits aircrew to wear normal flying clothes without the constriction of additional outer garments. It is designed to allow the user to wear undergarments to reduce sweat pollution of the activated charcoal used in the material utilised. The inner coverall, which can be worn under immersion suits and other garments, is a one-piece close-fitting garment made from anti-

NBC coverall for aircrew

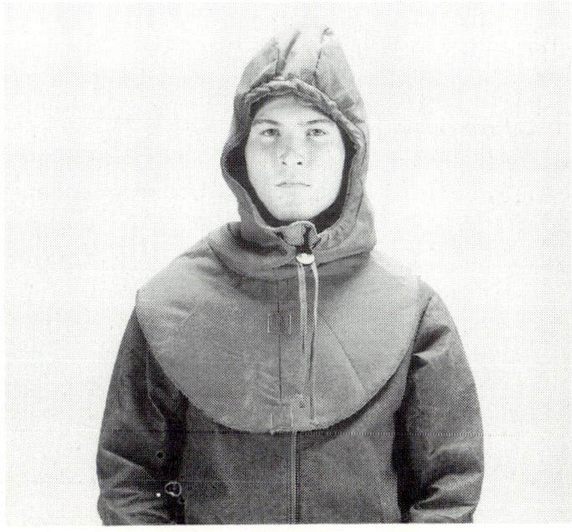

NBC hood for aircrew

NBC socks for aircrew

gas fabric and has a two-way front zip fastener. Pleating at the knee and elbow gives complete freedom of movement and elasticated stirrups allow a flying suit and boots to be pulled on quickly without risk of rucking. The garment can be worn around the clock and besides reducing fire hazards and protecting against burns will, when worn with normal flying clothing, give protection against all known chemical warfare agents. Under chemical warfare conditions the inner coverall can protect the wearer for up to 24 hours after the outer clothing has been contaminated. The garment can be worn for up to 50 hours in non-NBC conditions. Nine sizes are available. Weight of the inner coverall is 0.55 kg.

The NBC hood is made of the same anti-gas fabric as the inner coverall and is worn over Type G aircrew helmets and the AR5 respirator system (qv) or over leather helmets and S6 or S10 respirators. Only one size is available. Weight of the hood is 0.12 kg.

The NBC socks also use anti-gas fabric and are worn over the user's normal woollen socks but under the NBC inner coverall, flying suit and aircrew boots. They are available in eight sizes from 253 mm to 308 mm toe to heel. Weight is 0.08 kg.

All items are individually vacuum packed in envelopes with the added protection of an outer polythene bag.

STATUS
In production. In service with the British Armed Forces.

MANUFACTURER
Remploy Limited, 415 Edgware Road, Cricklewood, London NW2 6LR, United Kingdom.

Protective Overboots NBC

DESCRIPTION
The service model of the NBC overboot was the Mark 3. Since 1986 it has been replaced by the Mark 4 which is manufactured from high quality butyl rubber. It has a combat durability of approximately three weeks and at any one time chemical protection in excess of 24 hours is easily achieved. The materials can be readily decontaminated with standard decontaminants. Improvements introduced with the Mark 4 include an increase of 33 per cent in the thickness of the material used in the upper part of the boot. A better fitting than previously possible is assured by the choice of three sizes (sizes 3 to 5, 6 to 9 and 10 to 13) instead of the former one, and an integrally moulded heel. An improved tread pattern on the sole improves traction and stability.

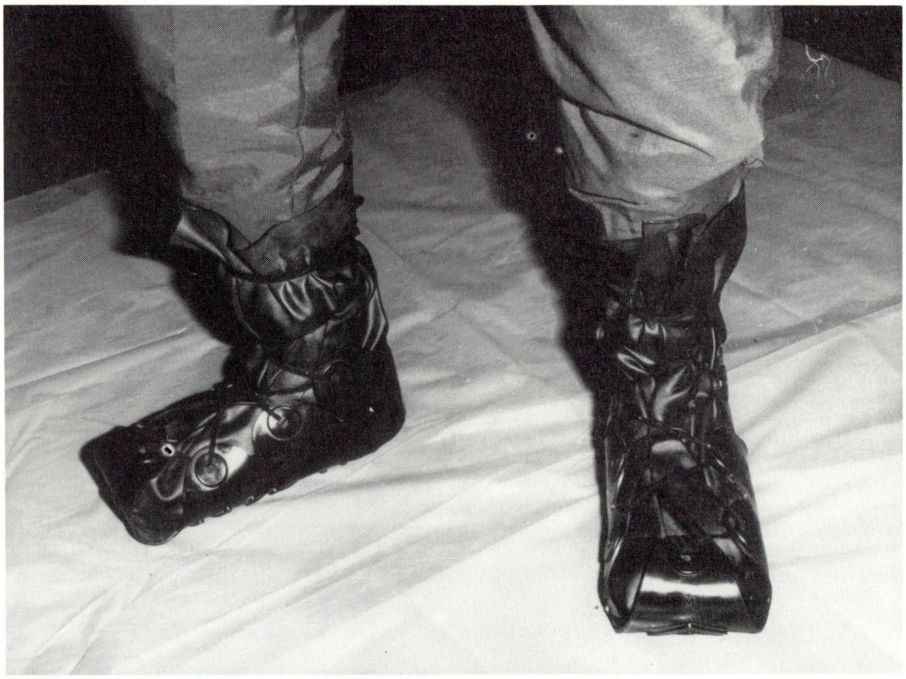

Mark 4 NBC protective overboots

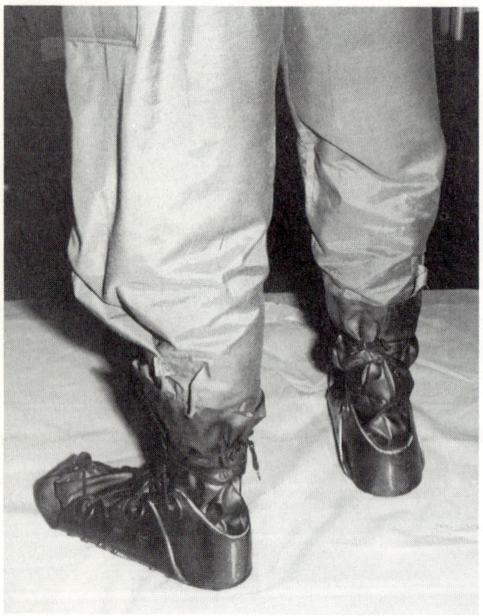

Mark 4 NBC protective overboots

A special version of the NBC overboot has been produced for use with skis and snowshoes. Known as the Overboot NBC Ski/March No 1 Mark 1 it is available in two sizes (6 to 9 and 10 to 13).

STATUS
In production. In service with the British Armed Forces.

MANUFACTURERS
Protection Division, Avon Industrial Polymers Limited, Bradford-on-Avon, Wiltshire BA15 1AA, United Kingdom.
Butyl Products Limited, 11 Radford Crescent, Billericay, Essex CM12 0DW, United Kingdom.
The Leyland and Birmingham Rubber Company Limited, Golden Hill Lane, Leyland, Preston, Lancashire PR5 1UB, United Kingdom.

Facelet NBC L1A1

DESCRIPTION
The Facelet NBC L1A1 was developed by the Chemical Defence Establishment (CDE) at Porton Down, Wiltshire, to provide personnel with interim protection against surprise attack by chemical weapons including nerve agents. It entered service with the British Armed Forces during 1983.

The Facelet NBC L1A1 provides oronasal protection and is donned as a precautionary measure when the wearer is on combat alert. It has been designed to exert a minimal physiological load, with the result that it can be worn over long periods and under all operational conditions. Each mask is provided with an adjustable harness which enables one size to fit the whole service population. Other features of the Facelet NBC L1A1 include its very low breathing resistance, high level of speech transmission and compatability with optical sights and other equipment. It provides a degree of protection at times when the full face respirator is not being worn, including periods of rest and sleep. Once a chemical attack is confirmed the Facelet can be removed quickly and the full face respirator donned. It is small and robust and can be easily carried in a serviceman's combat kit.

The main material used for the Facelet NBC L1A1 is activated charcoal cloth. Facelets are usually issued in vacuum-packed polythene packs with each pack containing three Facelets and one harness.

STATUS
In production. In service with the British and Dutch Armed Forces.

MANUFACTURERS
Charcoal Cloth Limited, East Wing, Bridgewater Lodge, 160 Bridge Road, Maidenhead, Berkshire SL6 8DG, United Kingdom.
Bondina NBC Materials, PO Box 3, Greetland, Halifax, West Yorkshire HX4 8NJ, United Kingdom.
Siebe Gorman and Company Limited, Avondale Way, Cwmbran, Gwent NP4 1YR, United Kingdom.
Remploy Limited, 415 Edgware Road, Cricklewood, London NW2 6LR, United Kingdom.

Charcoal Cloth Facelet NBC L1A1 in use

Suit Protective NBC – Decontamination No 1 Mark 1

DESCRIPTION
This is a two-piece translucent pvc suit consisting of a smock with integral hood, and overtrousers. It is intended for use by personnel decontaminating equipment and areas, and is worn over the normal NBC clothing and equipment, including the S6 and S10 respirators. The suit is disposable and should be worn for one hour only. The suit is supplied in three sizes.

STATUS
In production. In service with the British Armed Forces.

MANUFACTURER
Remploy Limited, 415 Edgware Road, Cricklewood, London NW2 6LR, United Kingdom.

Decontamination suit in use

Beaufort Maximum Protection NBC Suit

DESCRIPTION
The Beaufort NBC suit is a one-piece garment designed to provide maximum NBC protection and for rapid and easy donning in highly saturated areas where warnings are minimal, if given at all.

The one-piece suit is designed to fit over existing combat clothing. It has an integral hood and mitts, and overboots with Velcro tabs around each. There is a centrally placed sealed zip fastening which runs from crotch to chin closing the face seal around the appropriate respirator. The overboots are fitted with anti-slip ribbed soles. A belt is fitted for waist adjustment.

The fabric used for the garment is a combination of nylon textiles combined with a specially synthesised rubber compound which is designed to resist penetration and absorbtion of all known war gases and agents. It is designed to give a soft and flexible feel for comfort and drape and has a smooth regular matt finish to facilitate the run-off of liquids and to prevent beading. The fabric will not give off rubbery odours and will not block or stick to itself. It is also flame resistant and will not melt or drip in fire and can be treated or boiled in water for up to 3 hours without adverse effects or structural damage.

The fabric is totally impermeable to water up to 150 millibars and will not crack or blister at temperatures of up to +80°C; the minimum operational temperature is −40°C. It has good resistance to daylight ultra-violet radiation and can be treated to resist infra-red detection equipment.

STATUS
In production.

MANUFACTURER
Beaufort Air-Sea Equipment Limited, Beaufort Road, Birkenhead, Merseyside L41 1HQ, United Kingdom.

Beaufort maximum protection NBC suit

Maximum Protection NBC Suit

DESCRIPTION
This one-piece boiler suit is made from butyl coated fabric and is designed to provide maximum protection against NBC agents during either reconnaissance or decontamination operations. It has an integral hood with a drawstring, a Velcro fly along the front and the wrists and ankles are designed to be gathered by elastic seals and may be tucked into NBC gloves and overboots.

STATUS
In production.

MANUFACTURER
James North and Sons Limited, PO Box 3, Market Street, Hyde, Cheshire SK14 1RL, United Kingdom.

Maximum protection NBC suit

NBC Poncho

DESCRIPTION
This garment is designed to provide protection over and above that available from existing NBC suits. Manufactured from butyl coated fabric, there is an integral hood attached to a long poncho-type garment with large loose sleeves. It is worn over the usual NBC clothing and may double as a waterproof. There is also an internal charcoal cloth kilt/skirt. A Velcro fly front is provided and, in addition, there is a large front access zip under the poncho. If required a number of these garments can be joined together to provide a small shelter. The garments/suits may be provided in a number of colours.

STATUS
In production.

MANUFACTURERS
James North and Sons Limited, PO Box 3, Market Street, Hyde, Cheshire SK14 1RL, United Kingdom.
J & S Franklin Limited, Franklin House, 151 Strand, London WC2R 1HL, United Kingdom.

NBC poncho in sand colour scheme

Complete NBC Kit, Civilian, Heavy-duty

DESCRIPTION
This kit is intended for use by civilian and para-military authorities, including civil defence units, and can be produced to any size. The kit is stored and carried in a suitcase and consists of a heavy-duty washable and waterproof oversuit, an inner two-piece NBC protective garment, a full-face respirator with two canisters, a set of inner cotton and outer rubber gauntlets, a mask demisting pack, a pair of industrial rubber boots with steel toecaps, a roll of repair tape, decontamination powder and an instruction handbook.

The kit can be used as protection against either NBC or industrial hazards. Details of many of the above items can be found elsewhere in this section.

STATUS
In production.

MANUFACTURER
Civil Defence Supplies, Wellingore, Lincolnshire LN5 0JF, United Kingdom.

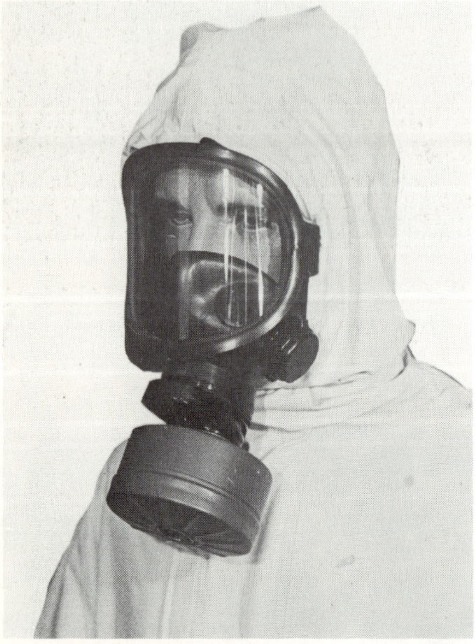

Full face respirator

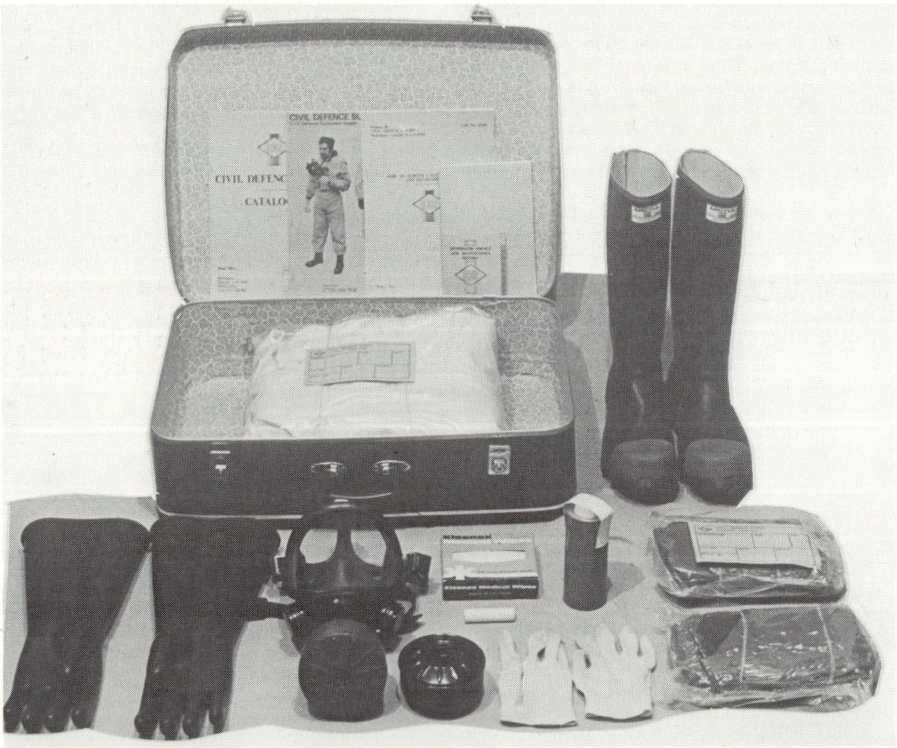

Complete individual NBC civilian kit

Complete NBC Kit, Civilian, Lightweight

DESCRIPTION
Intended primarily for civilian users as protection against NBC or industrial hazard, this kit can also be used by civil defence authorities. It consists of a complete lightweight kit of garments and equipment in a suitcase and can be produced in any appropriate size. It consists of a washable and waterproof lightweight oversuit, an inner NBC two-piece protective garment, a full-face respirator with one filter canister, a pair of industrial rubber boots with steel toecaps, a set of inner cotton and outer rubber garments, a mask demisting pack, decontamination powder and an instruction handbook.

Details of many of the above items can be found elsewhere in this section.

STATUS
In production.

MANUFACTURER
Civil Defence Supplies, Wellingore, Lincolnshire LN5 0JF, United Kingdom.

Lightweight civilian NBC suits in different sizes

Heavy-duty Outer Suit

DESCRIPTION
This is a one-piece heavyweight oversuit that is chemical resistant and flame retardant. Manufactured from pvc-coated knitted nylon the suit has extra wear patches and double cuffs and leggings. The integral hood can be adjusted to suit a variety of respirators. It is available in large and medium sizes. Intended primarily for civilian use, it can also be used as NBC protection and by civil defence authorities, or by rescue or other such emergency squads.

A lightweight version is also produced.

STATUS
In production.

MANUFACTURER
Civil Defence Supplies, Wellingore, Lincolnshire LN5 0JF, United Kingdom.

Combat suit protected against NBC agents in use

UNITED STATES OF AMERICA

Mask, Chemical-biological: Field, ABC-M17, ABC-M17A1 and ABC-M17A2

DESCRIPTION
These masks are the standard issue for all the US armed forces and have been in use since the early 1960s. The M17A1 differs from the M17 by having a self-contained drinking system and a resuscitation capability. The mask is well-designed with good vision and a voicemitter. The filter elements in the cheeks on each side are easily replaced without tools although the mask has to be removed for the changing operation. The masks are issued in an M15 carrier with two filter elements and an M1 CB mask waterproofing bag. The M17A1 mask is carried in an M15A1 carrier which also contains an M1 water canteen cap. The mask, which is available in three sizes, is kept in place by a six-strap head harness. When worn the mask can be covered by the Hood, Mask, M6A2 which can be held in position with underarm straps. The hood can be issued in camouflage colour schemes and has openings for the eyepieces and voicemitter and has a zip fastening in the front.

A version of this mask with a non-essential component omitted is known as the M17A2. In March 1985 the US Army Munitions and Chemical Command at Rock Island, Illinois, issued a $19 037 500 contract for 186 862 M17A2 masks with final deliveries completed in June 1986.

The Czechoslovak Model M-10 mask design (qv) appears to be a close copy of the basic M17.

STATUS
In service with the US Army, Navy, Air Force and Marine Corps. Available for export, and has been exported to nations such as Australia and Panama.

MANUFACTURER
Mine Safety Appliances Company, 600 Penn Center Boulevard, PO Box 426, Pittsburgh, Pennsylvania, USA.

Mask, Chemical-biological: Field, ABC-M17

Mask, Chemical-biological: MCU/2P

DESCRIPTION
The development of a new chemical-biological mask to replace the M17 series of masks has been under way for some years. At one time two possible replacement models were under development, the XM29 and the XM30. The XM29 was a one-piece mask but due to engineering difficulties with its flexible silicone lens it was passed over in favour of the XM30, a two-piece design which was redesignated the MCU/2P by the US Air Force. On the MCU/2P the lens is manufactured from moulded urethane. The face-piece is made from moulded elastomer with the periphery turned in to form a face seal. The lens is bonded into the face-piece. The mask uses an external filter canister developed in Canada which makes it suitable for right- and left-handed personnel. The design also has facilities for hose connections to chest-mounted filter systems, AFV or aircraft air systems, there are fittings to cater for drinking systems and mouth-to-mouth resuscitation, and there is a dual voicemitter. The mask is manufactured in three sizes.

The design of the XM30 (MCU/2P) was undertaken by the US Army Armament Research and Development Command's Chemical Systems Laboratory at Aberdeen Proving Ground, Maryland (now closed). For various reasons the US Army decided not to adopt the XM30 (MCU/2P) but the mask was accepted for service by the US Air Force who ordered an initial batch of 10 000 to be delivered during 1983 and 1984.

STATUS
In production for the US Air Force.

Mask, Chemical-biological: MCU/2P

Mask, Chemical-biological: Special Purpose, M9 and M9A1

DESCRIPTION
The only difference between the M9 and M9A1 is the carrier bag: the M9 uses the C15R1 and the M9A1 the M11 carrier. Other items are the M9 face-piece assembly and the M11 canister. The face masks are issued in three sizes and the filter canister can be fitted to either the left or the right. When worn the mask is held in place by a head harness. An optional extra for this mask is the winterising kit, CB mask, M1 which consists of a hood, insulating lenses, anti-snowglare lenses, a cheek pad and a plastic bag containing four spare nosecup valve discs. The kit is contained within the hood which is folded into a packet weighing approximately 0.32 kg.

The M9A1 is licence-produced by the Daewoo Corporation in South Korea as the KM9A1 (see separate entry in this section). The M9 was also produced in Sweden by Trelleborg AB and Forsheda AB as the Skyddsmask 51. The Finish KEMIRA M76 was also derived from the M9.

STATUS
In service with the US Army and Marine Corps.

Mask, Chemical-biological: Tank, M25 and ABC-M25A1

DESCRIPTION
These masks are virtually identical to the earlier ABC-M14A2. The M25 is a modification of the M14A2 and the ABC-M25A1 is virtually the same item. In all other respects the two masks follow the same lines as the ABC-M14A2.

STATUS
In service with the US Army and Marine Corps.

M40 and M42 Chemical-Biological Masks

DESCRIPTION
Following the non-acceptance of the XM30 protective mask (now the MCU/2P (qv)), during 1983 the US Army Chemical Systems Laboratory (later part of the now-closed US Army Chemical Research, Development and Engineering Center) produced prototypes of a new mask with a 'low risk' design incorporating the best features of many existing masks. This evolved into the XM40 multi-purpose chemical-biological mask which retained many features of the XM30 design plus the hard lens of the M17 series of masks. Other improvements included a periphery turned inwards to protect against leakage, a side transmitter for telephone communications, provision for a radio microphone and an external screw-on filter canister. Other design features include a silicon rubber facepiece and a six-point adjustment head harness. The mask is compatible with combat spectacles.

Three companies became involved in the XM40 programme resulting in three designations, XM40, XM41 and XM42. The three companies were Scott Aviation of Hebron, Ohio, ILC Dover Inc of Dover, Delaware, and Avon Industrial Polymers Limited of the United Kingdom (the latter submitted a version of its NBC S10 respirator known as the US10 which was later excluded from the programme for 'unsuitability'). Other companies became involved later, including Mine Safety Appliances of Murrysville, Pennsylvania. Testing began during fiscal year 1984 and the original schedule was that the first delivery should have been made in July 1985. However, by late 1986 the programme was over 15 months behind schedule and a request for proposals (RFP) for production had not been issued. By 1987 development of the XM40 was complete.

In July 1987 it was announced that Scott Aviation, a division of the Figgie International Company of Lancaster, New York, had been awarded a $22 155 979 firm-fixed-price contract to produce 115 750 chemical-biological masks with repair parts to support initial fielding. This would have been the first of two increments (to be completed by December 1989) that were intended to eventual produce 300 000 masks for a total contract value of $51 823 181. The masks involved were the M40, for infantry, and the M42 for combat vehicle crewmen. The original intention was that the M40 and M42 will eventually replace M9A1, M17-series and M25A1 masks in service with the US Army and Marine Corps from the summer of 1988 onwards.

In April 1988 it was announced that the contract awarded to Scott Aviation would expire after only one year of the contract involved. Scott Aviation did not deliver the first of the new masks on schedule and lobbying from the other companies involved in the bidding for the 1987 contract claimed that the contract award was unfair. Scott Aviation will be allowed to bid for the new contract along with the other original manufacturers.

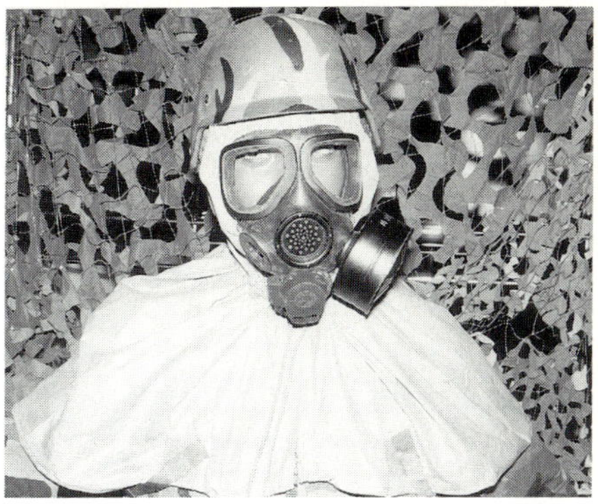
M40 chemical-biological protective mask (T J Gander)

STATUS
Ready for production.

MANUFACTURER
Scott Aviation, a division of Figgie International Company, Hebron, Ohio 43025, USA.

M43 Protective Mask, Aircraft AH-64

DESCRIPTION
The M43 (originally the XM43) Protective Mask, Aircraft AH-64 was specially developed for aircrew flying the AH-64A Apache two-seat attack helicopter. The mask consists of a form fitting facepiece with lenses fitted close to the eyes, an integrally attached hood and a skull-type suspension system. The mask provides AH-64A crew members with the required respiratory/skin protection from chemical and biological agents, toxins and radioactive fallout particles.

STATUS
Type classified for limited production. In service with the US Army.

DEVELOPMENT AGENCY
US Army Chemical Research, Development and Engineering Center, Aberdeen Proving Ground, Maryland, USA (now closed).

MANUFACTURER
ILC Dover Inc, Dover, Delaware, USA.

M43 Protective Mask, Aircraft AH-64

MF-11 Military Respirator

DESCRIPTION
The MF-11 military respirator is a light weight gas mask providing protection for the respiratory organs, eyes and facial skin. It consists of a facepiece with an integral drinking device, integrated exhaust valve and voicemitter, a fully adjustable six-point head harness and a gas-particulate filter canister providing protection against aerosol and gas/vapour NBC threats.

The MF-11 is available in three sizes (small, medium and large) and is made from specially compounded natural rubber for a gas-tight, non-irritating facial fit and increased resistance to chemical agent permeation. The lenses are manufactured from optical plastic and are surface treated for high strength and abrasion resistance. The triangular lens shape provides a wide visual field and good optical properties with low distortion.

An air guide is provided to direct inlet air along the interior surface of the lenses to prevent fogging. Air then enters a nosecup and is expired through a low resistance exhaust valve. The exhaust valve is integrated with a butyl-rubber membrane voicemitter. This assembly can be easily removed for cleaning or replacement. An optional valve is available to operate the mask under positive pressure from an external blower.

The MF-11 is equipped with a drinking tube

Component parts of the MF-11 military respirator

MF-11 military respirator in use

mounted on the right cheek position and can be used with a standard water canteen. The complete water access system consists of a water inlet valve assembly, a suction tube, a canteen cap and a connecting pipe. The system is provided with a check valve and screw caps to prevent the ingress of contamination.

The gas-particulate filter canister uses NATO threads and can be removed and replaced in seconds while still wearing the mask. The activated carbon used in the canister is impregnated with catalysts to provide extended protection against low molecular weight acid gases such as hydrogen cyanide (AC) and cyanogen chloride (CK). The mask/canister combination has a low air flow resistance.

Weight of the MF-11 with the canister and drinking system components is 0.7 kg. The canister weighs 0.2 kg.

The MF-11 is based on the Chinese M-85 (see page 24).

STATUS
In production.

MANUFACTURER
KT Corporation, 300 Long Beach Boulevard, Stratford, Connecticut 06497, USA.

NBC Mask for Dependents

DESCRIPTION
During early 1987 the US Army Laboratory Command's Human Engineering Laboratory at Aberdeen Proving Grounds, Maryland, announced that they had developed an NBC mask for use by the families and dependents of American personnel, and other civilians, in the event of a chemical attack in Europe or elsewhere. The mask has been adapted from a commercially produced off-the-shelf design and was originally designed by special operations forces who went on to re-define their operational requirements.

The mask is stated to be inexpensive and takes up half the space of a standard mask. It consists of a durable hood with an integral lens, a nasal breathing compartment, a neck seal and a filter canister. It provides protection for the wearer's head, face, shoulders and respiratory system against 'the full range of chemical and biological agents' and may be worn regardless of long hair, beards, moustaches or eye glasses. Weight is approximately 0.9 kg.

STATUS
Ready for production.

DEVELOPMENT AGENCY
US Army Laboratory Command, Human Engineering Laboratory, Aberdeen Proving Ground, Maryland 21005, USA.

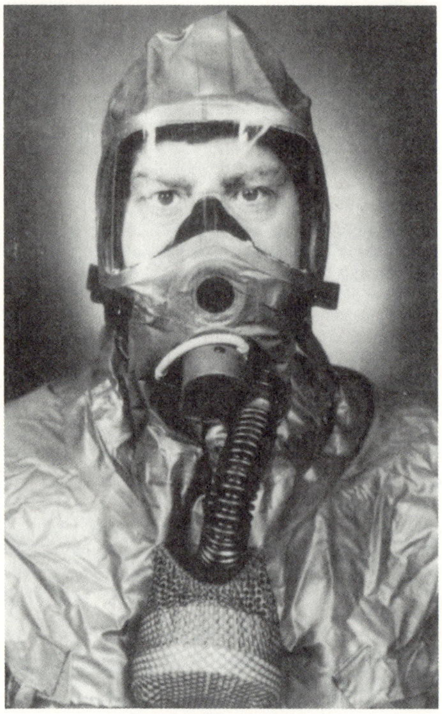

NBC mask for dependents

Chemical Protective Outfit

DESCRIPTION
The standard American protective outfit is issued in one complete package. It consists of a protective suit, gloves, overboots and a hood and is intended to be worn with the M17 and M40-series masks. The suit consists of a jacket and trousers, both made from charcoal-activated cloth with a nylon/cotton twill outer layer which can be either camouflaged to standard patterns or olive green. The suit is packed in a hermetically sealed polypropylene bag which has a shelf life of five years; the pack weighs approximately 1.7 kg. The gloves, overboots and hood are all manufactured from butyl rubber. If required, the suit and other garments can be issued in one pack which also contains the NBC face mask. The suit is made in eight sizes from XXX-small up to XX-large.

STATUS
In production. In service with the US armed forces and some other nations.

MANUFACTURER
Winfield International Limited, Empire State Building, 350 Fifth Avenue, New York, New York 10118, USA.

Chemical protective outfit

Saratoga System Protective Clothing

DESCRIPTION
The Saratoga System was originally developed by Blucher GmbH of the Federal Republic of Germany in close co-operation with the American and Israeli governments and industries and some European nations. The Saratoga System is a family of different materials using spherical activated carbon adsorbers providing good chemical agent protection and user comfort. Saratoga System garments can be washed and re-used repeatedly.

The Saratoga System uses spherical adsorbers which, due to their symmetry, have a higher strength than any other form of activated carbon. This allows higher quantities of activated carbon per unit area than other filters while offering the best possible flow conditions for adsorbtion.

During the manufacturing process the outer surface of the activated carbon spheres becomes extremely hard but there are enough pores to ensure access to the softer core, thus ensuring high adsorbtion efficiency with good wearing properties. Due to a special treatment of the inner surface of the carbon spheres, nitrogen-containing groups are formed which bind water only slightly. This treatment leaves Saratoga System materials practically insensitive to humidity and perspiration and allows repeated washings.

In the Saratoga System the carrier material is not fixed, thereby permitting optimum adaptation to specific problems by using the appropriate fibre or textile constructions. The special process of fixing the adsorbers on the carrier ensures maximum adherence with good hydrolytic stability, together with good fastness to rubbing abrasion. As the adhesive covers only a small part of the adsorbers, the larger surface area provides for a greater number of impact sites leading to improved adsorbtion.

The Saratoga System has been adapted to address a broad range of specific military applications by adjusting the fibre or texture construction. Spherical adsorber distribution can vary from 50 to 200 grammes per square metre.

Protective garments that are available using the Saratoga System include the following:
Army ground crew ensemble. Issued in monopack form this ensemble provides good liquid and vapour chemical agent protection with an overall weight of less than 550 grammes per square metre. In the construction used for this and other similar garments all the various layers are integrated into one textile material. The outer shell has a weight of 254 grammes per square metre, the carbon layer is 180 grammes per square metre while the liner is 57 grammes per square metre. As with all other Saratoga System-based garments the ensemble is washable and reusable.
Aircrew coverall. This is issued in monopack form and weighs less than 380 grammes per square metre. The shell weight is 140 grammes per square metre, the carbon layer 120 grammes per square metre and the liner 60 grammes per square metre. The aircrew coverall is flame resistant.
Light combat coverall or air force ground crew ensemble. The construction of this ensemble differs from the others described here in that the carbon layer adheres to the shell and the liner is sewn-in during manufacture. Issued in twin-pack form the weight is less than 340 grammes per square metre. Shell weight is 170 grammes per square metre, the carbon layer is 90 grammes per square metre and the liner 57 grammes per square metre.
Undercoverall for army ground crew, aircrew and fire fighters. This is worn as a pyjama undergarment and provides protection not only against chemical agent vapours but is also flame resistant. It weighs less than 360 grammes per square metre with the shell and liner both being 100 grammes per square metre. The carbon layer is 80 grammes per square metre.

STATUS
In production. In service with some NATO armed forces and under final evaluation for the US armed forces.

MANUFACTURER
Saratoga USA Inc, 350 Fifth Avenue, Suite 6608, New York City, New York 10118, USA.

Saratoga undercoverall pyjama with integral head protection

Saratoga CWU 66/P integrated chemical defence protection flight coverall

Spherical activated carbon adsorbers for use in various Saratoga System NBC protective garments

Brunswick Chemical Protective Gloves

DESCRIPTION
Brunswick Defense produces a series of chemical protective gloves made from butyl. They are produced in three thicknesses to suit the level of digital dexterity required for various military tasks. The thickest glove material is 0.635 mm thick, the next 0.355 mm thick and the thinnest 0.178 mm. They are produced in extra small, small, medium, large and extra large sizes. The maximum resistance to Mustard (HD) agent is 120 minutes for the 0.178 mm glove, 240 minutes for the 0.355 mm glove and 360 minutes for the 0.635 mm. Resistance to GB nerve agent is 150, 300 and 450 minutes respectively.

STATUS
Production complete. In service with the US Army.

MANUFACTURER
Brunswick Defense, One Brunswick Plaza, Skokie, Illinois 60077, USA.

YUGOSLAVIA

Protective Mask M1

DESCRIPTION
The Protective Mask M1 is used by all arms of the Yugoslav armed forces and some civilian organisations for protection against war gases and toxic fumes. It can be used under all conditions other than when the concentration of oxygen in the atmosphere is less than 16 per cent.

The complete equipment consists of a facepiece, a filter canister, a carrying bag and some accessories. The moulded rubber facepiece consists of the facepiece itself, two triplex vision panels, a nose cup body with inlet valves, a head harness, a tab assembly and valve sub-assemblies. The filter canister (TU-51) consists of the canister body with an anti-aerosol insert inside a charcoal cover together with a plug. The carrying bag may use either a shoulder or belt strap and also carries the accessories which are mainly a flannel and soap rag for cleaning and decontamination.

The M1 mask is available in three sizes, small (M), medium (S) and large (V). The individual sizes are determined by measuring the head volume of the user with the volume being determined horizontally by measuring the user's forehead above the eyebrows and ears and vertically by measuring from the top of the head before the ears to under the chin. Sizes can be determined from the following table:

Size	Vertical	Horizontal
small (M)	610–620 mm	550–560 mm
medium (S)	630–650 mm	570–590 mm
large (V)	660–670 mm	600–620 mm

STATUS
Production complete. In service with the Yugoslav armed forces and some other nations.

MANUFACTURER
Industry 'Miloje Zakic', 37000 Krusevac, Yugoslavia. Office: Street Marsala Tolbuhina 79/1V, Belgrade, Yugoslavia.

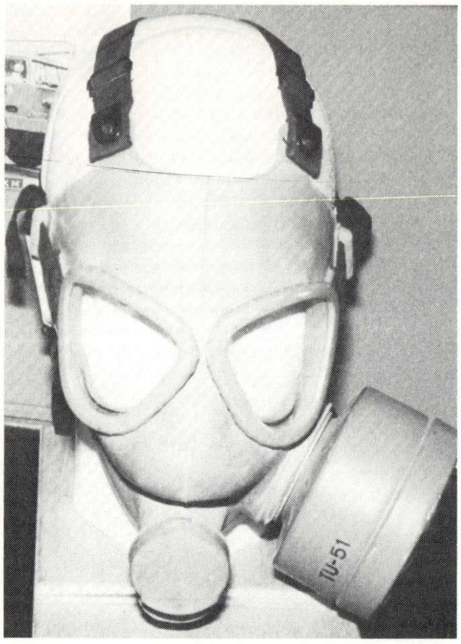

Yugoslav protective mask M1 (T J Gander)

Protective Mask M2

DESCRIPTION
The Protective Mask M2 has replaced the M1 version (see previous entry) in production and may be regarded as a progressive development of the earlier model. The M2 mask differs from the M1 mainly in the provision of a revised nose unit and a revised lens demisting system.

The complete M2 protective mask comprises the facepiece, a filter canister, a cleaning kit and a carrying bag. The M2 mask is available in three sizes, small, medium and large, and is compatible with protective clothing and steel helmets of Yugoslav manufacture.

A fully airtight fit against the face is produced by adjusting the elastic straps of the head harness. The user can maintain the mask using the cleaning kit supplied with the mask and which is normally contained inside the carrying bag. Individual damaged components can be replaced using ancillary tools that are supplied with the mask.

The M2 mask alone weighs approximately 0.8 kg. It will remain operational over a temperature range of from -30 to $+50$°C.

STATUS
In production.

MANUFACTURER
Federal Directorate of Supply and Procurement (SDPR), 9 Nemanjina Street, 11105 Belgrade, Yugoslavia.

Protective mask M2

NBC Protective Suit M3

DESCRIPTION
The main type of NBC protective suit used by the Yugoslav armed forces is the M3, a two-piece suit manufactured from polyester fabric rubberised on both sides plus an extra layer of butyl rubber. The suit consists of a blouse with an integral hood and trousers. All parts are sewn and glued together in such a way that no seams are exposed. There are extra inserts at the blouse and trouser waist, arm and leg endings and hood opening to ensure a good seal with other clothing and in areas such as the hood opening rubber bands are also provided.

The M3 suit is provided in three sizes, as follows:

Small (M)	height up to 1.67 metres
Medium (S)	height from 1.68 to 1.79 metres
Large (V)	height over 1.8 metres

Repairs can be made by glueing on rubber patches and the inserts can be replaced when damaged or frayed.

Gloves used with the M3 suit are the protective rubber gloves M3. These are butyl rubber five-fingered gloves with no textile insert so that talcum powder has to be scattered into the glove before wearing. They are available in two sizes, small (M) and large (V) and have a shelf life of ten years.

STATUS
In production. In service with the Yugoslav armed forces.

MANUFACTURER
Federal Directorate of Supply and Procurement (SDPR), 9 Nemanjina Street, 11105 Belgrade, Yugoslavia.

NBC Protective Suit M3 in use together with Protective Mask M1

NBC Protective Overall M5

DESCRIPTION
The NBC Protective Overall M5 is a one-piece garment designed for use by NBC decontamination and reconnaissance personnel. It is manufactured from flame resistant rubberised fabric and is provided with an integral hood and elastic sealing closures at the hood opening and at the wrist and ankle. All seams are impermeable. A front zip fastening is provided for easy donning and doffing and a draw string can be used around the waist.

The M5 overall is available in three sizes (small, medium and large) and weighs approximately 3 kg regardless of size.

STATUS
In production. In service with the Yugoslav armed forces.

MANUFACTURER
Federal Directorate of Supply and Procurement (SDPR), 9 Nemanjina Street, 11105 Belgrade, Yugoslavia.

NBC protective overall M5

NBC Protective Overboots M1

DESCRIPTION
The NBC Protective Overboots M1 are manufactured from flameproof rubberised fabrics. The upper parts of the overboots are thinner that the bottom parts and have straps sewn on them for securing them to the legs. The lower parts of the overboots have tie-wraps to secure them to the ankles and they are sewn to the uppers. The soles are ribbed for improved traction on smooth surfaces.

Only one size of overboot is available and they are finished in an olive grey colour. Weight is approximately 0.55 kg.

STATUS
In production. In service with the Yugoslav armed forces.

MANUFACTURER
Federal Directorate of Supply and Procurement (SDPR), 9 Nemanjina Street, 11105 Belgrade, Yugoslavia.

NBC protective overboots M1

NBC Protective Mask for Horses M1

DESCRIPTION
The NBC Protective Mask for Horses M1 is used to provide respiratory protection against NBC agents for draught, pack and riding horses and mules. The complete mask consists of a facepiece covering the animal's head (covering the nose, eyes and ears) and secured around the neck, two filter canisters, a cleaning kit and a carrying bag. The carrying bag may be carried on a pack or back saddle.

Once fitted the mask may be placed in one of two conditions, preparatory or protective. The state can be altered by the animal's handler by altering the position of a lever on the inhalation valve; the process takes about five seconds. It is claimed that once the mask is fitted the animal retains 80 per cent vision.

The mask, also known as the MZK M-1, is available in two sizes. The smaller size (S) is for animals with muzzles up to 470 mm in circumference; this version weighs 3.25 kg. The larger size (L) weighs 3.55 kg and is for animals with a muzzle circumference greater than 470 mm.

STATUS
In production. In service with the Yugoslav armed forces.

MANUFACTURER
Federal Directorate of Supply and Procurement (SDPR), 9 Nemanjina Street, 11105 Belgrade, Yugoslavia.

M1 protective mask for horses in use

DECONTAMINATION KITS AND EQUIPMENT

BRAZIL

ENGESA EE-25 (4 × 4) NBC Decontamination Truck

DESCRIPTION
This vehicle was designed and produced specifically for the NBC decontamination of troops, roads and vehicles and uses the 4 × 4 chassis of the ENGESA EE-25 2500 kg truck. For vehicle details refer to *Jane's Military Logistics 1989* pages 380 and 381.

The load-carrying area of the EE-25 truck has a 3000-litre capacity water tank integral with the platform, a 200-litre fuel tank for the heater, a decontaminant solution tank with a 160-litre capacity, a heater with a fuel consumption of 12 litres an hour, a decontamination solution dispenser with a normal mix of 1 litre of decontaminant to every 20 litres of water (graduated from zero to 30 litres an hour), a water pump with a flow of 48 litres a minute, 16 showers, two 100 mm diameter hoses, a road or terrain decontamination set and canvas covers for the shower areas.

STATUS
Production as required. In service with a North African country, believed to be Libya.

MANUFACTURER
ENGESA Engenheiros Espacializados SA, Avenida Tucunaré 125/211, Caixa Postal 152/154, 06400 Barueri, SP, Brazil.

ENGESA EE-25 (4 × 4) NBC decontamination truck

ENGESA EE-25 (4 × 4) NBC decontamination truck

CANADA

Decontamination Mitt

DESCRIPTION
This decontamination mitt is designed to meet the need for a ready means of individual decontamination and is used for the rapid removal of liquid agents from skin or other surfaces. The pouch-type mitt encloses the whole hand and wrist and there is a thumb opening on both sides. The mitt is large enough to wear over protective gloves.

The mitt is made of an open-weave material with a Velcro fastener at the wrist. Pockets on both sides of the mitt contain a natural bentonite powder similar to fuller's earth. In use the powder is dusted onto the contaminated surface through the material of the mitt.

The mitt is expendable and is packaged individually in a sealed bag made of laminated material.

STATUS
In production.

MANUFACTURER
Sheldon M Kasman Limited, 8 Milner Avenue, Scarborough, Ontario, Canada M1S 3P8.

CHINA, PEOPLE'S REPUBLIC

M-73-1 Decontamination Vehicle

DESCRIPTION
The M-73-1 decontamination vehicle is based on the chassis of the CA-30 (6 × 6) 2500 kg truck. The vehicle carries at the rear a 2.5 cubic metre tank to carry decontaminating agents that are dispensed over large areas of terrain through spray bars at the rear of the vehicle. The rate of dispensing is controlled by two operators seated at the rear of the tank and over the spray bars when the vehicle is operating. For travelling purposes the two operators are seated in an enlarged crew cab behind the driver and vehicle commander.

The M-73-1 has a gross weight of 9300 kg. Vehicle length is 6.86 metres, width 2.4 metres and height 2.42 metres.

STATUS
In production. In service with the Chinese armed forces.

DEVELOPMENT AGENCY
Research Institute for Chemical Defence, PO Box 1044, Beijing, China.

M-73-1 decontamination vehicle

M-82 Personnel Shower Vehicle

DESCRIPTION
The M-82 personnel shower vehicle is based on the chassis of the CA-30 (6 × 6) 2500 kg truck with an enclosed box body. When in use for personnel decontamination the box body slides to the rear where it is supported on folding legs; this provides more internal space for preparatory disrobing and donning fresh clothing after showering. Extra compartments enclosed by canvas covers are then erected on either side of the rear body. Access to the shower section is via ladders through one of three entrances at the rear and egress is through side exits towards the front of the canvas extensions. The M-82 carries a water pump and heater with water supplied to the system through a flexible hose normally carried on the front of the box body.

STATUS
In production. In service with the Chinese armed forces.

DEVELOPMENT AGENCY
Research Institute for Chemical Defence, PO Box 1044, Beijing, China.

M-82 personnel shower vehicle

M-82 personnel shower vehicle

Decontaminating Apparatus, Tank, M-84

DESCRIPTION
This portable decontamination apparatus is a rechargeable cylinder containing 1.5 litres of a decontamination agent known as T-191. It is carried on a variety of vehicles for local 'first aid' decontamination.

Weight of the M-84 is 3.2 kg. It is 380 mm high and 108 mm in diameter.

STATUS
In production. In service with the Chinese armed forces.

DEVELOPMENT AGENCY
Research Institute for Chemical Defence, PO Box 1044, Beijing, China.

Decontaminating Apparatus, Tank, M-84

CZECHOSLOVAKIA

Decontamination Apparatus, Truck-mounted, Model TZ-74

DESCRIPTION
Based on the chassis of the Tatra 148 PPR 15 (6 × 6) truck, this equipment is a progressive development of the Soviet Model TMS-65. It follows the same general lines in that it employs a small gas turbine to spray decontaminating agents over vehicles and equipment. As with the Soviet Model TMS-65, the gas turbine is mounted on the rear of the vehicle with the operator's cabin on the right-hand side. The rest of the vehicle rear is taken up with tanks for the decontaminating liquids and fuel for the gas turbine. The gas turbine, which is designated Type M 701 c-500, can be traversed through 120 degrees, elev-

Decontamination Apparatus, Truck-mounted, Model TZ-74

ated 30 degrees and depressed 20 degrees. It can dispense liquids at the rate of 900 litres an hour. A fuel tank for the gas turbine holds 2000 litres. The main decontaminant tank has a capacity of 5000 litres. Fully loaded the Model TZ-74 weighs 22 000 kg and empty 14 000 kg. Length is 9.5 metres, width 2.5 metres and height 2.95 metres. It has an operational crew of two. As well as being used for equipment decontamination, it can be used to create smoke screens.

A training system known as the TTZ-74-B has been developed for use with this system. Images of equipment to be 'decontaminated' are projected onto a screen placed in front of the jet exhaust. A sound system simulates the noise of the equipment's operation and other vehicle sounds and an electronic system records the operator's reactions to the ranges of the equipment being 'decontaminated'. Engine faults can be fed into the system and the training system may be used under cover or in the field.

STATUS
In production. In service with the Czechoslovak and East German armies.

MANUFACTURER
Truck chassis: Tatra, Narodni Podnik, Koprinice, Czechoslovakia.

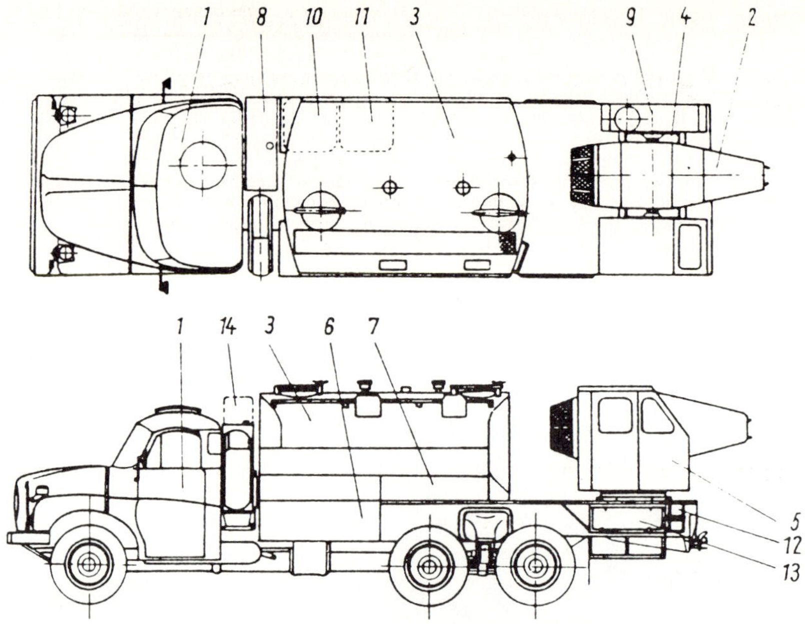

Model TZ-74 truck-mounted decontamination apparatus **(1)** *basic Tatra 148 chassis* **(2)** *turbine* **(3)** *decontaminant tanks* **(4)** *turntable* **(5)** *operator's cabin* **(6)** *hydraulics compartment* **(7)** *water heater* **(8)** *heater fuel tank* **(9)** *smoke screen equipment* **(10)** *hydraulic fuel reservoir* **(11)** *fuel tank* **(12)** *jerrican* **(13)** *vehicle service hatch* **(14)** *camouflage net stowage*

FRANCE

Chemical Decontamination Glove

DESCRIPTION
This glove is intended for personal decontamination of individual users' skin, clothing and equipment. It is a simple fingerless olive green glove made from natural fibres. One side has decontaminating powder (fuller's earth or a similar substance) behind a fine mesh and the other is a sponge cloth. The powder is patted onto the affected surface and then wiped off with the sponge. The gloves are issued in sealed plastic bags measuring120 × 200 mm and 20 mm thick. Opened and ready for use the glove measures 160 × 250 mm. Each one weighs 127 grammes, ± 5 grammes.

STATUS
Production complete. In service with the French armed forces.

MANUFACTURER
Société Industrielle de Réalisations et d'Etudes de Conditionnement (SIREC), Chemin d'en Bas, 77102 Villenroy, France.

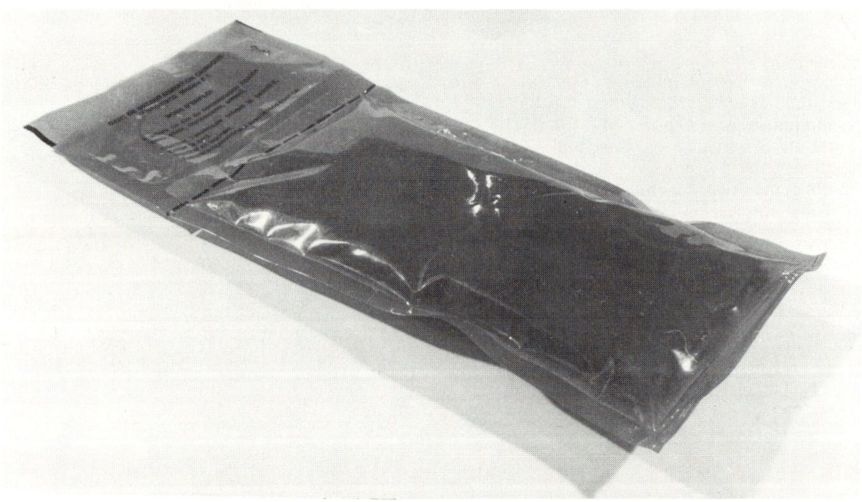

Chemical decontaminant glove in plastic envelope as issued

Decontamination Appliance, Emergency, 2.5-litre

DESCRIPTION
This apparatus is carried on vehicles for emergency decontamination of the carrier vehicle and, to a lesser extent, weapons and other equipment. Normally the apparatus is carried on a special 400 mm long rack. Once filled with a decontaminating solution, the apparatus is not brought under pressure until immediately before it is required. A commercial 'Sparklets' CO_2 bulb is used to pressurise the contents within 30 seconds which enables the contents to be sprayed for up to 40 seconds approximately. The tank has a total capacity of about 2.5 litres but only 1.6 litres can be effectively used, enough to cover an area of approximately 5 square metres. The appliance weighs 2.7 kg and is 385 mm long with a diameter of 110 mm.

STATUS
In production. In service with the French Army.

MANUFACTURER
Groupement Industriel des Armements Terrestres (GIAT), 10 place Georges Clémenceau, 92211 Saint-Cloud, France.

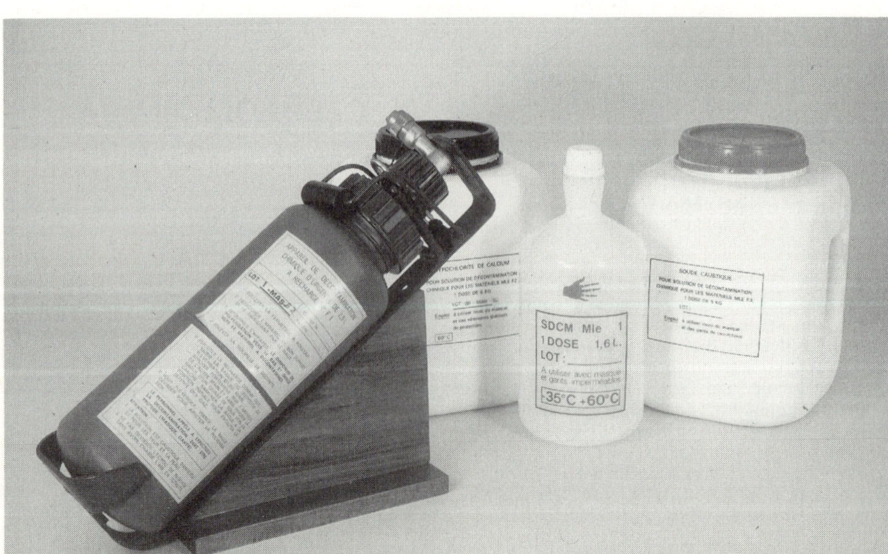

2.5-litre emergency decontamination appliance

GERMANY, EAST

Individual Decontamination Kit Model EP-60

DESCRIPTION
The EP-60 kit is carried in a plastic bag and is used for decontaminating clothing and small items of personal equipment. The plastic bag contains three tubes, numbered 1, 2, and 3, two wooden and one metal spatulas, cotton pads and instructions. The procedure is to remove as much as possible of the contaminating agent with the spatulas and the cotton pads. If the agent is identified the contents of the appropriate tube only need be applied: tube 1 contains nerve agent decontaminant, tube 2 is for use against blister agents and tube 3 is for use against both biological and radiological agents. If the nature of the contamination is not known the ointment is applied direct from the tubes in numerical order. The ointment is removed after ten minutes using the cotton pads. Decontamination is completed by washing the area in water. Complete, the EP-60 kit weighs about 150 grammes.

STATUS
Probably no longer in production. In service with the East German forces.

MANUFACTURER
State factories.

Individual Decontamination Kit, Model EP-62

DESCRIPTION
The EP-62 kit is used for personal decontamination of skin, clothing, individual weapons and equipment. It is contained in a plastic or light alloy case and consists of two plastic bottles, instructions and absorbent pads. The two bottles contain ointment, one blue for use against nerve agents and the other white for use against blister agents. If the nature of any contamination is not known the blue bottle contents are used first. The agent is first removed as far as possible with the absorbent pads and the ointment is then applied thickly. It is left for ten minutes and then removed with more pads.

STATUS
Probably still in production. In service with the East German forces.

MANUFACTURER
State factories.

Large Decontamination Kit, Model GES-10

DESCRIPTION
This kit is used for decontaminating vehicles and other large equipment, and is carried as a standard item on each vehicle. The Model GES-10 is intended as 'first-aid' decontamination to make the vehicle usable until a full cleansing can be carried out by a fully-equipped decontamination unit. The kit comprises a metal tray, with two 5-litre water cans and an accessories container which can be split into two halves to act as further trays. The kit contains two scrubbing brushes, two ring-shaped brushes, a flat short-handled brush, rags, three 50 ml bottles of Mersolat D decontaminant and three cardboard boxes each containing 0.5 kg of calcium hypochlorite. In use any large areas of contamination are wiped clean with the rags, and mud and debris are removed with a scraper. The standard decontaminant, EF1-III, is produced by mixing 0.5 kg of calcium hypochlorite in one of the cans and then pouring the solution into one or more of the trays from where it is applied with the various brushes. In a nuclear radiation environment, one bottle of Mersolat D (a detergent) is added to a can of water and the vehicle is scrubbed. After either procedure the vehicle is dried with rags from the kit.

STATUS
Probably no longer in production in the above form. In service with the East German and some other Warsaw Pact forces.

MANUFACTURER
State factories.

Decontamination Apparatus, Backpack, Models TEG-57A and TEG-57B

DESCRIPTION
These two back-pack equipments have replaced the earlier TEG-10. They consist of two 5-litre brass tanks and the required pumping gear. Both are issued in wooden cases, and mounted on the back frame are the two tanks, a hand-operated air pump, a pressure gauge, control valves, rubber hose, and a spray pipe with a nozzle to which a brush can be fitted. The Model TEG 57B differs from the earlier Model TEG-57A in having a redesigned frame with no base plate and an integral carrying handle (the even earlier Model TEG-10 used only a single tank).

In service the decontamination solutions to be used are mixed and strained before loading the tanks. The tanks can be filled with a different solution in each, or the same solution in both: a three-way valve selects either tank or, alternatively, it can be used to dispense a mixture of the two tank contents. Both tanks are pressurised using the pumping lever to about 6 kg/cm^2, and the discharge rate can be varied by fitting one of four nozzle sizes: 2, 1.2, 0.8 and 0.6 mm, and with a 10-litre filling the spraying times are 4.5, 10, 24 and 45 minutes respectively.

There are four types of decontamination fluids known to be available for use with the TEG-57. They are EAF-1 (used when radioactive fall-out is present), FF-7, EF-8 and EAF-1. The latter consists of a 2 to 6.5 per cent aqueous solution of calcium hypochlorite.

SPECIFICATIONS
Weights:
(empty) 11 kg
(full) 21 kg
(full, with frame) 27 kg
Height: 415 mm
Max volume: 2 × 6.7 litres
Max capacity: 2 × 5 litres
Hose length: 1.75 m
Hose diameter: 20 mm

STATUS
It is probable that neither model remains in production. In service with the East German armed forces and some other armed forces.

MANUFACTURER
State factories.

Decontamination and Deactivation Apparatus, Model EEA-61

DESCRIPTION
The Model EEA-61 is built into some East German vehicles fitted with air brakes. It is an integral part of the vehicle and consists of two 40-litre tanks, two hose lines, two spray pipes with nozzles and round brushes, and some other fittings. Each tank is normally filled with 36 litres of decontamination solution with a different solution in each tank. Depending on the nature of the contaminant, the appropriate tank is selected and the tank is then pressurised from the vehicle's air brake system. The solution is driven through the hose system to the nozzles where the brushes assist in clearing the contaminant.

STATUS
Probably still in production. In service with the East German forces.

MANUFACTURER
State factories.

Decontamination Apparatus, Truck-mounted, Model GEW-1

DESCRIPTION
This equipment is mounted on the G-5 (6 × 6) 3500 kg truck chassis and is used for decontaminating vehicles, weapons, equipment, buildings and terrain. It consists of a 4500-litre tank, a pump driven from a power take-off from the main engine, hoses, nozzles and various accessories. The pump can discharge up to 1500 litres of decontaminating solution every minute at a pressure of 8 kg/cm^2. Six 9.5 mm or three 42 mm hoses can be used to suit the task in hand, or two spray nozzles can be fitted to the front or rear of the vehicle to clear areas of road or terrain. One filling of the tank is enough to decontaminate 16 vehicles or a strip 5 metres wide and 1000 to 1500 metres long using the front-mounted nozzles. Using the nozzles mounted at the rear a strip 3 metres wide and 2500 metres long can be cleansed. The Model GEW-1 can be used for fire-fighting.

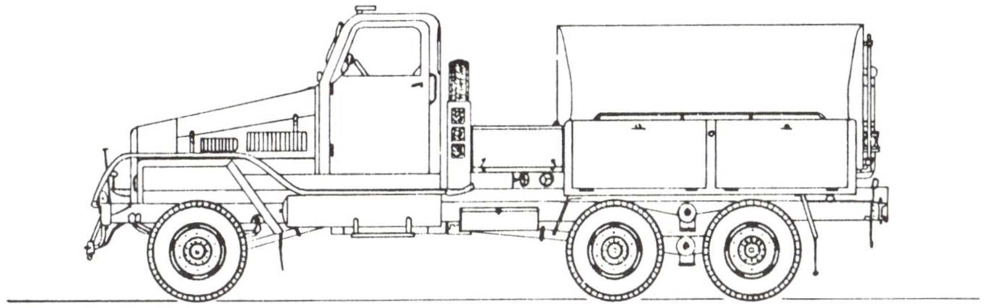
GEW-1 decontamination vehicle

STATUS
Production complete. Probably no longer in full service with the East German forces but still held in reserve.

MANUFACTURER
State factories.

Decontamination Apparatus, Truck-mounted, Model GEW-2

DESCRIPTION
The Model GEW-2 is used to dispense solid, dry decontaminants over roads and terrain. It is mounted on a G-5 (6 × 6) 3500 kg truck chassis and consists mainly of a covered-hopper body into which the decontaminating material is loaded. A hydraulically-driven spreader disc at the rear, the speed of which can be varied, dispenses the decontaminating material. The spreading width can be varied from 3 to 6 metres. The floor of the hopper is also hydraulically driven to the rear. The hopper body has a capacity of 3000 kg. The Model GEW-2 can also be used for salting or sanding roads.

STATUS
Production complete. In limited service with the East German forces.

MANUFACTURER
State factories.

Decontamination Apparatus, Truck-mounted, Model GEW-3

DESCRIPTION
The Model GEW-3 is mounted on a G-5 (6 × 6) 3500 kg truck chassis and replaced the earlier Model GEW-1 in East German service. It is used to decontaminate vehicles, buildings, equipment and terrain and can also be used as a water or fuel carrier, for fire-fighting and as a mobile or static smoke generator. It consists of a 3100-litre tank, a diesel-engine driven pump, hoses, nozzles, fittings, valves and accessories. The tank has internal baffles, two access and filling manholes, and is corrosion-proofed internally. The diesel engine for the pump is between the cab and the tank and its exhaust pipe passes through the tank to warm the contents in winter, as does the vehicle engine exhaust pipe. The pump is at the rear of the tank and the pump and diesel engine controls are all inside the vehicle cab. Various hoses can be fitted to the equipment to suit the decontamination role and spray nozzles can be fitted to the front and rear of the vehicle for terrain or road cleansing. At the rear of the equipment there is a special fitting with five nozzles which is used to disperse smoke fumes produced by the dissemination of chlorosulphonic acid.

Model GEW-3 truck-mounted, decontamination apparatus

STATUS
Production complete. In service with the East German forces.

MANUFACTURER
State factories.

Decontamination Apparatus, Model EA 65

DESCRIPTION
This equipment is a portable decontamination system using pre-pressurised and pre-filled tanks. The tanks are carried to their point of use by specially-equipped LO 1800 A (4 × 4) 1800 kg trucks which can carry 16 tanks each. The trucks are fitted with racks for the tanks, an air compressor, and a loading and unloading ramp. The corrosion-resistant steel tanks are mounted in tubular steel frames, and each has a pressure gauge, outlet hose, spray pipe, nozzle and nozzle brush. The tanks are pre-filled with an appropriate decontamination solution and pressurised on the truck to a maximum working pressure of 6 kg/cm^2. Tank capacity is approximately 40 litres, and the weight loaded 85 kg. The cylinders can be used to decontaminate vehicles, weapons, equipment and general areas. Once empty the tanks can be collected, refilled and repressurised.

STATUS
Probably still in production. In service with the East German forces.

MANUFACTURER
State factories.

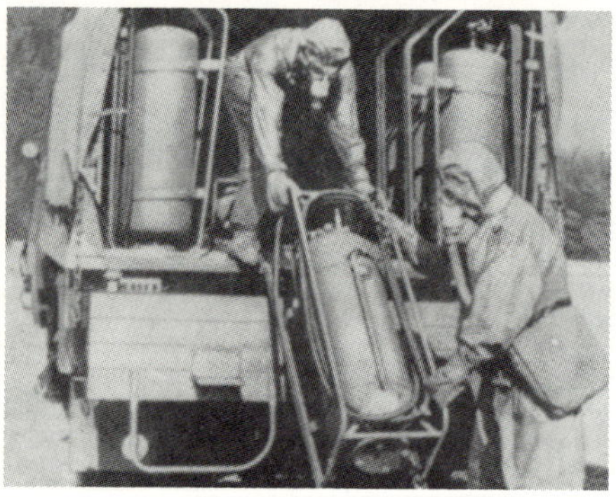

Unloading EA 65 decontamination tanks

Decontamination Apparatus, Truck-mounted, Model MOE

DESCRIPTION
The Model MOE is not intended mainly as decontamination equipment; it was produced primarily for water-damping roadways and runways under construction. However, it can be used for decontamination, and it can also be used as a water-carrier or a water-pumping station. Mounted on the G-5 (6 × 6) 3500 kg truck chassis, the Model MOE can be fitted with a variety of hoses and fittings but consists mainly of a 3750-litre tank, an air-cooled four-cylinder petrol engine which drives a centrifugal pump, spray booms and various other accessories. The pump can discharge 600 litres a minute which is normally passed through the extensible spray booms, but other fittings to suit particular tasks can be employed. When extended the booms cover an area 7.3 metres wide.

STATUS
Production complete. In reserve with the East German forces.

MANUFACTURER
State factories.

Decontamination Shower Apparatus, Vehicle-mounted, Model DA-2S

DESCRIPTION
The Model DA-2S replaced the earlier Model DA-2 in East German service but in its turn is being replaced by the later Model DA-66. It can be used as a normal field shower unit as well as for decontamination. It consists of two units, one, which is carried on a G-5 (6 × 6) 3500 kg truck chassis, consisting of an oil-fired boiler, a heat exchanger, hoses, plumbing, pipes, pumps and controls and the other, the shower unit, which is carried on a towed trailer. The shower trailer has 12 shower heads, but on the move it also carries the associated dressing and undressing tents, canvas-covered passage-ways, duckboards, stoves and various other items. The oil-fired boiler heats the water, using a heat exchanger, to 70°C from cold in about 50 minutes. Before the water reaches the shower heads it is mixed with cold water to produce a working temperature of 35°C. The oil burner-blower and the shower pumps are driven by electric motors: the water is delivered to the shower heads at a pressure of 1.5 kg/cm^2. After use the shower water is collected from a single point for disposal. Tents, heated if necessary, complete the equipment.

STATUS
Production complete. Still in service with the East German forces.

MANUFACTURER
State factories.

Decontamination Shower Apparatus, Truck-mounted, Model DA-66

DESCRIPTION
The Model DA-66 is much lighter than the earlier Model DA-2S, but like it is carried in two units. One unit is carried on the chassis of an LO 1800 A (4 × 4) 1800 kg truck which carries the equipment's water-heating unit and a pump. A trailer carries a compartmented tent which is divided into undressing, shower and dressing sections. This trailer has a single axle. The Model DA-66 is replacing the Model DA-2S.

STATUS
Production probably complete. In service with the East German forces.

MANUFACTURER
State factories.

Decontamination Apparatus, Truck-mounted, Model EW-1 Decontamination Apparatus, Skid-mounted, Model S-4

DESCRIPTION
These two equipments are the same, but when mounted on a truck it is designated the Model EW-1 and on a ground skid-mount it is the Model S-4. They are used to decontaminate vehicles, weapons, field fortifications and equipment. Both units have two 700-litre drum tanks, a double two-piston pump, an air-cooled petrol engine, two jet pumps, hose reels and hoses, nozzles and accessories. The tanks can be filled with either decontaminants or water through their 380 mm filling apertures. The apertures are fitted with screens, and once a small quantity of liquid has been poured into the tanks it can be circulated by the piston pump to operate the jet pumps. The pump discharges through seven hoses which are stowed on two reels, three on the left reel and four on the right. The piston pump can draw from either tank separately or both together so different decontaminants can be poured into each tank. The equipment is three metres long, 1.7 metres wide, 1.5 metres high and weighs about 2100 kg.

STATUS
Production complete. In service with the East German forces and civilian defence organisations.

MANUFACTURER
State factories.

Decontamination Apparatus, Truck-mounted, Model EW-2 Decontamination Apparatus, Skid-mounted, Model S-6

DESCRIPTION
The Model EW-2 and the Model S-6 are the same equipment but the Model EW-2 is mounted on a truck and the Model S-6 on a ground skid. They differ from the earlier Models EW-1 and S-4 in several respects. The Models EW-2 and S-6 both have rectangular 700-litre tanks but each has its own air-cooled petrol engine and its own three-piston pump. There are also two jet pumps, and there are six hoses fitted to two reels. The pumps can be used either to fill the tanks or discharge the tank contents through the hoses, and each pump can draw or fill both or either of the tanks. The equipment is 2.9 metres long, 1.7 metres wide and 1.5 metres high and weighs approximately 2230 kg.

STATUS
Production complete. In service with the East German forces.

MANUFACTURER
State factories.

Decontamination Pumping Apparatus, Trailer-mounted, Model TS-8

DESCRIPTION
The Model TS-8 pumping apparatus is the military version of a civilian fire-fighting appliance and is used for a variety of decontamination purposes. In its basic form it can be used for fire fighting or to pump water over equipment or vehicles to remove radio-active particles, but for chemical and biological cleansing it is more often used to pump water for decontamination stations (two or more units can be connected in series if necessary). The unit consists of a rotary pump driven by a ZW 1103 1.1-litre air-cooled two-cycle engine developing 28 hp at 3500 rpm, together with the associated pressure hoses, pipes and fittings. The unit is mounted on a frame in a box-type body carried on a TAS 8/8 single-axle trailer. The pumping unit, which weighs 155 kg, can deliver 800 litres a minute to a pressure head of 80 metres.

STATUS
Production complete. In service with the East German forces.

MANUFACTURER
State factories.

Decontamination pumping apparatus, trailer-mounted, Model TS-8

GERMANY, WEST

Kärcher Decojet Decontamination System

DESCRIPTION
The Kärcher Decojet is a mobile, self-contained decontamination system integrated in a single light metal frame carried on a Mercedes-Benz or Peugeot P4 4 × 4 light vehicle. It is used as a quick reaction 'first aid' decontamination system close to the front line or as a company level self decontamination measure. It can also be used in conjunction with the Kärcher Decocontain system at locations such as air bases for decontaminating vehicles, aircraft and their weapon systems.

The Decojet system can be used to decontaminate personnel using a pulsating shower apparatus which is part of the system; provide a pre-wash for material decontamination; provide post treatment for material decontamination with hot steam (140°C); provide the main decontamination treatment for material using an injector for prepared decontamination solutions (such as calcium hypochlorite and water); for deradiation and disinfection with specific agents, decontamination of clothing and equipment in an evaporating container; for terrain decontamination; and for daily high pressure and steam cleaning purposes.

The complete Decojet system consists of a Kärcher multi-purpose decontamination system (MPDS – see entry in this section) module with a cold and hot water circuit using a two-spray-lance operating principle; a 435-litre water tank and run-back pipe preheating the water in the tank; an additional 200-litre water tank with an injector system for mixing and applying decontamination solutions; hose reels for pre-wash and post treatment; a two-stage personnel shower with an injection system for adding chemicals to the water jets; a boiling vessel for clothing and equipment decontamination; decontamination chemicals such as calcium hypochlorite for material decontamination, RM 21 chemicals for personnel decontamination, and RM 54 for deradiation and material decontamination. The system is completed by various spray lances, tools and other accessories.

Kärcher Decojet providing showers for personnel decontamination

Kärcher Decojet being used to decontaminate trucks

All operations are controlled from a central operating panel at the rear of the Decojet unit.

SPECIFICATIONS
Weight: 1055 kg
Length: 1.653 m
Width: 1.41 m
Height: 1.24 m
Water tank capacity:
(main) 435 litres
(additional) 200 litres

STATUS
In production. In service with France, various countries in the Middle East, and Portugal.

MANUFACTURER
Alfred Kärcher GmbH & Co, Alfred-Kärcher-Strasse 28-40, D-7057 Winnenden, Federal Republic of Germany.

Kärcher Decojet ready for transport on Mercedes-Benz (4 × 4) light vehicle

Kärcher Decojet-trailer Decontamination System

DESCRIPTION
The Kärcher Decojet-trailer is a mobile decontamination system mounted on a 4-tonne capacity single-axle trailer. Units carried on the trailer include a Kärcher MPS 3200 motor pump set (see entry in this section), a Kärcher C8-DADS (Direct Application Decontamination System – see entry in this section), and a Kärcher Multi-Purpose Decontamination System (MPDS – see entry in this section). The trailer can be used for a range of decontamination tasks such as deradiation, detoxification and disinfection of material, clothing and equipment, personnel, terrain and buildings. The system can also be used for fire fighting, having a flow rate of up to 11 000 litres of water per hour, and for general maintenance and cleaning purposes.

With the help of the Decojet-trailer system a complete decontamination site can be established within a very short time. The site could contain up to 15 stations for the complete decontamination of personnel, equipment and material, all of which could be carried out simultaneously. The 15 stations would proceed as follows:

Station 1: personnel briefing and preparation prior to cleansing
Stations 2 to 5 deal with material decontamination
Station 2: prewash of material to remove dirt
Station 3: main treatment including the application of cleansing emulsion incorporating decontaminating chemicals
Station 4: waiting period for the emulsion to react
Station 5: post treatment by washing off the decontaminating chemical emulsion using a 140°C steam jet
Stations 6 to 15 deal with personnel and equipment decontamination
Station 6: preparation
Station 7: pre-decontamination of NBC protective clothing with aqueous C8 solvent
Station 8: removal of NBC protective clothing
Station 9: removal of all other clothing
Station 10: decontamination of NBC clothing using dry steam at temperature of 210°C
Stations 11 and 12: cleansing NBC protective masks in aqueous C8 solvent, plus any medical first aid
Station 13: pulsation showers for personnel
Station 14: personnel drying
Station 15: issue of new NBC protective clothing and equipment.

All accessories and equipment for such a decontamination site are carried on the Decojet-trailer. All decontamination chemicals, as well as 1000 litres of water, can be carried on the trailer so a degree of independent operation is possible for a certain period of time. The pumps of the modules mounted on the trailer extract water from sources such as rivers or streams up to a maximum suction height of 5 metres. The trailer also carries equipment for dry steam production (with temperatures up to 210°C) and application.

A single Decojet-trailer can decontaminate four or five tanks or heavy trucks each hour (up to eight if they are chemical agent resistant) or 150 soldiers. The system can also decontaminate 15 to 20 sets of clothing and personal equipment per hour.

SPECIFICATIONS
Weight: 3000 kg
Length: 5.16 m
Width: 2.45 m
Height: 2.46 m
Ground clearance: 0.38 m
Water tank capacity: 1000 litres

STATUS
In production. In service with Australia, Austria and Portugal and with other countries in North Africa and the Middle East.

MANUFACTURER
Alfred Kärcher GmbH & Co, Alfred-Kärcher-Strasse 28-40, D-7057 Winnenden, Federal Republic of Germany.

Kärcher Decojet-trailer decontamination system

Decojet-trailer decontamination system in operation cleansing heavy trucks

Kärcher Decocontain Decontamination System

DESCRIPTION
With the Kärcher Decocontain decontamination system all the necessary equipment to meet most military decontamination requirements is integrated in a standard 20-foot/6.096-metre ISO container. Tasks that can be carried out using the Decocontain system include complete deradiation; disinfection and detoxification of material such as trucks, armoured vehicles and aircraft; complete decontamination of clothing and equipment; full decontamination of personnel; terrain decontamination; and water recycling. The system can also be used as a fire fighting and water supply system, for field showers and for general maintenance and cleaning work using the integrated high pressure steam jet cleaners. The system can also be used for de-waxing and de-icing material, and

steaming explosive from various forms of munitions.

By placing the various equipment and components inside a standard container the Decocontain system can be readily moved by standard handling systems and transport methods to wherever it is required. The container is fitted with interchangeable corner fittings and is normally located on two pairs of skids for locking the container onto a carrier vehicle and for loading by crane or helicopter. Various container options are available including an NBC filter ventilation system with air locks and equipment designed to produce an internal over-pressure. Features such as EMP protection, heat insulation and fire protection can be added to suit customer requirements.

A typical Decocontain interior for personnel decontamination could have eight main areas. They are an entrance compartment; an entry cell; a pre-shower with an integral WC; a rinsing post-shower; a sterile cell; a departure sluice, a decontamination cell for clothing and equipment; and a service compartment.

The following are typical operating capacities for a Decocontain system:

Decontaminating tanks or heavy trucks – 4 to 5 per hour

Personnel decontamination in a contaminated area – 20 to 30 persons per hour

Personnel decontamination in a non-contaminated area – approximately 100 person per hour

Decontamination of clothing and equipment – approximately 15 to 20 sets per hour

Water recycling – approximately 700 litres per hour.

Kärcher Decocontain decontamination system in use

STATUS
In production. In service with Spain.

MANUFACTURER
Alfred Kärcher GmbH & Co, Alfred-Kärcher-Strasse 28-40, D-7057 Winnenden, Federal Republic of Germany.

Kärcher Decont Tent

DESCRIPTION
The Kärcher Decont Tent is a tent system designed for decontamination tasks. The tent with its inflatable tubular frame can be pitched in five minutes using a manual air pump supplied with the tent. Once set up the tent can be used for weather protection when decontaminating personnel, in which case a field shower system is set up inside the tent, or for the decontamination of clothing and equipment. In the latter case saturated steam (at temperatures of up to 210°C) or hot gas is fed into the tent using either the Kärcher Multi-Purpose Decontamination System (MPDS – see entry in this section) or a similar module.

The tent tubular frame is reinforced over the floor area and has a safety overflow valve. The tarpaulin used in the tent construction is resistant to chemical warfare agents and can be decontaminated. It has a watertight inner lining, and a canvas groundsheet doubles as a cover for the folded tent when it is being tranported or stored. The tent seams are sewn and treated with a high temperature resistant adhesive. Two hose connections are provided at opposite corners for waste water disposal. Once erected the tent is fastened securely by four ground loops and four guy lines.

SPECIFICATIONS
Weight: approx 43 kg
Length: (internal) 2 m
Width: (internal) 2 m
Height: (internal) 2.15 m
Temperature resistance:
(outer tent) −30 to +80°C
(inner tent) −30 to +140°C

STATUS
In production. In service with several countries.

MANUFACTURER
Alfred Kärcher GmbH & Co, Alfred-Kärcher-Strasse 28-40, D-7057 Winnenden, Federal Republic of Germany.

Kärcher Decont Tent in use

Kärcher Decont Tent ready for transport

Kärcher C8-DADS Direct Application Decontamination System

DESCRIPTION
The Kärcher C8-DADS direct application decontamination system was designed for the whole range of decontamination tasks, including deradiation, detoxification and disinfection of terrain and large items of military hardware such as armoured vehicles and aircraft.

The system is driven by an air-cooled diesel engine which drives a high pressure (75 bar) water pump, a solvent pump for liquid agents, and two hydraulic pumps. The diesel engine may be started by hand or by an electric starter. Once started an automatically regulated hydraulic system guarantees the correct mixing of the various decontamination chemicals. The heavy duty high pressure pump can draw water up to a height of 5 metres. All important parts of the system are constructed using corrosion-resistant materials.

With the C8-DADS system optimum detoxification is accomplished using the Munster emulsion, once surface dirt has been removed using cold water at high pressure. The Munster emulsion, developed by the German Research and Science Centre in Munster, is a composition of a special calcium hypochlorite known as C8, tetrachlorethylene (which may be added to act as a solvent), a phase transfer catalyst and water. When used with the C8-DADS system the emulsion is mixed automatically under the control of a single switch. The emulsion is then applied under pressure using a special spray lance. For tanks and large vehicles with chemical-resistant paintwork approximately 150 litres of emulsion are required. For vehicles with non-resistant paintwork 300 litres are needed. The emulsion is applied in two coats.

Optimum deradiation (10 mrad/h) is guaranteed by a hot foam treatment. For this two deradiation powders are mixed within the system and best results are achieved at a water temperature of 80°C. Hot water is provided by the Kärcher Multi-Purpose

Decontamination System (MPDS – see next entry.

Disinfection of biological and bacteriological warfare agents is carried out with the help of aqueous C8 solution.

When decontaminating terrain the decontaminating liquids may be sprayed using nozzles mounted on vehicles. The basic C8-DADS module can mix up to 254 kg of C8 with water (providing 1300 litres of aqueous solution). To increase this quantity additional pumps systems can be deployed.

The C8-DADS system may be carried on a truck with the components mounted in a tubular steel frame and can be dismounted for use. Accessories include 20-litre cans for fuel and chemicals, a 25 metre application hose, hand spray guns, a spray lance 1.04 metres long, various suction hoses, emulsion application lances with lengths of 1.2 and 2 metres, other spray lances, and nozzles and spare parts.

SPECIFICATIONS
Weight: 320 kg
Length: 0.96 m
Width: 0.85 m
Engine: 4-stroke diesel developing 13 hp at 3000 rpm
Electrical system: 12 V

STATUS
In production. In service with Australia, Austria, Egypt, France, West Germany, Taiwan, Thailand, NATO headquarters and some nations in the Near East. The system has been tested by the US Army.

MANUFACTURER
Alfred Kärcher GmbH & Co, Alfred-Kärcher-Strasse 28-40, D-7057 Winnenden, Federal Republic of Germany.

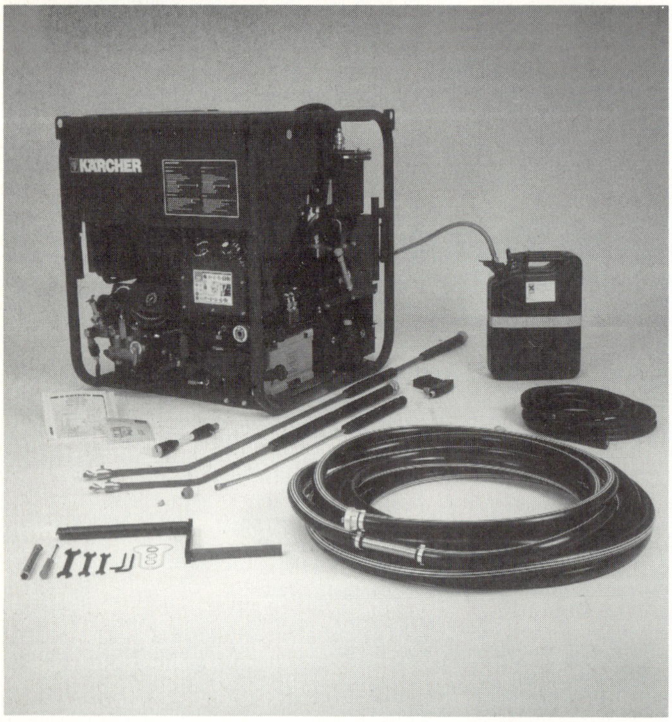

Truck-mounted Kärcher C8-DADS Direct Application Decontamination System in use spraying Munster emulsion onto Leopard 1 MBT

Kärcher C8-DADS Direct Application Decontamination System

Kärcher MPDS Multi-Purpose Decontamination System

DESCRIPTION
The Kärcher MPDS Multi-Purpose Decontamination System was developed following the Kärcher HDS 1200 BK high pressure steam cleaning system (see next entry) and is a self-contained high pressure (60 bar) system that can be operated using cold water, hot water (80°C), hot steam (140°C) or dry steam (210°C).

The MPDS is equipped with a diesel engine and a diesel burner. The fuel supply is taken from a NATO jerrican. The system uses a double-walled high performance burner with the flame being electronically controlled via a photocell. There is a high pressure water pump with a maximum suction height of 5 metres to pump water for the system from streams and rivers – the pump is sea water resistant. Liquid chemicals can be added to the system via the high pressure pump up to a rate of 60 litres per hour. For optimum winter starts the diesel engine, oil and air are preheated. All functions can be controlled via an electrical control panel with all operations pre-selected using a single main switch. The engine can be started by hand or by an electric starter powered by a maintenance-free 12-volt battery. All components are located in a light metal frame.

Detoxification and disinfection operations are carried out using hot steam (140°C) and a steam spray nozzle. Liquid chemicals are added via the high pressure pump and with the introduction of an injector accessory kit more agressive chemicals such as calcium hypochlorite can be used. Processes such as decontaminating personal equipment, structural interiors, NBC protective clothing and respirators, are carried out using dry steam (210°C).

For deradiation a special foam nozzle and a rotating brush can be used. Using chemicals such as RM 54 foam cleaner and a calcium inhibitor it is possible to carry out an effective foam treatment.

The MPDS can also be used to supply field showers and for many maintenance and general cleaning tasks. It can also be use to steam explosive from munitions and for de-icing aircraft and missile systems. A sand-blasting set is available for removing corrosion from equipment.

Kärcher MPDS Multi-Purpose Decontamination System in use

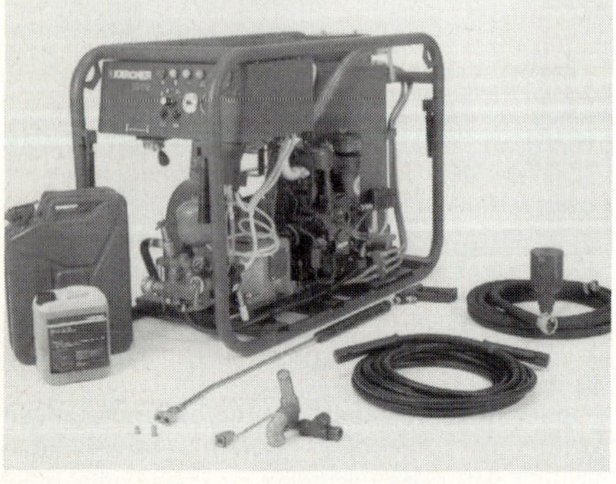

Kärcher MPDS Multi-Purpose Decontamination System

SPECIFICATIONS
Weight: 220 kg
Length: 1.25 m
Width: 0.575 m
Height: 0.85 m
Engine: 4-stroke diesel developing 5.6 hp at 3000 rpm
Operating temperature: -30 to $+60$°C

STATUS
In production. In service with Austria, Australia, NATO headquarters, Portugal and countries in North Africa and the Middle East.

MANUFACTURER
Alfred Kärcher GmbH & Co, Alfred-Kärcher-Strasse 28-40, D-7057 Winnenden, Federal Republic of Germany.

Kärcher HDS 1200 BK and EK High Pressure Steam Jet Cleaner Units

DESCRIPTION
The Kärcher HDS 1200 BK and EK high pressure steam jet cleaner units can be used for a variety of cleaning, maintenance and decontamination tasks. The units can produce cold or hot water, steam or dry steam. The main difference between the two units is that the HDS 1200 BK is powered by a two-stroke petrol engine while the HDS 1200 EK is powered by a 4 kW electric motor. The HDS 1200 BK is equipped with an electric starter and a manual starter.

When operating, a pressure switch trips an electromagnetic coupling on the pump whenever a spray gun trigger is squeezed. The drive system is disengaged when the trigger is released. On the HDS 1200 EK the electric motor is switched on as the spray gun trigger is squeezed and switched off as it is released. These drive systems ensure that the system pump operates only when it is needed to save energy and extend service life. The system uses a high pressure (up to 50 bar) pump producing enough suction to raise water up to 5 metres from streams and rivers. The pump feed rate is between 240 and 1200 litres per hour and decontamination or other chemicals can be introduced via the pump at a rate of up to 60 litres per hour. The heater burner fuel consumption is controlled via a closed circuit system: diesel fuel is injected when the burner is on and returned to the fuel tank when the burner is turned off. A continuous duty ignition system is employed. An integral electronic water-softening unit is provided to prevent scale from forming on the heater coil. The softener unit is adjustable to suit local water conditions.

There are two independent safety systems. A thermostat and low water cut-out prevents thermal overload and a pressure switch and safety valve prevent over-pressure.

Kärcher HDS 1200 EK (left) and HDS 1200 BK (right) high pressure steam jet cleaner units

All components are mounted on skids with fold-down wheels. The entire unit is enclosed in a corrosion-resistant casing.

As well as being used for general decontamination and cleaning tasks these units can be adapted for de-icing aircraft and missile systems, for cleaning fuel tanks and for steaming explosive from munitions. The units can also be used to supply water for field showers.

SPECIFICATIONS
Weight:
(HDS 1200 BK) 300 kg
(HDS 1200 EK) 280 kg
Length: 1.45 m
Width: 0.75 m
Height: (without exhaust pipe) 1.11 m
Pressure range: 15 to 50 bar

Kärcher HDS 1200 BK high pressure steam jet cleaner in use

STATUS
In production. In service with the West German armed forces and 40 armed forces world-wide.

MANUFACTURER
Alfred Kärcher GmbH & Co, Alfred-Kärcher-Strasse 28-40, D-7057 Winnenden, Federal Republic of Germany.

Kärcher Decontamination Accessories

In addition to the decontamination units and systems covered in the previous entries, Alfred Kärcher GmbH & Co also produce the following associated units that can be used as part of various decontamination processes.

MPS 3200 Motor Pump Set
The MPS 3200 motor pump set can be used as part of any decontamination system but is normally used for the pre-wash stage to remove dirt prior to the application of decontamination chemicals. It allows the operation of two spray lances at one time with a water flow rate of 1600 litres per hour for each lance. If required one of the lances can be replaced by a shower unit. The MPS 3200 can also be used for fire fighting in which case the flow rate can be increased to 11 000 litres per hour by a high capacity injector. The high pressure pump is sea and waste water resistant and can raise water up to a maximum suction height of 5 metres. All components are integrated in a light metal frame and all function are controlled automatically.

The MPS 3200 weighs 280 kg and dimensions are 0.96 × 0.7 × 1 metre.

Kärcher MPS 3200 motor pump set

Field Shower Unit
This is a two-stage shower unit for personnel decontamination. It can be supplied with water by the Kärcher MPDS Multi-Purpose Decontamination System jet or the HDS 1200 BK/EK high pressure steam cleaner unit. The five shower nozzles operate to two sides so one side may be used to deliver water with a decontaminant chemical such as RM 21 (which does not irritate the skin) while the other side can be used for rinsing with clean water. Shower water is prepared by an injector where heated water and cold water are blended to produce comfortably warm shower water. A heat-compensation reservoir in the shower base regulates the temperature which

is set by a thermostat. Pulsating nozzles are used.

The weight of the shower unit is approximately 48 kg and it is 2 metres high.

Collapsible Water Tank

The Kärcher collapsible water tank has a maximum capacity of 2500 litres and can be used to supply fresh or drinking water to a decontamination system or unit. It weighs 26 kg when empty and when full measures 2.3 × 1.5 × 0.7 metres.

STATUS

All the above units are in production. The MPS 3200 is in service with a NATO headquarters, Australia, Austria, Portugal and countries in North Africa and the Middle East. The Shower Unit and Collapsible Water Tank are in service in several countries.

MANUFACTURER

Alfred Kärcher GmbH & Co, Alfred-Kärcher-Strasse 28-40, D-7057 Winnenden, Federal Republic of Germany.

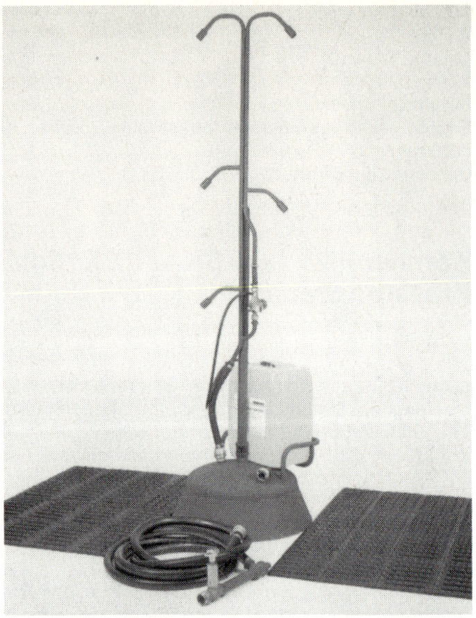

Kärcher field shower unit

Kärcher Decontamination and Cleaning Agents

RM 21

A liquid, slightly alkaline agent used for personnel decontamination as it does not irritate the skin. Also used for cleaning painted surfaces and normal cleaning surfaces. A derivative is RM 21 DsK, a mild liquid agent used for cleaning weapon systems and specifically developed to act as an additive foam emulsifier. When used with tetrachloroethylene, RM 21 DsK can be used for chemical agent decontamination.

RM 31

A liquid alkaline agent used for cleaning unpainted surfaces.

RM 32

A liquid alkaline cleaner with a disinfectant additive used for the cleaning of industrial and sanitary areas.

RM 35

A highly active disinfecting agent for the cleaning of industrial and sanitary areas.

RM 54

A foam cleaner for cleaning sensitive weapon systems, also used for nuclear decontamination.

Calcium inhibitor

A cleaning agent used to prevent the calcifying of high pressure steam jet cleaning equipment.

Calcium hypochlorite – C8

A mixture of powder and granulate with available chlorine (64 per cent by weight). The aqueous C8 solvent is used for decontamination purposes and is a component of the Munster emulsion used with the Kärcher C8-DADS decontamination system (qv).

C8 emulsion component

A mixture of tetrachloroethylene and emulsifier PTC 2000. It is a component of the Munster emulsion.

Emulsion training simulator

Used in place of C8 and other decontamination agents for training purposes.

STATUS

All the above agents are in production and are in service world-wide.

MANUFACTURER

Alfred Kärcher GmbH & Co, Alfred-Kärcher-Strasse 28-40, D-7057 Winnenden, Federal Republic of Germany.

Kärcher Hot Air Generator FB 60 E

DESCRIPTION

The Kärcher Hot Air Generator FB 60 E was designed for the hot air decontamination of indoor areas as well as the decontamination of NBC protective clothing and other equipment. It can also be used for the ventilation of NBC decontamination stations.

Decontamination of interiors is carried out using the FB 60 E and an interior decontamination extension set with special NBC filters. Vehicles, aircraft and vessels can be decontaminated on a purely physical basis by heat application with a remote thermostat controlling the required temperature. Clothing and equipment can be effectively decontaminated using the hot gas/hot steam process. For example, NBC protective garments may be hung in the Kärcher Decont tent (see entry in this section) and treated with hot air and a small quantity of hot steam.

The FN 60 E can also be used as a hot air generator to heat cabins, cockpits, tents or shelters, and during summer it can be used as a fan to cool these installations.

All the main components of the FB 60 E are installed in a torsionally stiff tubular aluminium frame with pick-up points for forklifts or air transport. It can be easily moved by one man using the integral handles and wheels.

The main fan and fuel pump are driven by a central 220-volt electric motor. Fresh air is drawn in via a radial fan and is then heated by a burner and heat exchanger. Optimised combustion is achieved by

Kärcher hot air generator FB 60 E

Kärcher hot air generator FB 60 E with accessories

the specially-shaped burner sphere. The burner sphere and heat exchanger are made of high temperature-resistant material, both of which are gas-tight welded to ensure that no combustion gases are mixed with the heated fresh air.

Due to its two integrated fuel heaters the FB 60 E is able to work in temperatures down to −30°C. The generator can operate using diesel, diesel-petrol and kerosine fuels.

SPECIFICATIONS
Weight: 200 kg
Length: 1.72 m
Width: 0.74 m
Height: 0.91 m
Nominal burner heating capacity: 60 kW
Nominal radiator air flow: 3000 m^3/h
Fuel consumption: 7.5 litres/h

STATUS
In production. In service with the Australian, West German and US armed forces.

MANUFACTURER
Alfred Kärcher GmbH & Co, Alfred-Kärcher-Strasse 28-40, D-7057 Winnenden, Federal Republic of Germany.

OWR DEKON Decontamination System

OWR DEKON 2000 decontamination system mounted on MAN 20.280 DFAEG (6 × 6) 10 000 kg truck

DESCRIPTION
The OWR DEKON decontamination system is a mobile system based on 6 × 6 and 8 × 8 trucks and consists of five basic components, each of which has its own carrying vehicle. The trucks used for the system are the MAN 20.280 DFAEG (6 × 6) 10 000 kg truck or the MAN 10 000 kg (8 × 8) and 7000 kg (6 × 6) high mobility tactical trucks.

The first component is the DETECT 1000, normally carried on an 8 × 8 chassis. It has three main compartments, the first of which is the working area or laboratory fitted with an air-conditioned filtered over-pressure system. It contains instruments for detecting, measuring and marking NBC-contaminated areas and also has communications equipment for reporting findings. The area has space for two operators but can accommodate up to a further six persons for travelling. The second area is the decontamination lock accessible from the outside by a hinged ladder. A shower basin is installed near the ground with a direct operable shower head and drainage for contaminated liquids. This cubicle is also used as a reception area for contaminated clothing and equipment. There is access to the laboratory area. The third area, the decontamination centre, houses all necessary equipment including a 600-litre tank, the hot water plant, storage for decontaminating agents and associated accessories. A spreader and spray bar are fitted at the rear of the vehicle for area distribution of decontaminants. Warning devices such as horns and lamps may be fitted. The driver's cab has partial armour protection.

The DEKON 2000 is a decontamination disaster protection vehicle and is designed to be operated independently of the rest of the system if necessary. It can be used for personnel and equipment decontamination, area decontamination and for producing toxic-free drinking water. The equipment carried on the DEKON 2000 includes a shower lock, wide jet spray tubes, spray shower heads, water brushes with adjustable telescopic shafts, mixing injectors for liquid and powder decontaminants, a water spray device under the vehicle platform, a spreading arrangement and a foam gun, a water purifying plant, two 1500-litre drinking-water tanks, a heat-insulated special tank for the supply of 2500 litres of water, a water circulation heater, a water supply electric pump, mixing devices, an electric heater, a water supply electric pump, mixing devices, an electrical control panel and 380/220-volt and 24-volt generators. All devices and equipment are carried on a flat working area accessible by a ladder at the rear or side. Normally they are covered by a canvas tilt which can be raised when working.

The third component is the SHOWER 3000 which can supply cleansing showers for up to 3500 personnel within a 24-hour period. Personnel walk into the showers from one side and leave by ladders on the other side. The equipment consists of a water tank, an electrical generator, a water circulation heater, shower cubicles with ten showers each, a water pump and switch gear. Associated with the equipment is a collapsible 5000-litre water tank along with clothing bags, towels, decontaminating agents etc.

The fourth component is the WASH 4000, a field laundry with a capacity for processing approximately 100 to 150 kg of clothing every hour. This is enough for 150 persons. The WASH 4000 is equipped with two washing machines, one centrifuge, one rotary dryer, an electrical generator, a 3500-litre water tank, a water pump and various auxiliary appliances.

The fifth component is the MOBILE WORKSHOP 5000. This is a van containing a workshop and a supply of spares capable of maintaining the complete DEKON system in the field. It contains its own small decontamination cubicle for personnel and equipment, and radio equipment may be fitted.

All vehicles in the DEKON system can be fitted with the Decontamination Emulsion Direct Application System (DEDAS) which has an output of more than 100 litres per minute of decontamination solution.

STATUS
In production.

MANUFACTURER
Odenwaldwerke Rittersbach OWR GmbH, D-6957 Elztal-Rittersbach, Federal Republic of Germany.

OWR DEKON 2000 decontamination system fitted with spray bars for area decontamination

OWR SHOWER 3000 for personnel decontamination mounted on MAN 20.280 DFAEG (6 × 6) 10 000 kg truck

OWR DEKON Trailer 6000

Complete OWR DEKON Trailer 6000 with shower unit in use

DESCRIPTION
The OWR DEKON Trailer 6000 is intended for the decontamination of personnel, equipment and terrain but it can also be used as a water or foam fire extinguishing equipment, as a mobile disinfecting equipment or as a mobile shower station. It is carried on a single-axle trailer that can be towed by any vehicle with a 3000 kg capacity and upwards.

On the move the trailer is enclosed by a framed cover that can be removed to act either as a dressing station or a weather cover for a shower installation. The trailer has its own 1000-litre water tank and carries several further 1000-litre foldable water tanks. There is an integral heater that can heat 3600 litres of water by 28°C every hour for use by the shower units or for other purposes. The canvas cover can enclose a shower unit with eight shower heads which is sufficient to process about 100 persons an hour. Power to pump the water is provided by an engine on the trailer that also provides power for a generator to supply the circulation heater and for possible lighting units. The trailer also has facilities for mixing decontamination and other agents with the water, and by using the mixing injector and a projection tube, the trailer can be used to produce foam for decontamination or fire fighting. Also carried on the trailer can be a work bench, fire extinguishers, entrenching tools, spare parts and tools, first-aid kits and extra supplies of decontaminating and other agents.

STATUS
In production.

MANUFACTURER
Odenwaldwerke Rittersbach OWR GmbH, D-6957 Elztal-Rittersbach, Federal Republic of Germany.

NBC Decontamination Truck

DESCRIPTION
The standard Bundeswehr NBC decontamination truck is a MAN (6 × 6) 7000 kg truck which has been adapted for the role by the addition of special mounting points on the load area. The trucks may carry either decontaminating equipment for use at main centres or equipment for terrain, equipment and clothing decontamination in the field. Both equipments are low-pressure types. The field version equipment includes a 1.5 cubic metre tank, a continuous-flow heater and special fittings for the field decontamination role.

STATUS
Production probably complete. In service with the West Germany Army.

MANUFACTURER (decontamination equipment)
MINIMAX, Urach, Federal Republic of Germany.

MAN 7000 kg (6 × 6) truck converted for NBC decontamination

HUNGARY

Decontamination Apparatus, Truck-mounted

DESCRIPTION
This equipment is mounted on a Csepel D-346 (4 × 4) 3000 kg truck chassis and consists of a tank with a capacity of approximately 2000 litres, a heater, a pump, nozzles, hoses, fittings and various accessories. Used for decontaminating vehicles, equipment, weapons and buildings, the equipment has stowage boxes along each side for the various fittings, and the heater is just behind the cab with a smoke flue on the left-hand side.

STATUS
Production complete. In service only with the Hungarian armed forces.

MANUFACTURER
State factories.

ISRAEL

M-812 Portable Spray Tank for NBC Decontamination

DESCRIPTION
The M-812 portable spray tank for NBC decontamination is carried by one man using a shoulder harness. It has a capacity for 10 litres of decontaminating agent and delivers the agent using compressed air produced by using a vertical hand-operated pump on top of the tank; a pressure relief system is provided. The tank is connected by a hose to a spray gun that can deliver the agent in either a fine or heavy spray and a fan or circular pattern. A hard nylon bristle brush can also be fitted. Spares are carried in a container located under the tank.

The tank is zinc-coated and constructed using electric seam welding. It is 600 mm high and has a diameter of 195 mm.

UZI & UZI also produce a hand-held decontamination tank based on the design of the American M11 decontaminating apparatus (see entry under *United States* in this section). This tank has a capacity of 1.3 litres.

STATUS
In production. In service with the Israeli armed forces.

MANUFACTURER
UZI & UZI Metal and Hydraulic Works Limited, PO Box 693, Kiryat Shmona, Israel.

Portable Decontamination Apparatus

DESCRIPTION
The design of this portable decontamination apparatus appears to be based on that of the American M11 (see entry under *United States* in this section). The apparatus has a capacity of 1.26 litres of DS-2 decontaminating agent which is stated to be sufficient to cover an area of approximately 12.5 square metres at an effective spray range of 1.8 to 2.4 metres. The agent is sprayed under pressure from a high-pressure nitrogen-filled cylinder fitted close to the spray head at the top of the tank. The spray rate is controlled by a thumb lever. To prevent accidental discharge a lead-sealed captive safety pin is provided. After use the apparatus may be re-filled and re-pressurised.

The apparatus can be carried in a special vehicle rack. Weight of the apparatus full is 3 kg (1.36 kg empty).

STATUS
In production.

MANUFACTURER
Electra Mikun Limited, PO Box 1044, Hod Hasharon 45110, Israel.

ITALY

Cristanini Sanijet C 921 Decontamination System

DESCRIPTION
The Cristanini Sanijet C 921 decontamination system is a self-contained unit that is available in two forms. One is the Sanijet C 921 D 18/50 which has an electrical system operating on 50 Hertz while the D 18/60 operates on 60 Hertz.

The system is powered by an air-cooled diesel engine which drives a water pump that enables the system to operate by delivering cold or hot water (up to 95°C), steam (120°C) or dry steam (190°C). The pump has a capacity to raise water up to 5 metres from streams and rivers via a suction hose. The engine uses pre-heating devices to assist starting under cold weather conditions. Once in the unit the water can be heated or super-heated by a coil burner assembly with a 2 kVA alternator feeding the burner and other auxiliary electrical equipment. The engine also provides a 24 volt dc supply for recharging a dry battery that can be used to start the engine and there is a 24 volt dc socket that can be used to supply other auxiliary equipment, such as lighting, or for connecting into another 24-volt dc supply in case of battery failure.

Once the water is heated it can be mixed with various decontamination chemicals and sprayed using spray lances or similar devices. The system has provision to vary the chemical-water ratio as required. All operations are controlled from a waterproof control panel which also houses the pressure and temperature adjustment controls. Safety devices against over-pressure and over-temperature are provided.

The Sanijet C 921 system can be used with various accessories such as a water softener, a foam emulsion production unit, various types of field shower unit and devices intended to steam explosive from various types of munitions.

Cristanini Sanijet C 921 decontamination system in use decontaminating a M109 155 mm self-propelled howitzer

SPECIFICATIONS
Weight: 230 kg
Length: 0.85 m
Width: 0.85 m
Height: 0.8 m
Engine: air-cooled diesel developing 10 hp
Capacity:
(hot water) 90 bar, 14 litres/min at 95°C
(steam) 20 bar, 550 kg/h at 120°C
(super-heated steam) 20 bar, 300 kg/h at 190°C

STATUS
In production. In service with Italian armed forces.

MANUFACTURER
Cristanini SpA, 37010 Rivoli (Verona), Italy.

Cristanini Sanijet 3000/3 Containerised Decontamination System

DESCRIPTION
The Cristanini Sanijet 3000/3 containerised decontamination system makes use of a single 20-foot/6.096- metre ISO standard container that encloses a fully self-contained unit for the decontamination of personnel, vehicles and other equipment, and clothing.

The container used for the Sanijet 3000/3 system is RFI and EMP resistant and uses an NBC filtration system to produce an internal over-pressure. Also provided is an air-conditioning unit with a capacity of 1000 cubic metres an hour and capable of working in temperature conditions from −32 to +52°C. An integral water tank has a total capacity of 2000 litres and is supplied by a pump and filter system that can collect water from various sources. Electrical power for the complete system is provided by a 39 kVA generator.

The container is divided into a number of compartments with a central compartment acting as a control room. Other compartments are used for showers, garment decontamination and access/exit compartments. One entire end of the container is occupied by an engine compartment housing detergent tanks, a compressor, the main engine, the electrical generator and a water boiler. The engine room has its own ventilation and access hatches.

For personnel decontamination, entry to the container is via an end hatch. Once inside personnel remove their clothing with the NBC mask being the last item to be removed; the masks are then discarded through a hatch to outside the container. The interior of the entrance hall where disrobing takes place is kept at an over-pressure of 25 mbar. Personnel then enter a shower section where there is a differential pressurisation of 30 mbar. The shower unit has a flow of 18 litres a minute with a selection of hot water from 20 to 40°C. It is possible to add detergents to the shower water flow. After the initial shower a further shower is provided for rinsing. Once the showers are completed personnel pass to the main control compartment for drying and dressing. They then leave through a pressurised exit compartment which has two airtight doors.

For decontaminating vehicles and equipment, an internal compartment houses spray lances and other accessories. These include two cold water lances, each operated at a pressure of 110 bar and with a capacity of 18 litres a minute. One of the lances may be replaced by a hot water lance with a similar flow rate and operating pressure. Also provided is a foam application lance operating at a pressure of 35 bar and with a flow rate of 6 litres a minute, and a vapour lance operating at 16 bar and with a flow rate of 10 litres a minute. The compartment housing the lances also contains hose reels that can be unwound for external use.

Garments can be decontaminated in a seperate compartment which is equipped with racks, clothes hangers, bags and a sump for waste water collection. Decontaminating vapour is fed into the chamber at a pressure of 16 bar, with a flow rate of up to 10 litres a minute and at a temperature between 120 and 140°C.

STATUS
In production.

MANUFACTURER
Cristanini SpA, 37010 Rivoli (Verona), Italy.

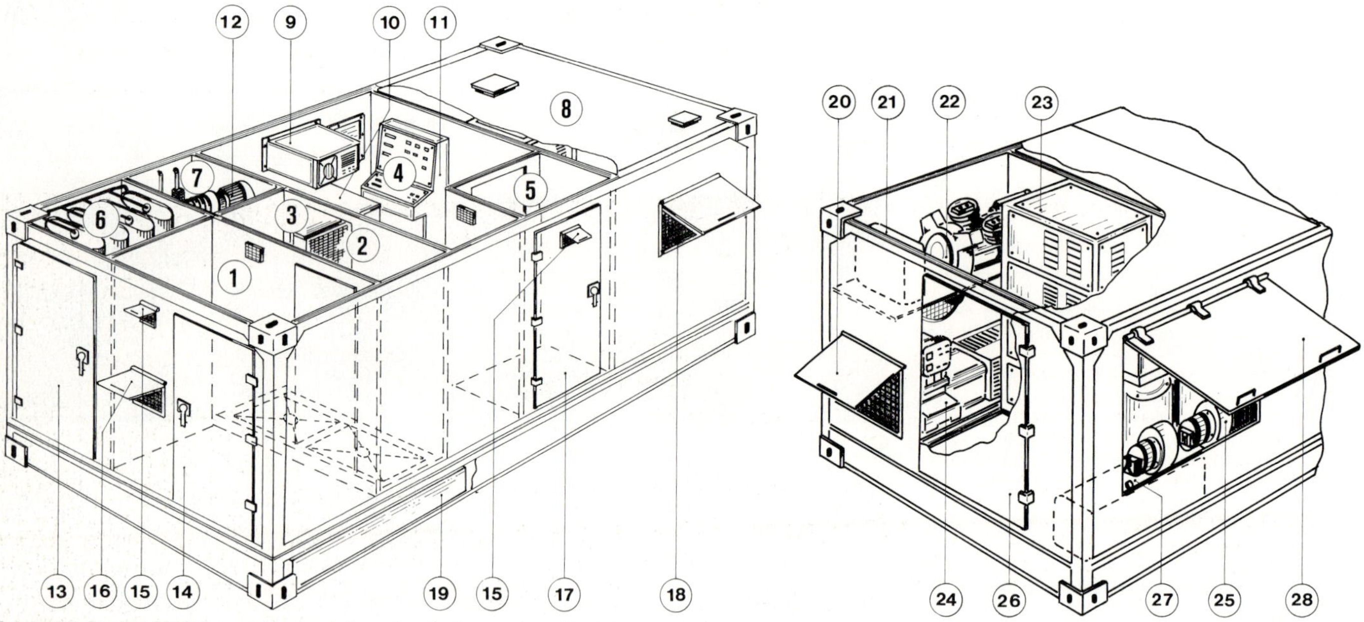

Cristanini Sanijet 3000/3 containerised decontamination system **(1)** *entrance compartment* **(2)** *decontamination shower* **(3)** *rinsing shower* **(4)** *main control compartment* **(5)** *pressurised exit compartment* **(6)** *garment decontamination compartment* **(7)** *lances and accessories* **(8)** *engine compartment* **(9)** *NBC pressurisation filter unit* **(10)** *air conditioning unit* **(11)** *internal control panel* **(12)** *hose reels* **(13)** *entrance hatch to garment decontamination compartment* **(14)** *entrance hatch* **(15)** *differential pressure valves* **(16)** *hatch for mask disposal* **(17)** *exit hatch* **(18)** *air vents for engine compartment* **(19)** *water cistern* **(20)** *engine ventilation hatch* **(21)** *detergent tank* **(22)** *compressor unit* **(23)** *modular high pressure pump unit* **(24)** *generator unit* **(25)** *boiler unit* **(26)** *entrance hatch to engine compartment* **(27)** *fuel tank* **(28)** *side hatch*

Tirrena Small Decontamination Set SDS T 155

DESCRIPTION
Most of the chemical solutions used to decontaminate vehicles and equipment after exposure to chemical warfare agents are corrosive to varying degrees. In practice this means that decontaminants have to be mixed or otherwise prepared immediately before use or their dispenser/containers will become corroded to the point of uselessness. This has several disadvantages, not only in a time-scale, but there will be certain areas or sets of circumstances in which the water essential to produce the working solutions will be either unavailable or scarce.

For these circumstances the Small Decontamination Set SDS T 155 produced by Tirrena SpA was developed. It is a small fire extinguisher-type decontaminant dispenser designed to be carried on a vehicle for the immediate first-aid decontamination of the vehicle or other equipment. The SDS T 155 dispenser can be carried fully loaded for long periods, as it is made entirely of AISA-304L stainless steel which is proof against chemical corrosion. The equipment can be loaded at a remote point and carried in its special bracket until required. The DS-2 decontaminant is propelled from the container by compressed nitrogen which is supplied with the equipment in a special loading container which weighs 300 grammes and has a diameter of 40 mm. The 95 cm^3 of nitrogen stored in the loading container pressurises the contents of the dispenser to a level of 20 kg/cm^2, and the loading container is itself reloadable. The dispenser is tested to a pressure of 55 kg/cm^2. The dispenser cylinder has a maximum capacity of 1.85 litres but the normal working capacity is 1.5 litres. This is sufficient to decontaminate area of from 6 to 8 square metres at a spray range of 2 metres.

SPECIFICATIONS
Weight:
(full) 4.3 kg
(empty) 2.8 kg
Height: 310 mm
Diameter: 125 mm
Working pressure: 20 kg/cm^2
Testing pressure: 55 kg/cm^2
Contents:
(max) 1.85 litres
(working) 1.5 litres
Loading container
Weight: 0.3 kg
Diameter: 40 mm
Contents: 0.95 litre
Loading pressure: 150 kg/cm^2
Testing pressure: 400 kg/cm^2

STATUS
In production. Over 35 000 produced for the Italian Army.

Tirrena decontamination set SDS T 155

MANUFACTURER
Tirrena SpA, Via del Quirinale 22, 00187 Rome, Italy.

JAPAN

Japanese NBC Systems

Chemical protection (defence) vehicle based on chassis of Type SU 60 APC (K. Nogi)

Decontamination vehicle based on 6 × 6 truck chassis (K. Nogi)

NORWAY

LDS NBC-Sanator III Lightweight Decontamination System

DESCRIPTION

The Sanator lightweight decontamination system consists of an air-cooled engine, a water pump and a heater. The basic unit weighs 163 kg and is compactly assembled (including instrumentation) in an aluminium frame for mobility. Fuels for the engine and heater are supplied from standard 5-gallon/20-litre fuel cans which are attached to the frame, making the apparatus independent of outside power. The basic unit is supplemented by a 65 kg accessory kit containing hoses, cleaning wands, a decontamination agent injector and personnel shower hardware. A rubberised fabric, self-supporting, collapsible tank with a 6000 litre capacity is an auxiliary item – a 12 000-litre tank and other sizes of tank are also available. The Sanator provides a hot water-/steam capability at low water consumption rates and is capable of drawing water from a natural source 10 metres away and 3 metres below pump level. The shower attachment can decontaminate up to 12 personnel at one time and can produce up to 80 litres of heated water per minute. The extendable trigger wands can supply a total flow of up to 22 litres a minute. The 7.3 hp two-stroke pump engine has a fuel consumption of 3 litres per hour while the heater multi-fuel consumption is 10 to 35 litres per hour. Booster pumps can be incorporated into the system as necessary.

The basic Sanator unit measures 1020 × 590 × 860 mm. The standard accessory pack measures 1060 × 520 × 390 mm and weighs 61 kg.

In addition to producing hot water for NBC decontamination the Sanator can also be used for hot or cold drinking water supply, for water storage purposes and for supplying water to field hospitals. The Sanator is also in use by humanitarian aid and civil construction organisations.

The Sanator is licence-produced in the United States as the M17 Lightweight Decontamination System, Sanator. For details refer to entry under *United States* in this section.

Sanator lightweight decontamination equipment in use with a Royal Norwegian Air Force F-16A fighter

STATUS

In production. In service with the Norwegian and Swedish armed forces and with the United States armed forces (licence production).

MANUFACTURER

Karl H. Høje & Co AS, Storgt 37, 0182 Oslo 1, Norway.

POLAND

Truck-mounted Vehicle Decontamination Apparatus

DESCRIPTION

This truck-mounted vehicle decontamination apparatus was developed along the same lines as the Czechoslovak TZ-74 and the Soviet TMS-65 in that it uses a jet engine to spray decontaminants over vehicles that drive past the efflux. This Polish equipment is based on the chassis of the Star 266 (6 × 6) 3500 kg truck and mounts the jet engine at the rear behind a decontaminant tank. It would appear that the jet engine is fixed with no variation in traverse but some degree of elevation possible. The jet engine is controlled from a transparent housing mounted on the right-hand side of the vehicle cab roof. This housing accommodates the operator and his bank of controls. As the engine is fixed facing to the rear, vehicles to be decontaminated have to drive past the jet exhaust.

STATUS

In service with the Polish armed forces.

MANUFACTURER

State factories.

Decontamination Apparatus, Truck-mounted, Model IRS

DESCRIPTION
The Model IRS is used to decontaminate large areas of terrain, large pieces of equipment, vehicles and buildings. It can also be used as a water carrier, to provide heated water for showers or for fire fighting. The equipment is mounted on a Star 66 (6 × 6) 3500 kg truck and consists of a 2500-litre tank, a pump driven by the vehicle engine, a hand pump, a heater, various hoses, pipes and nozzles, and shower equipment. The tank is fitted with internal baffles and is internally corrosion-proofed. The inlet manhole is screened and the system can be used to discharge solid or liquid decontaminants. Liquid solutions are mixed by the internal cycling of the pump which is driven by a power take-off, and when discharging the pump can deliver 600 litres a minute at a working pressure of 4 kg/cm^2. The tank contents are heated by a fuel-fired heater fed from the vehicle fuel tank by compressed air from the vehicle air brakes. The heater can warm 2000 litres of solution an hour to a temperature of 70°C, which is then maintained. The various accessories are stowed along the sides of the tank in cases and drums of decontaminant are stowed over the cases. The equipment consists of fourteen 10 mm hoses, each with a spray pipe, nozzle and nozzle brush. For cleansing large areas or buildings (or fire fighting) three 20-metre long hoses, 25 mm in diameter, can be fitted. The shower unit has eight heads. For terrain cleansing, fan-shaped nozzles can be fitted to the front and rear of the vehicle. The total weight of the Model IRS is 9560 kg.

IRS decontamination vehicle based on Star 66 (6 × 6) 3500 kg truck chassis

STATUS
No longer in production. In service with the Polish armed forces.

MANUFACTURER
State factories.

Decontamination Apparatus, Model UDU

DESCRIPTION
This equipment is carried to its place of use by a truck and placed on the ground. It is operated by two men, and is used to remove radioactive particles from clothing, tents, tarpaulins and similar equipment. The equipment has a small internal combustion engine which powers a system of beaters and a vacuum system. The contaminated clothing is fed into the equipment where it is beaten and the particles are then removed by suction to a special container for disposal. The Model UDU can process up to 120 pieces of clothing an hour. The weight of the equipment is 270 kg.

STATUS
Production complete. In service with the Polish armed forces.

MANUFACTURER
State factories.

UNION OF SOVIET SOCIALIST REPUBLICS

Individual Decontamination Kit, IPP

DESCRIPTION
The IPP kit has largely replaced the earlier IPP-3 kit and contains decontaminants capable of dealing with nerve agents as well as blister and biological agents. The kit is normally carried in the protective mask carrier and is contained in a plastic case. The main items inside the case are a glass phial and a plastic phial which contains a glass ampoule. Gauze pads are also included, and another small compartment contains four anti-smoke gauze-wrapped ampoules. The glass phial contains anti-nerve gas decontaminant, and the plastic phial decontaminant for use against blister gases. The plastic container holds alcohol and the glass ampoule chloramine-B powder: crushing the ampoule mixes the two, and the resultant mixture can be used once the plastic phial has been punctured with the metal point in the case lid. The gauze pads are used to apply the solution to the affected areas. Enough mixture is supplied to cover 500 cm^2. In use the nerve gas decontaminant would be spread first from the glass phial, followed by the mixture from the plastic phial. The four anti-smoke ampoules can be removed quickly from the kit case by fitted drawstrings. They are crushed and placed inside the wearer's protective mask, and the resultant inhalations can overcome the effects of most irritant smoke. The ampoules contain a mixture of chloroform, ethanol, ethyl ether and ammonia water.

STATUS
In production. In service with the Soviet and other Warsaw Pact armed forces.

MANUFACTURER
State factories.

Individual Decontamination Kit, IPP-3

DESCRIPTION
The Model IPP-3 is essentially the same as the later IPP kit but it does not contain any nerve gas decontaminant. It has been largely replaced by the IPP but many kits are still kept in reserve stocks. The Model IPP-3 is contained in a plastic case and is normally carried in the protective mask carrier. The case holds two plastic phials for use against blister agents, each phial containing a glass ampoule filled with chloramine-B powder. The plastic phial also contains alcohol and crushing the plastic phial mixes the powder with the alcohol. A metal spike in the case lid is used to pierce the phial and the solution can then be used with the gauze pads contained in the kit. The kit also contains four crushable anti-smoke ampoules which are the same as the ones in the IPP kit and are used in the same manner.

STATUS
No longer in production. Still in reserve and training service with the Soviet and Warsaw Pact armed forces.

MANUFACTURER
State factories.

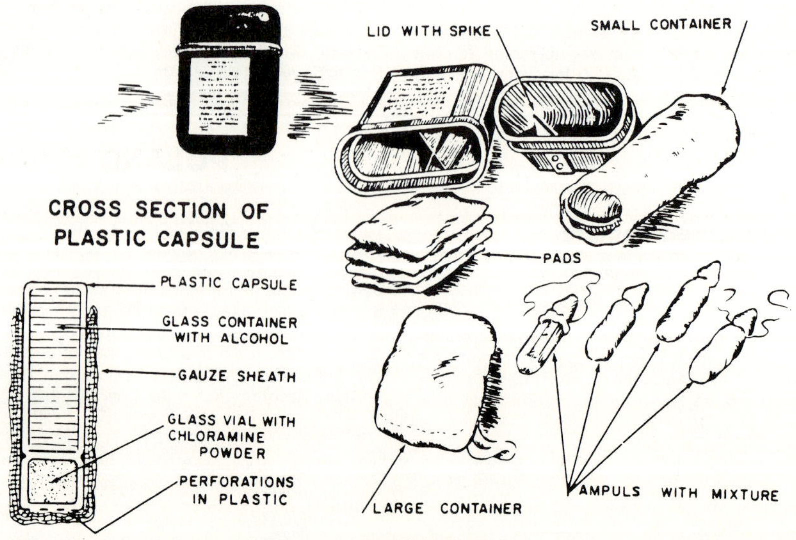

Model IPP-3 individual decontamination kit

Decontamination Packet, Model DPS

DESCRIPTION

This packet is used for the personal decontamination of clothing once a contaminated area has been left. The packet is made from clear plastic and is issued sealed. Inside are an instruction sheet and a fabric dusting bag which is used to dust the agent-absorbing brown powder over all outside clothing and headgear. Once applied the dust is rubbed into the clothing with either the bag or rubber gloves. The dust is then shaken or brushed from the clothing before the protective headgear is removed.

STATUS

Probably still in production. In service with the Soviet and Warsaw Pact armed forces.

MANUFACTURER

State factories.

Decontamination Kit, Model PKhS

DESCRIPTION

The Model PKhS kit is a personnel and clothing decontamination set for areas which are too large to be covered effectively by an IPP kit. It is contained in a plywood case and consists of three 0.5-litre bottles, two large and two small packets of decontaminant agent, two packets containing ten gauze pads each, a mixing dish and a wooden stirrer. Two of the bottles are sealed with white wax and the other has a red wax seal. The powdered contents of one large and one small decontaminant agent packet are emptied into the dish and mixed with the contents of one of the white-wax sealed bottles and the resultant solution can be freshly used as an anti-blister agent decontaminant. The contents of the red wax sealed bottle can be used direct from the bottle as a nerve agent decontaminant. Whichever solution is used it is normally applied using the gauze pads. The carrying case, which is supplied with a carrying sling, is 305 mm wide, 305 mm high and 90 mm deep.

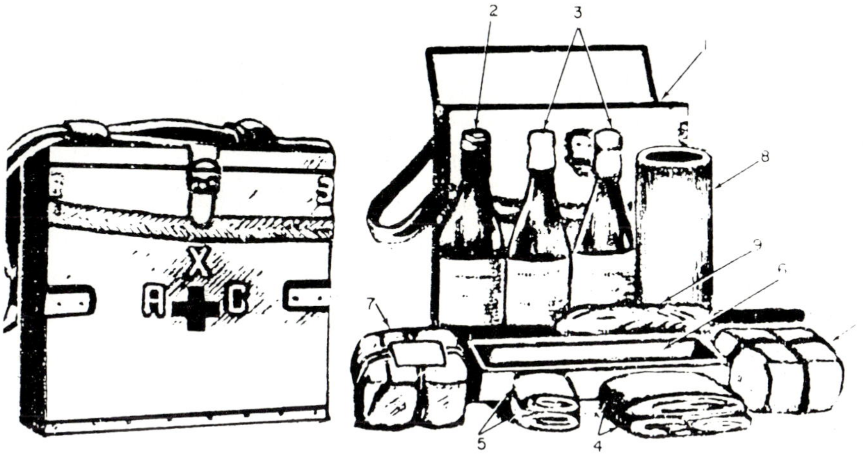

Model PKhS decontamination kit **(1)** *carrying case* **(2, 3)** *solvents* **(4, 5)** *powdered decontaminants* **(6)** *mixing dish* **(7)** *gauze pads* **(8)** *container for powdered decontaminants* **(9)** *wooden stirrer*

STATUS

In production. In service with the Soviet and Warsaw Pact armed forces.

MANUFACTURER

State factories.

Personal Weapons Decontamination Kit, Model IDP

DESCRIPTION

The Model IDP kit is contained in a drab olive metal case and although primarily issued for decontaminating personal weapons, there is enough decontaminant in the kit for use on small crew-served weapons. The kit consists of five cotton swabs and two ampoules. One of the ampoules is marked with a red tip and is intended for use with blister agents. Its position in the case is marked by a '1' embossed in the case wall. The other ampoule (No 2) has a black tip and is intended for use against nerve agents. The weapon is first wiped with a swab to clear as much of the agent as possible. The appropriate ampoule is then opened and the decontaminant solution applied, after which the weapon is wiped dry and lightly oiled. Each ampoule contains 82 ml of solution. The case is 130 mm long, 80 mm wide and 40 mm deep, and the entire kit weighs 0.305 kg. Printed instructions are glued to the side of the case.

STATUS

In production. In service with the Soviet and Warsaw Pact armed forces.

MANUFACTURER

State factories.

Artillery Decontamination Kit, Model ADK

DESCRIPTION

The Model ADK kit is issued to the crews of artillery pieces and large-calibre mortars for decontaminating their weapons and is effective against both blister and nerve agents. The kit is contained in a metal case which holds four 1-litre cans of decontamination solution, two smaller plastic containers, two application brushes with handle extensions, two metal scrapers, about 0.5 kg of cotton wool, a 150 mm long roll of sealing tape, and four cork gaskets. Instructions are inside the case lid and four rubber blocks keep the cans in place. Two of the four cans are embossed with the number '1' and have red lids: they are for the decontamination of blister agents and V-type nerve agents. The two black-lidded cans are embossed with the number '2' and are for use against G-type nerve agents. The scrapers are used to remove mud and dirt from the equipment as the solutions, normally two, are being mixed.

The first solution is made by emptying one 80-gramme packet of DT-6 decontaminant powder into a red-lidded can which contains 1 litre of dichloroethane and shaking for about 5 to 10 minutes. All possible surfaces are wiped with the cotton wool, and then the mixture is applied with the brushes supplied. The black-lidded cans contain the ready-mixed No 2 aqueous solution of 2 per cent sodium hydroxide, 5 per cent monoethanolamine and 20 per cent ammonia. Even if there are no G-type nerve agents present, the solution No 2 has to be used to remove the corrosive solution No 1 from unpainted metal surfaces. When any contaminating agent is unidentified the solutions are used in numerical order. After use the kit can be refilled with solutions and powder by an ADM-48 decontamination vehicle and the spare gaskets and tape supplied with the kit are used to reseal the kit ready for re-use. The scrapers can also be used to loosen the solution container caps.

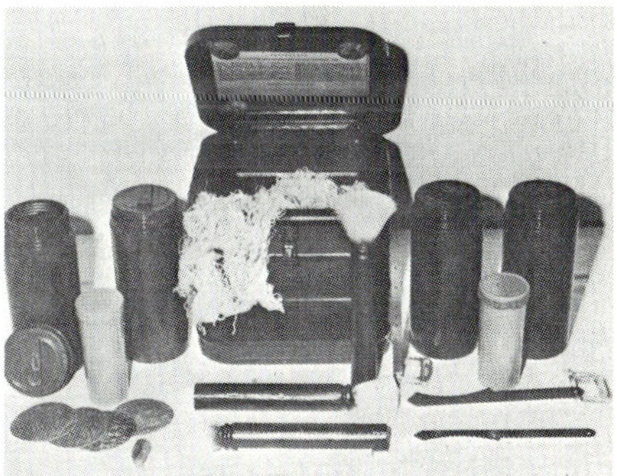

ADK artillery decontamination kit

STATUS

In production. In service with the Soviet and Warsaw Pact armed forces.

MANUFACTURER

State factories.

Machine Gun/Mortar Decontamination Kit, Model PM-DK

DESCRIPTION
The kit Model PM-DK follows much the same lines as the larger artillery kit model ADK and the solutions involved and their use are the same. The difference lies in the smaller sizes of the items concerned. Only two solution cans are contained in the metal case, one of the No 1 and one of the No 2. There is only one application brush which fits directly onto the solution can in use and a cleaning wire is supplied to clean the hollow handle of the brush. Each solution can contains 250 ml. Like the larger artillery kit, the Model PM-DK can be refilled and reused and spare gaskets are supplied to reseal the solution cans. There are two types of carrying case, both with carrying slings. One has straight sides with rounded corners and the other has curved, body-contoured sides.

STATUS
In production. In service with the Soviet and Warsaw Pact armed forces.

MANUFACTURER
State factories.

Model PM-DK machine gun/mortar decontamination kit

Decontamination Apparatus, Back-pack, Model RDP-3

DESCRIPTION
This back-pack equipment has been largely replaced in front-line service by the later Model RDP-4V but still remains in reserve and civilian defence organisation use. It follows the same general lines as the later Model RDP-4V but differs in that the pumping handle is on the right and the nozzle is guided by the left hand. A design fault is that the pump is immersed in the corrosive decontaminating liquids in action and thus the pump life is shortened and maintenance is difficult: this fault was rectified on the later Model RDP-4V. The Model RDP-3 tank contains approximately 12 litres of decontaminant and when full weighs around 20 kg (7.2 kg empty). Pumping at a rate of 25 to 30 strokes a minute gives a discharge rate of about 0.8 litre a minute and the tank contains enough liquid to cover approximately 5.1 square metres. The Model RDP-3 is 40 mm high, 350 mm long and 180 mm wide.

STATUS
No longer in production. Probably still in reserve and training use with Soviet and Warsaw Pact armed forces and some civilian defence organisations.

MANUFACTURER
State factories.

Decontamination Apparatus, Back-pack, Model RDP-4V

DESCRIPTION
This hand-operated back-pack equipment is used to decontaminate vehicles, equipment, weapons, small buildings and small areas of ground. The back-pack is held in place by a pair of straps at the waist and shoulders. It consists mainly of a tank with a large filling aperture, a filter screen, a clamp-on pressure cap, a piston-type air pump operated by the left hand, outlet tube and hose, and a spray pipe and nozzle to which a brush may be attached. A wrench and a small canister containing tools are fitted to the bottom of the tank. The equipment can be used to spray decontamination solution No 1 or No 2, depending on the nature of the contamination. The pump handle is normally clipped in the vertical but in use it is horizontal and operated by the left hand: a pumping rate of 25 to 30 strokes a minute gives a discharge flow of about 0.7 litre a minute. When full, the Model RDP-4V weighs approximately 17 kg and empty 8.5 kg. It is 355 mm high, 290 mm long and 190 mm wide.

STATUS
In production. In service with the Soviet and Warsaw Pact armed forces.

MANUFACTURER
State factories.

Model RDP-4V back-pack decontamination apparatus

Decontamination Kit, Model IDK-1

DESCRIPTION
The Model IDK-1 kit is normally carried in a vehicle cab and is used to decontaminate the vehicle. The kit, which is designed to be used in conjunction with a standard 20-litre jerrican, consists of a spray pipe with a nozzle and brush, a cap for the 20-litre can, a hose, an aspirator attachment and various accessories. The kit can be used in two ways: one is to mix either the decontamination solution No 1 or No 2 and strain it into the 20-litre can which is then pressurised and kept so using a hand pump, or the aspirator is used when fitted to the vehicle air brake system.

STATUS
In production. In service with the Soviet armed forces.

MANUFACTURER
State factories.

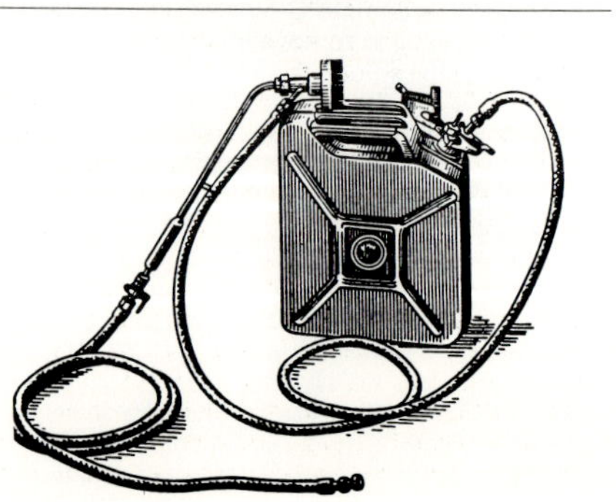

Model IDK-1 decontamination kit

Decontamination Kit, Models DK-4 and EEA-64

DESCRIPTION
These decontamination kits are essentially similar, and use a vehicle exhaust system as the pressure source. Each vehicle so equipped has provision over the exhaust for fitting a special cap with a branch valve fitted anterior to the vehicle's silencer system. To this branch valve is fitted a jet pump which uses the exhaust pressure to provide suction to power a pump. The pump drives decontaminating solution or powder from a standard 20-litre jerrican through a hose and eventually through a nozzle which may be fitted with a brush. Alternatively, the system can be driven in reverse so that the pump provides suction from the nozzle to remove radioactive fallout particles. When used in this manner the standard 20-litre jerrican can be used as the dust receptacle, but it is more usual to pass the exhaust dust through water in a specially-dug hole or channel.

The East German version of the Model DK-4 is known as the Model EEA-64. Polish versions have several designations including IZS, EZS, EZCS-34 and EZCS-54.

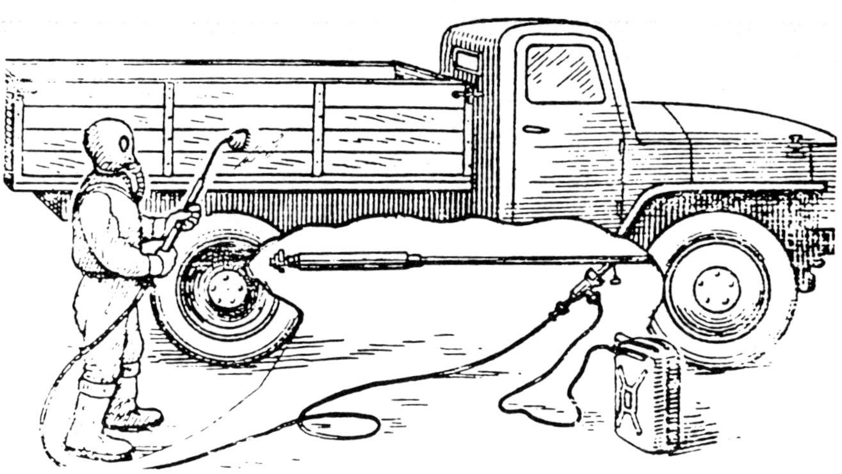

Model DK-4 decontamination kit in use

STATUS
In production. In service with the Soviet, East German and Polish armed forces.

MANUFACTURER
State factories.

Decontamination System, Portable, Model DKV

DESCRIPTION
This portable system is used to decontaminate vehicles, and is composed of 78 cylindrical tanks carried on a specially-equipped truck and trailer. Each cylindrical tank is pre-loaded with decontamination solution No 1 or 2, and each tank can be fitted with either one or two spray pipes. The tanks are delivered to the users where they can be pressurised by either the user vehicle's air brake system or by a separate compressor. The spray pipes are usually fitted with a brush and the contents of two cylinders are usually sufficient to decontaminate a truck; an MBT takes three cylinders. Each cylinder can take up to 30 litres. Several cylinders can be fitted to a single compressed air source, and the cylinders can be collected and refilled.

STATUS
In production. In service with the Soviet and Warsaw Pact armed forces.

MANUFACTURER
State factories.

Decontamination Apparatus, Trailer-mounted, Model DDP

DESCRIPTION
This trailer-mounted equipment follows the same general lines as the truck-mounted Model DDA-53 series, but has slightly less capacity. Otherwise the two equipments are essentially similar and the same boiler is used. However, the trailer-mounted Model DDP has only one steam chamber with a capacity of 1.4 cubic metres, even though the boiler can provide enough water for about 80 showers an hour. The equipment is mounted on a 1-AP-1 or 1-AP-5 trailer and the complete equipment weighs approximately 2050 kg.

STATUS
Probably no longer in production. In service with the Soviet and Warsaw Pact armed forces.

MANUFACTURER
Various state factories.

Decontamination Apparatus, Truck-mounted, Models DDA-53, DDA-53A, DDA-53B and DDA-66

DESCRIPTION
These equipments are used by decontamination units and by medical units for sterilisation, disinfecting and disinfestation. With chemical units the Model DDA-53 series is used for steam-decontamination of chemically- and biologically-contaminated clothing and small items of equipment, and the apparatus can also provide hot water for showers. Each equipment consists of a vehicle-borne system with two steam chambers, a vertical boiler (the RI-3), a fuel oil tank, a water pump, a formaldehyde tank, a 12-head shower unit and all the associated hoses and fittings. The system does not have its own water tank, so water has to be provided by another tanker vehicle or a stand tank if no natural source is available. The water boiler which produces steam or hot water contains 250 litres and is normally fired by fuel oil, although wood may be used with a loss in heating efficiency. The fuel oil tank holds 55 litres which is enough for 8 to 10 hours' operation. Pipes connect the boiler output to steam chambers each of which can contain 25 to 30 summer uniforms, 20 winter uniforms or 12 sheepskin jackets. Decontaminants can be added to the steam if necessary. The system can thus decontaminate up to 80 uniforms an hour in summer and up to 48 in winter. When used for showers the system is normally used in conjunction with tents which are carried on other cargo trucks, and enough water can be provided for up to 100 showers an hour in summer and 70 to 72 in winter.

DDA-53A truck-mounted, decontamination apparatus based on GAZ-63 (4 × 4) 1500 kg truck chassis

There are four models of the DDA-53 (sometimes known as the ADA). The Model DDA-53 is mounted on the GAZ-51 (4 × 2) truck chassis, the Model DDA-53A on the GAZ-63 (4 × 4) chassis and the Model DDA-53B on a ZIL-130 (4 × 2) 4000 kg truck chassis. The DDA-53B differs also in having the boiler and steam chambers enclosed in a metal body. The Model DDA-66 is mounted on a GAZ-66 (4 × 4) 4000 kg truck chassis.

STATUS
Probably no longer in production. In service with the Soviet and Warsaw Pact armed forces.

MANUFACTURER
Various state factories.

Decontamination Apparatus, Truck-mounted, ADM-48D

DESCRIPTION

This truck-mounted equipment is used for large-scale decontamination of vehicles, weapons, equipment and ground areas. It can be mounted on either a GAZ-51 or GAZ-63 truck chassis, and consists of two main units. Unit No 1 on the right side is filled with the standard decontaminant solution No 1 which is an 8 per cent solution of decontaminant DT-6 in dichloroethane, effective against blister and V-type nerve agents. As this solution is corrosive all the fittings on the unit are brass. Unit No 2, on the left, is filled with decontaminant solution No 2, an aqueous solution of 2 per cent sodium hydroxide, 5 per cent monoethanolamine and 20 per cent ammonia, effective against G-type agents. Both units consist of a 500-litre tank, a hand pump, an anti-surge tank, a pipe system, a metering unit (for rapid and measured refilling of PM-DK and ADK decontamination kits), six 12-metre hoses, six nozzles and brushes, four suction hoses, spare parts and various tools. Each unit can supply up to six spray lines, and when not in use the hoses are carried in boxes by the rear tailgate. A rectangular steel tank with a 65-litre capacity fitted by the tailgate contains dichloroethane, and six RDP-4V back-pack equipments are also carried on the vehicle for remote use. The ADM-48 is one of the basic equipments used by Warsaw Pact chemical units and its employment is widespread.

STATUS

Probably no longer in production. In service with the Soviet and Warsaw Pact armed forces and possibly with some Middle East states.

MANUFACTURER

State factories.

Decontamination Apparatus, Truck-mounted, Model TMS-65

DESCRIPTION

The Model TMS-65 is intended for the rapid decontamination of vehicles and towed weapons and equipment. It is mounted on a Ural-375 (6 × 6) 4000 kg truck chassis and the main operating component is a Model VK-1F modified gas turbine aircraft engine mounted on a turntable with the operator's cabin. A hydraulic system turns the turntable to either side of the vehicle and the operator's cabin is equipped with floodlights and a screen wiper. Fuel for the engine is contained in a 1500-litre tank between the turntable and the right rear of the cab. A tank on the left contains 1500 litres of decontaminant solution and another 4000 litres of solution are carried in a trailer towed behind the vehicle. Normally the Model TMS-65 is employed in pairs. Two equipments are placed about 50 metres apart and the engines are started. Decontamination solution is directed into the engine exhaust and so to wherever the operator directs the hot gas stream: the engine can be elevated or depressed as well as traversed. The vehicles to be decontaminated are driven between the two equipments and are cleansed in between 30 seconds to 3 minutes, depending on the nature and degree of contamination. For nuclear cleansing, water may be used. If space allows the two Model TMS-65 equipments are driven along a static column, which is stated to be particularly effective in winter.

STATUS

In production. In service with the Soviet and Warsaw Pact armed forces.

MANUFACTURER

State factories.

TMS-65 with turbojet covered by tarpaulin

Close-up of TMS-65 turbojet outlet showing storage tank to its rear

Model TMS-65 truck-mounted decontamination apparatus in operational use decontaminating a BDRM-2-rkh radiological-chemical reconnaissance vehicle

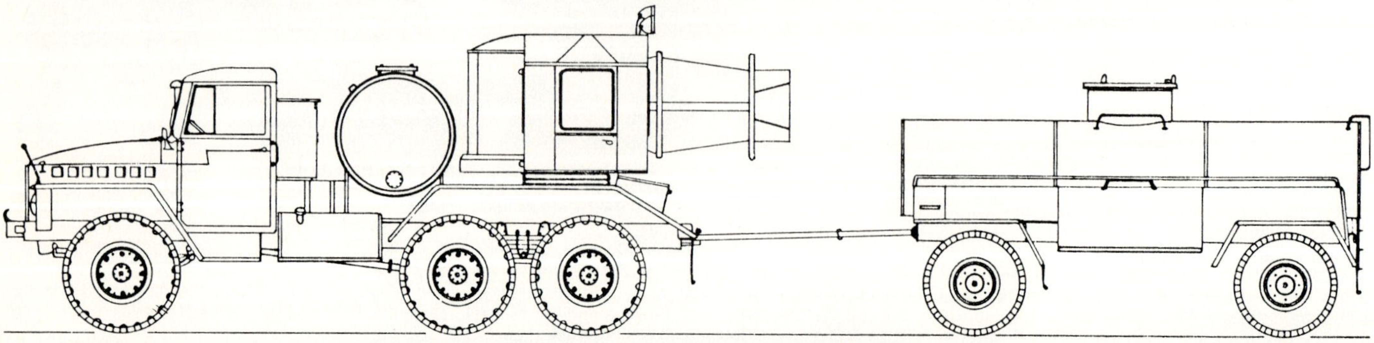

Model TMS-65 decontamination vehicle (Ural-375E) towing trailer

Decontamination Apparatus, Truck-mounted, Models ARS-12D and ARS-12U

DESCRIPTION
Apart from use as decontamination equipment the Models ARS-12D and ARS-12U can be used as water carriers, for fire fighting and to provide cold showers for various purposes. The two models are basically similar but the Model ARS-12D is mounted on a ZIL-151 (6 × 6) 2500 kg truck chassis and the Model ARS-12U on a ZIL-157 (6 × 6) 2500 kg truck chassis. Each consists of a 2500-litre tank divided by two baffles and with a large manhole for access and filling. Also provided is a depth gauge, a self-priming pump driven from the vehicle engine, a hand pump, a piping system, hoses, nozzles and other spares and accessories. The main pump drive shaft turns at 1400 to 1600 rpm and can turn the pump to deliver 300 to 400 litres a minute. The hand pump, operated at 45 strokes a minute, can deliver 4.5 to 5.5 litres a minute. Decontaminants are mixed in the tank as water is poured in and a thorough mixture is made by internally recycling the solution through the pump, otherwise the vehicle motion and the filling mixing is all that is necessary. Many different decontaminants may be used but the standard solutions are numbers 1 and 2. There are several administration methods. To clear roads or terrain a wide spreading nozzle (the DN-3) can be fitted directly to the main discharge pipe and the vehicle is then driven over the affected road or terrain. A full tank can then clear a strip 500 metres long and five metres wide. For decontaminating vehicles and other equipment up to eight 18-metre hoses can be fitted, each with spray pipes, nozzles and nozzle brushes. Up to four vehicles can be cleansed at one time and there is sufficient solution in a full tank to cleanse 12 MBTs, 13 APCs, 15 trucks or 45 artillery pieces. To decontaminate buildings (or for fire fighting) four 25 mm diameter hoses can be used. Racks on top of the tank hold drums of decontaminant and several RDP-4V back-pack equipments can be carried for remote use. The Model ARS-12D can be used to emit smoke screens by discharging chlorosulphonic acid.

ARS-12U in action using DN-3 spreader nozzle

STATUS
No longer in production. In service with the Soviet and Warsaw Pact forces, and may be in service with some Middle East states.

MANUFACTURER
State factories.

Decontamination Apparatus, Truck-mounted, Model ARS-14

DESCRIPTION
This equipment is a development of the earlier Model ARS-12U and it is expected that it will eventually replace the earlier models in service. The main change is the use of a ZIL-131 (6 × 6) 3500 kg truck chassis but the full extra tank capacity this provides cannot be used because extra drums of decontaminant are carried on specially-fitted racks. However, the extra drums enable the equipment to be used over a longer period without reloading. Some changes have also been made to the piping system in that the previously fixed outlet pipes are now carried separately and fitted to the equipment by hoses. The wide-spreading DN-3 nozzle is now fitted at the vehicle front as well as the rear, and the eight hoses are wound onto only four drums instead of the former eight; these drums are at the left rear. Other changes have been made to the general 'plumbing' but an innovation is an extra rubber hose which can be fitted to the vehicle exhaust in winter to thaw out any frozen parts of the equipment. Filling pistols can also be fitted to the system for loading or reloading decontamination kits.

Model ARS-14 truck-mounted decontamination apparatus

STATUS
In production. In service with the Soviet and East German armed forces.

MANUFACTURER
Various state factories.

Decontamination Station, Models AGV-3M and AGW-3M

DESCRIPTION

The Decontamination Station Model AGV-3M is used for steam decontamination of chemically- or biologically-contaminated clothing and small pieces of equipment. It consists of four special vehicles: one Model AGV-3M truck-mounted decontamination steam and hot air generator, two Model AGV-3M truck-mounted decontamination steam chamber equipments and one cargo truck carrying a drying tent, a shower tent, a collapsible water tank and accessories.

The steam and hot air generating unit is mounted on a ZIL-157 or ZIL-151 (6 × 6) truck chassis. It has a van-type body which contains an oil-fired heater, a 500-litre boiler with a super-heater, a petrol engine for driving the fuel oil and water pumps, a single-stage turbo-blower and a heat exchanger. The steam generated in the boiler is super-heated to between 160 and 200°C and then passed to the steam chamber vehicles. The turbo-blower supplies 350 cubic metres a minute through the heat exchanger at a pressure of 0.5 kg/cm^2, which is then supplied to the drying tent via a large diameter hose. The boiler also supplies hot water for the crew's shower unit. The steam chamber units are mounted on either a ZIL-130, ZIL-150 or ZIL-164 (4 × 2) truck chassis. Each unit has three 2 cubic metre pressure chambers and all the pipes and other fittings. Clothing is hung inside the steam chambers which are then pressurised by the steam from the steam generator vehicle. Extra decontaminants can be added to the steam if necessary. After a period in the steam chambers the clothing is taken to the drying tent. The cargo truck used with the station is usually a ZIL-130 or a ZIL-164. The station can process between 50 and 150 uniforms an hour, depending on the type of contamination.

The decontamination station Model AGV-3M has now replaced the earlier Model AGV-2.

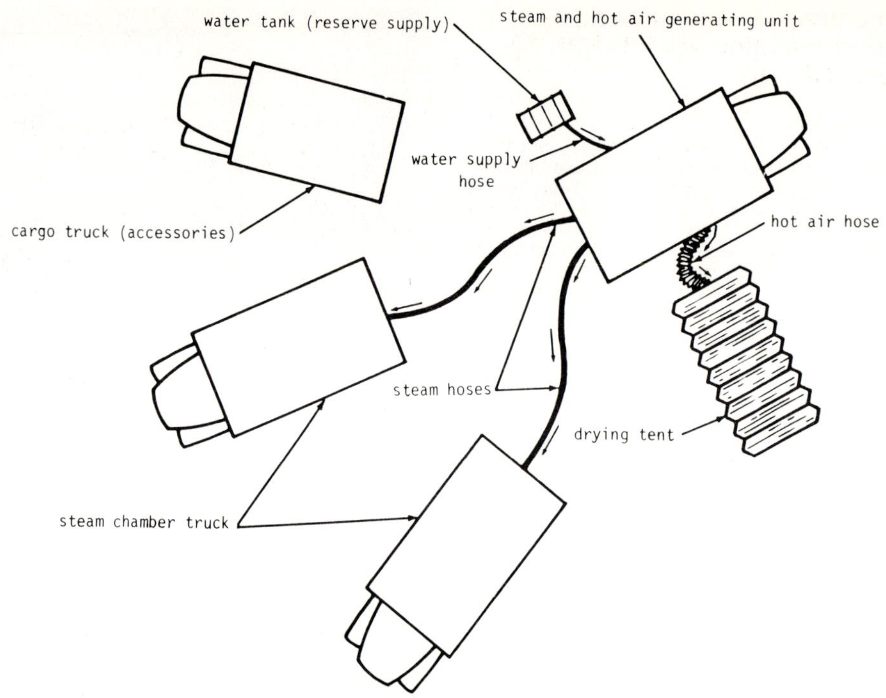

Layout of AGV-3M and AGW-3M decontamination system

In East German service the Model AGV-3M is known as the Model AGW-3M.

STATUS

Production completed. In service with the Soviet and Warsaw Pact armed forces. The earlier Model AGV-2 may remain in service with some civil defence organisations.

MANUFACTURER

State factories.

Decontamination Apparatus, Clothing, Truck-mounted, Models BU-2, BU-3, BU-4 and BU-4M

DESCRIPTION

All these equipments have the same function which is to decontaminate clothing, protective clothing, small canvas shelters and similar articles by boiling, after which the items can be conventionally laundered. All the equipments are carried on either GAZ-53, GAZ-63 or LO 1800 A trucks and are unloaded from the trucks by hand-operated cranes before use. The Model BU-2 has two 350-litre boilers which are heated either by a fire or by an integral steam coil. The equipment also has a collapsible water tank, or tanks, a hand pump, a manual drying press and a drying tent. Each boiler can hold 40 uniforms and up to 1000 uniforms can be processed in 24 hours. The Model BU-3 is essentially similar but has only one boiler. The Models BU-4 and BU-4M are improved versions of the Model BU-2 and differ from one another only in detail. Each equipment has two 450-litre boilers, a hand pump, one 1000- and one 1200-litre collapsible water tank, metal baskets, a hand drying press and a drying tent. The metal baskets each take ten uniforms and each boiler can accommodate four baskets. Extra decontaminants can be added to the boiling water if necessary. Normally these equipments are used in conjunction with a decontamination station Model AGV-3M.

STATUS

Production complete. In service with the Soviet and Warsaw Pact armed forces.

MANUFACTURER

State factories.

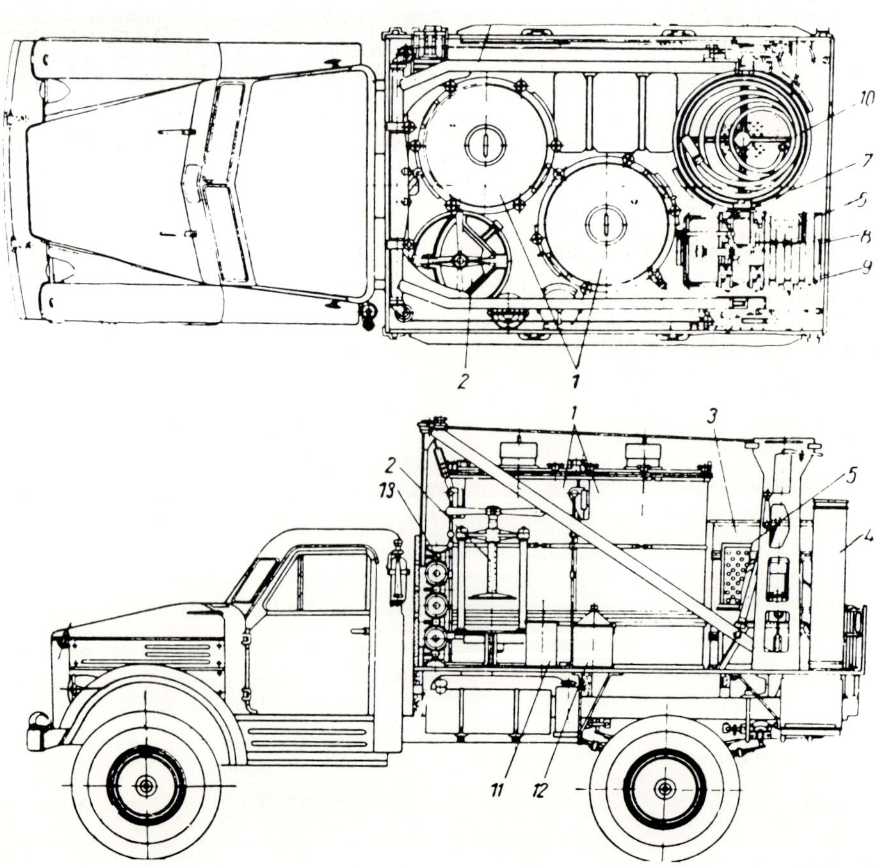

Model BU-4M truck-mounted decontamination apparatus **(1)** *decontamination boilers* **(2)** *drying press* **(3)** *water tank* **(4)** *chimney* **(5)** *baskets* **(6)** *crane* **(7)** *storage box* **(8)** *manual water pump* **(9)** *benches* **(10)** *hose* **(11)** *bucket* **(12)** *container* **(13)** *stakes and pegs for drying tent*

Dry-decontaminant Spreader, Truck-mounted, Models PDP-53 and PDM

DESCRIPTION
This spreader is usually fitted to the GAZ-51 or GAZ-63 (4 × 4) truck (both of which have a payload capacity of 1500 kg) and the ZIL-150 (4 × 2) 3500 kg truck. The equipment is used to spread dry decontaminant over roads or terrain and it is fitted to the back of the vehicle. The spreader itself is driven by a chain, driven in turn by a sprocket fitted to the right rear wheel hub. The chain drives a rotary spreader at the bottom of a hopper which is kept topped up by two men with shovels in the rear of the truck. The actual distribution of the spreader can be varied between 0.5 to 1 kg/m^2 on the Model PDP-53 but on the East German Model PDM it is fixed at 0.5 kg/m^2.

STATUS
Production complete. In service with the Soviet and East German armed forces.

MANUFACTURER
State factories.

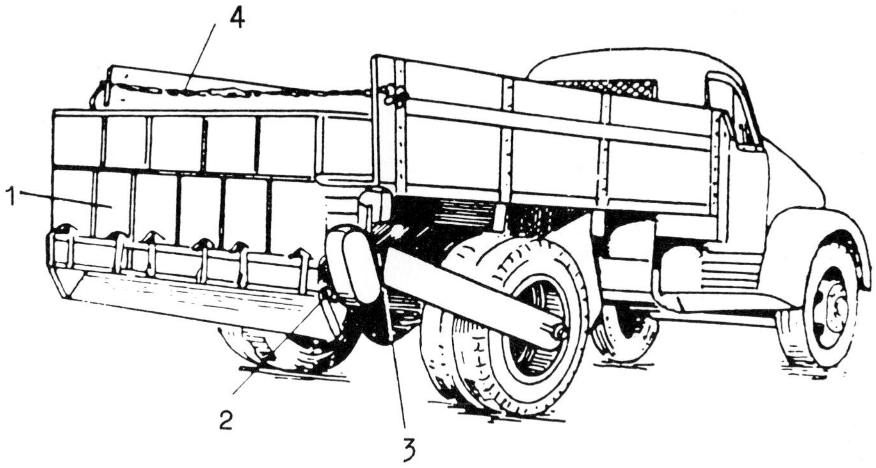

Model PDP-53 truck-mounted dry-decontaminant spreader **(1)** *hopper* **(2)** *sprocket cover* **(3)** *drive chain cover* **(4)** *safety chain*

UNITED KINGDOM

Decontamination Kit, Personal No 1 Mark 1

DESCRIPTION
This decontamination kit is normally issued to all field personnel and is carried as part of their standard personal equipment. It consists of a sealed clear plastic bag containing a set of instructions and four pads charged with fuller's earth. In use the plastic bag is torn open and one pad is taken out and placed over one hand like a glove. The pad is then banged over suspected areas of skin or clothing and equipment to dispense the fuller's earth which is then rubbed over using the pad to spread the powder.

Decontamination Kit, Personal No 1 Mark 1 in use

Decontamination Kit, Personal No 1 Mark 1

STATUS
In production. In service with the British Armed Forces.

MANUFACTURERS
Remploy Limited, 415 Edgware Road, Cricklewood, London NW2 6LR, United Kingdom.
J & S Franklin Limited, Franklin House, 151 Strand, London WC2R 1HL, United Kingdom.

Decontamination Kit, Personal No 2 Mark 1

DESCRIPTION
This personal decontamination kit consists of a flat polythene dispenser containing 113 grammes of fuller's earth. It is used for decontaminating the user's boots and personal equipment and would normally be used in conjunction with the Decontamination Kit Personal No 1 Mark 1.

STATUS
Production complete. In service with the British Armed Forces.

SUPPLIERS
Richmond Packaging, New Road, Winsford, Cheshire, United Kingdom.
J & S Franklin Limited, Franklin House, 151 Strand, London WC2R 1HL, United Kingdom.

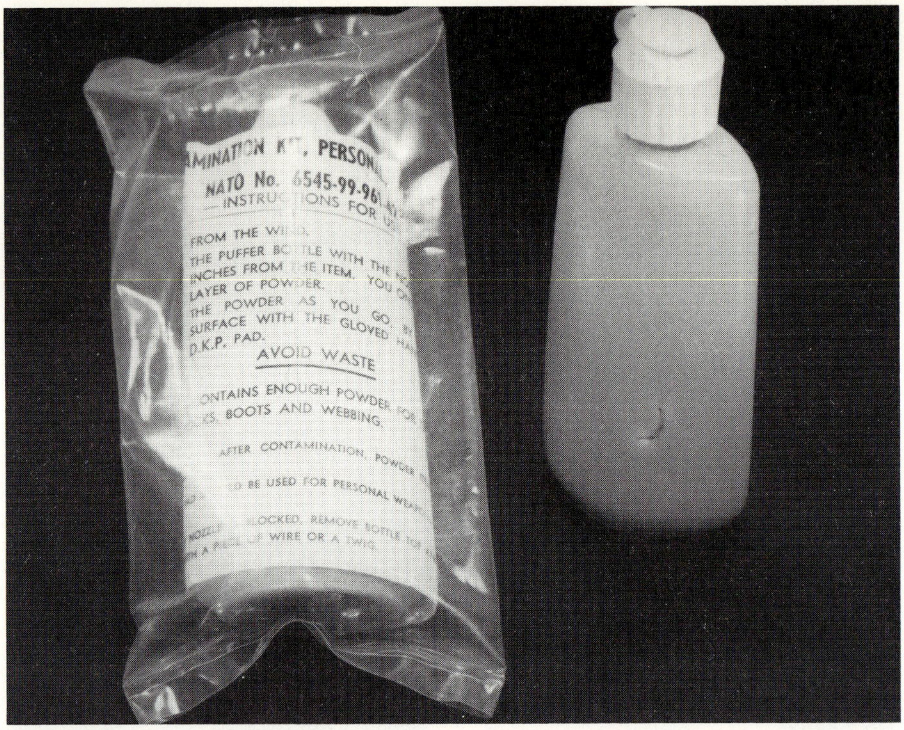

Decontamination Kit, Personal No 2 Mark 1

Decontamination Apparatus, Chemical Agent

DESCRIPTION
Intended for emergency front-line use, this kit is normally carried on vehicles and uses normal 9-litre fire extinguishers. The kit is supplied in an olive-green canvas tool roll containing two car-wash type brushes, extension pipes, fittings, a spanner, and an instruction sheet. In use the extinguisher body is charged with a water soluble decontaminant which is normally issued in sealed bags packed into tins. One tin is required for each loading of the extinguisher body. Once the fire extinguisher body has been charged it can be connected to the kit with the fittings issued and decontamination can begin.

STATUS
In service with the British Armed Forces.

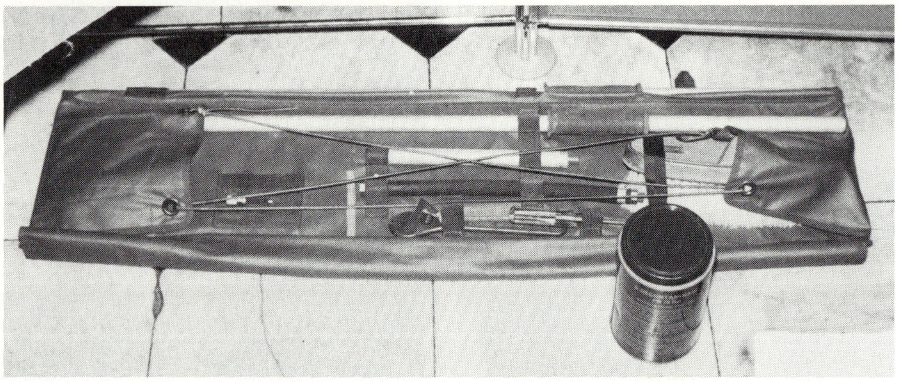

Decontamination Apparatus, Chemical Agent, tool roll, and can containing decontaminant (T J Gander)

Westland Dynamics Limited (WDL) Dual Purpose Decontamination and Large Area Screening System

DESCRIPTION
The WDL Dual Purpose Decontamination and Large Area Screening system was developed in conjunction with Engineered Air Systems Inc of St Louis, Missouri, in order to meet a US Army requirement. The decontamination part of the system provides high pressure heated water or hot air for the decontamination and cleaning of personnel and equipment, and provides for the addition and mixing of decontaminating fluids. The system can also be used to produce large area obscuration by disseminating smoke material.

The system may be mounted as a static installation or on a mobile platform such as a light vehicle or an armoured personnel carrier. The primary power source is the Weslake Aeromarine Engines

NGL WAEL 600G turbocombustor power pack

NGL Dual Purpose Decontamination and Large Area Screening System deployed in the decontamination mode

Limited 600G turbocombustor power pack. This rugged turbocharger-based gas turbine was developed for reliability, low maintenance and simplicity of operation. Control of the complete system is from a remotely-mounted control panel. In the system a turbo-expander air motor, powered by the bleed air from the turbocombustor, drives a hydraulic high pressure fluid pump and a low pressure vane type water pump. These units, together with all the associated pipes and other components are contained within the unit with hoses to interconnect with the water boiler which attaches to the exhaust pipe of the turbocombustor. The system is provided with two operator wands and pick-up and return lines to and from a water source. The pick-up line incorporates a filter. The wands have decontaminating agent connection and metering facilities.

The system can provide up to 25 litres of decontaminating fluid or hot water per minute at 70 bar. Dry weight of the total system is approximately 130 kg. Overall dimensions of the turbocombustor pack and decontamination unit are both approximately 840 × 420 × 208 mm. Fuel for the system may be diesel or most gas turbine fuels. The system is designed to operate off a 24 volt dc power supply. A 10 kW 28-volt dc generator can be incorporated into the turbocombustor system if required.

STATUS
In production.

MANUFACTURER
Westland Dynamics Limited, Yeovil, Somerset BA20 2YD, United Kingdom.

Portaflex 100 Decontamination Shower Unit

DESCRIPTION
The Portaflex 100 decontamination shower unit was originally developed for use by fire-fighting units involved in clearing hazardous chemical spillages but it also has many potential military applications.

The shower is a lightweight unit weighing 26 kg and was designed to be carried and assembled by one man. It can be assembled in about 30 seconds and erected at any convenient point. Once connected to a fire appliance or some form of mobile water pump the four flexible shower frame hoses used with the unit become rigid under a pressure of 3 bar plus and assume a configuration surrounding the object or person to be decontaminated. Water is delivered through 16 fog-producing water heads in a full head-to-foot 360-degree pattern. Water is supplied to the unit via a 2.5-inch (63.5 mm) male instantaneous coupling in the metal base frame. When not under pressure the four flexible shower frame hoses are held suspended from an overhead gantry arrangement.

The Portaflex 75 is similar but uses only two flexible shower frame hoses. It weighs 15 kg.

STATUS
In production.

MANUFACTURER
J.D.Hughes Fabrications Limited, Whitefield Road Industrial Estate, Bredbury, Stockport, Cheshire SK6 2SS, United Kingdom.

Portaflex 100 decontamination shower unit

UNITED STATES OF AMERICA

Decontaminating and Re-impregnating Kit, Individual, ABC-M13

DESCRIPTION
The ABC-M13 kit is issued to all field personnel and is intended for the emergency decontamination of skin, clothing and equipment and it also has a limited ability to re-impregnate items of clothing. The kit is contained in a plastic bag, which itself is contained within a plastic or light alloy case, and consists of a plastic bag containing a pad filled with skin-decontaminating powder, two cloth bags filled with decontaminating and re-impregnating powder, a capsule of dye, and a cutter. The kit weighs approximately 318 grammes and measures 476 × 730 × 105 mm.

The M13 kit is licence-produced in South Korea by the Daewoo Corporation and is known as the KM13.

STATUS
In service with the US Army.

Decontamination Kit, M258A1

DESCRIPTION
The Decontamination Kit, M258A1, is intended for use by individual personnel for the decontamination of skin areas, clothing, personal equipment and weapons. The original kit was the M258 carried in a plastic container that doubled as a solution container. Inside the container are two plastic bottles and a number of tissues. The bottles are marked Solution 1 and Solution 2 and are mixed before use. This type of kit can be used only once. Numbers of M258 kits are still held by the US armed forces for possible use.

The M258 kit is licence-produced in South Korea by the Daewoo Corporation and is known as the KM258.

The M258A1 kit is contained in foil envelopes. Each envelope contains a pre-impregnated towelette, and there are enough for three separate decontaminating operations. In April 1985 the US Army placed a $4.8 million contract for the production of M258A1 kits. The contract contained a 200 per cent option to be exercised at a later date.

STATUS
In production. In service with the South Korean armed forces and the US Army.

MANUFACTURER
Brunswick Defense, One Brunswick Plaza, Skokie, Illinois 60077, USA.

Decontaminating Apparatus, Portable, DS2, 1½-quart, ABC-M11

DESCRIPTION
This equipment is carried on a rack on nearly all American-produced military vehicles and is intended for the decontamination of vehicles or crew-served weapons. Normally the equipment is issued with the nitrogen pressure cartridge already filled and in place but the container has to be filled with DS2 decontaminating agent before the equipment is ready for use. When used, the handle safety pin seal wire is removed and the handle is lifted to puncture the pressure can. A thumb lever then controls the spray. The optimum spray range is about 2 metres and the equipment will cover approximately 42 square metres.

Expanded capacity versions holding either 1.5 or 2.5 litres are available.

The ABC-M11 is being replaced in US Army service by the M13 Portable Decontamination Apparatus (DAP) (see following entry).

SPECIFICATIONS
Internal volume: 1.42 litres
Agent capacity: (DS2) 1.26 litres
Weight filled: 3.008 kg
Weight empty: 1.36 kg
Diameter: (with handle) 139.7 mm
Height: 362 mm
Effective spray range: 1.83 to 2.44 m
Duration of one charge: (continuous) 30 s

STATUS
In service with the US Army, Israeli armed forces and some other nations.

MANUFACTURER
All Bann Enterprises Inc, Anaheim, California, USA.

MARKETING AGENCY
Tradeways Limited, 307-F Maple Avenue West, Vienna, Virginia 22180, USA.

M13 Portable Decontamination Apparatus (DAP)

DESCRIPTION

The M13 portable decontamination apparatus is a vehicle-mounted, man-portable, manually-operated unit for the decontamination of wheeled and tracked vehicles, towed artillery and crew-served weapons described as 'larger than 60 caliber', ie 0.60-inch/15 mm. Development began in 1979 with first deliveries commencing during 1985. The M13 will eventually replace the M11 apparatus in service.

The apparatus consists of a pre-filled disposable container holding 14 litres of DS2 decontaminating agent (70 per cent diethylenetriamine, 28 per cent ethylene glycol monomethyl ether, 2 per cent sodium hydroxide), an accessory container holder, a manual in-line pump, two wand sections (one optional), and a disposable synthetic filament polypropylene brush.

SPECIFICATIONS

Volume: 14 litres
Weight filled: 24.5 kg
Weight empty: 10.9 kg
Coverage per filling: 112 m^2
Length: 355.6 mm
Width: 168 mm
Height: 476 mm
Operating temperature: -31 to $+49$°C

M13 portable decontamination apparatus (DAP) in use (US Army)

STATUS

In production. In service with the US Army.

MARKETING AGENCY

Tradeways Limited, 307-F Maple Avenue West, Vienna, Virginia 22180, USA.

Decontaminating Apparatus, Power-driven, Skid-mounted: Multi-purpose, Non-integral, 500-gallon, ABC-M12A1

DESCRIPTION

This multi-purpose equipment consists of three main components, a pump unit assembly, a tank unit assembly and personnel shower assembly (combined) and an M2 600-gallon per hour liquid fuel water heater. The ABC-M12A1 can be used for a variety of purposes including decontamination, de-icing, fire-fighting with water or foam, pumping and showering personnel or equipment. When used for decontamination the pump unit (which has two hose reels, a dc generator, a petrol engine, a fuel tank and a priming tank) can be used to pump up to 190 litres of decontaminating fluid every minute from both hoses. The decontaminant most likely to be used is STB (supertropical bleach – chlorinated lime) but other agents may be used. The tank unit has a stainless steel tank and a hopper-blender for mixing decontaminating agents, as well as a fluid agitation system. This unit also has an integral shower assembly for 25 men. The M2 water heater can provide heated water at the rate of up to 2270 litres an hour. All three components are mounted on skids and accessories are provided for inter-connection of all three units. For transport the units are contained in three crates which together weigh 1903 kg.

STATUS

In service with the US Army, Air Force and Marine Corps.

M17 Lightweight Decontamination System, Sanator

DESCRIPTION

The M17 Lightweight Decontamination Equipment, Sanator is produced in the United States under licence from Karl H. Høje & Co. AS of Oslo, Norway. For full details of this equipment refer to entry under *Norway* in this section.

STATUS

In production. Type classified and in service with the US Army, Air Force and Marine Corps. Orders to date have exceeded 3500.

MANUFACTURER

Engineered Air Systems Inc, 1270 N. Price Road, St Louis, Missouri 63132, USA.

M17 Lightweight Decontamination System, Sanator

XM18 Truck-mounted Decontamination Apparatus

DESCRIPTION

The XM18 truck-mounted decontamination apparatus was originally known as the XM14 but the need for modifications resulted in a new designation. The XM18 is intended to be the replacement for the ABC-M12A1 apparatus (qv) and will follow the same overall form, but delays have meant that the apparatus will not be ready for service until June 1989 at the earliest.

STATUS

Development.

YUGOSLAVIA

NBC Decontamination Vehicle

DESCRIPTION
Few details are known regarding a NBC decontamination vehicle built onto the chassis of a TAM 150 T11 BV (6 × 6) 3000/5000 kg truck. The vehicle carries a tank for decontamination solution over the rear axles with what appears to be an independently-powered pump unit at the rear, and hose and reel units just behind the hard-topped cab. The pump, hose and reel units are contained inside housings with upward-opening flaps. There appears to be no provision on the vehicle for other than the usual crew of two, including the driver.

For full details of the TAM 150 T11 BV (6 × 6) 3000/5000 kg truck refer to *Jane's Military Logistics 1989* page 548.

STATUS
In service with the Yugoslav armed forces.

MANUFACTURER (vehicle only)
Tovama Avtomobilov Motorjev, Maribor, Slovenia, Yugoslavia.

Yugoslav decontamination vehicle on chassis of TAM 150 T11 BV (6 × 6) 3000/5000 kg truck

CHEMICAL AND BIOLOGICAL DETECTION AND IDENTIFICATION EQUIPMENT

CANADA

Detector Kit, Chemical Agent (C-2)

DESCRIPTION
This kit is contained in a vinyl-coated carrying case and may be used by one person after only a minimum of training. It has a variety of operational roles including determining the presence or absence of chemical agents after an attack; identifying chemical agents; collecting samples for later analysis; identifying when it is safe to unmask for long or short periods; testing for area contamination; monitoring the expected arrival of a vapour hazard; and testing for the presence of agents after decontamination operations. The kit can be used to detect and identify Tabun (GA), Sarin (GB), Soman (GD), V agent (VX), Mustard (H, HN and T), Phosgene oxime (CX), Hydrogen Cyanide (AC), Cyanogen chloride (CK) and Phosgene (CG).

The kit contains a booklet of Paper, Chemical Agent Detector, Liquid, 3-Way (see separate entry), 20 Detectors, Chemical Agent, Nerve Vapour (see separate entry), 30 plain detector tubes, 20 detector tubes with white bands for sampling, 3 bottles of chemical reagents, an air sampling pump, a container intended to hold ten detector tubes, an instruction card set, a pencil, water bottle, some anti-freeze solution, report cards and envelopes. The items are all stored in arranged compartments and pockets, and the kit is normally carried slung from a shoulder.

The kit is intended for issue down to small unit level.

SPECIFICATIONS
Weight: 1.4 kg
Length: 230 mm
Height: 150 mm
Width: 70 mm

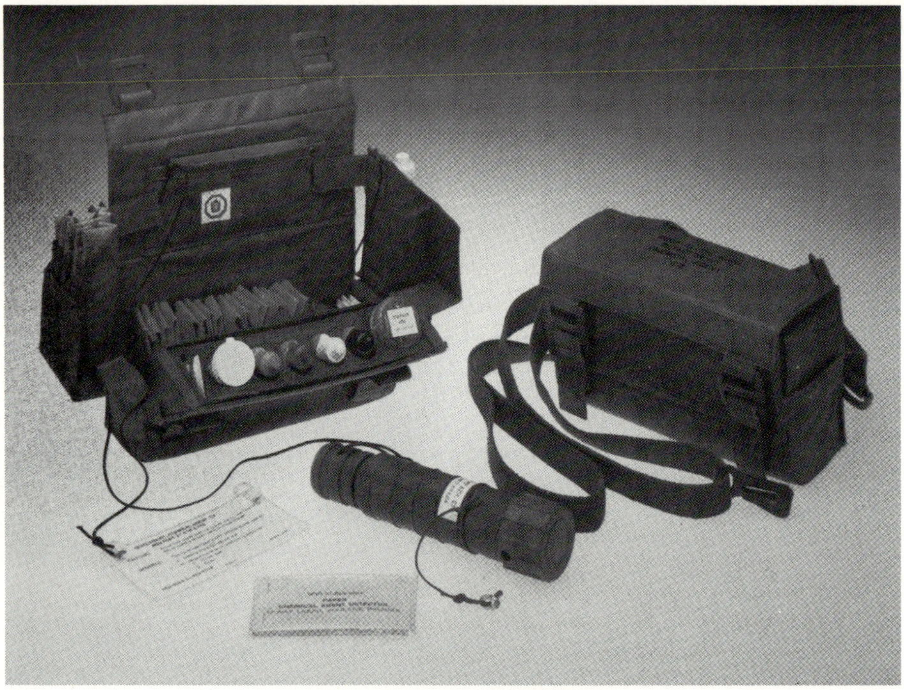

Detector Kit, Chemical Agent (C-2)

STATUS
In production. In service with the Canadian Armed Forces.

MANUFACTURER
Anachemia Canada, 500 Second Avenue, PO Box 147, Lachine (Montreal), Québec, Canada H8S 4A7.

Chemical Agent Liquid Detector Papers – 3-way, M-8 and M-9

DESCRIPTION
These chemical agent liquid detector papers were designed to meet the need for a simple and rapid method of detecting and differentiating between the three major groups of liquid chemical warfare agents and can detect G, V and H agents. The material used is basically a dye impregnated paper sensitive to liquid chemical agents. They are produced either in booklet form (3-way with backing adhesive or M-8 without) or in a dispenser roll (M-9). Provided with the booklets are colour comparison charts and instructions.

To use, a piece of the paper is detached from a booklet or dispenser roll. The paper is exposed to a surface suspected of contamination by a liquid chemical agent either by wiping the suspected surface with the paper or by attaching it to a surface or piece of equipment which may be exposed at some point. If the paper comes into contact with a liquid chemical agent it will change colour. The 3-way and M-8 papers exhibit one of three colours to signify the presence of G (shades from yellow to orange), V (very dark blue-green to light blue-green) or H agents (dark red). The M-9 paper develops coloured spots in the presence of any of the agents.

The 3-way papers are issued in booklets containing 12 sheets of paper each measuring 100 × 65 mm. M-8 papers have similar dimensions but are issued in booklets containing 25 sheets. The M-9 papers are issued in a roll 10 metres long contained in a dispenser box measuring 75 × 75 × 60 mm. The dispenser box has a serrated feed-out edge to cut lengths of paper as required and is issued in a sealed bag.

STATUS
In production. In service with the Canadian and US armed forces.

MANUFACTURER
Anachemia Canada, 500 Second Avenue, PO Box 147, Lachine (Montreal), Québec, Canada H8S 4A7.

Detector, Chemical Agent, Nerve Vapour

DESCRIPTION
Developed by the Canadian Armed Forces, this simple detector is intended for issue down to individual level and is intended to indicate if nerve agents are present or if it is possible to remove protective masks and clothing after a nerve agent attack. The detector is in two parts, both using clear plastic bases. On the main body is a disc of enzyme impregnated test paper, while a holder has a small sample of a chemically-impregnated test paper. In use the paper on the body is moistened. The paper on the holder is then pressed against it. If the test paper on the body changes colour to blue or green no nerve agent is present. If the test paper remains unchanged, nerve agent is present.

Each detector measures 55 × 25 × 2 mm and is packaged in an airtight foil wrap, together with an instruction sheet and a silica-gel air dryer pack. The individual units are packed 40 to an air tight, moisture-proof container measuring 95 × 95 × 51 mm.

STATUS
In production. In service with the Canadian Armed Forces.

MANUFACTURER
Anachemia Canada, 500 Second Avenue, PO Box 147, Lachine (Montreal), Québec, Canada H8S 4A7.

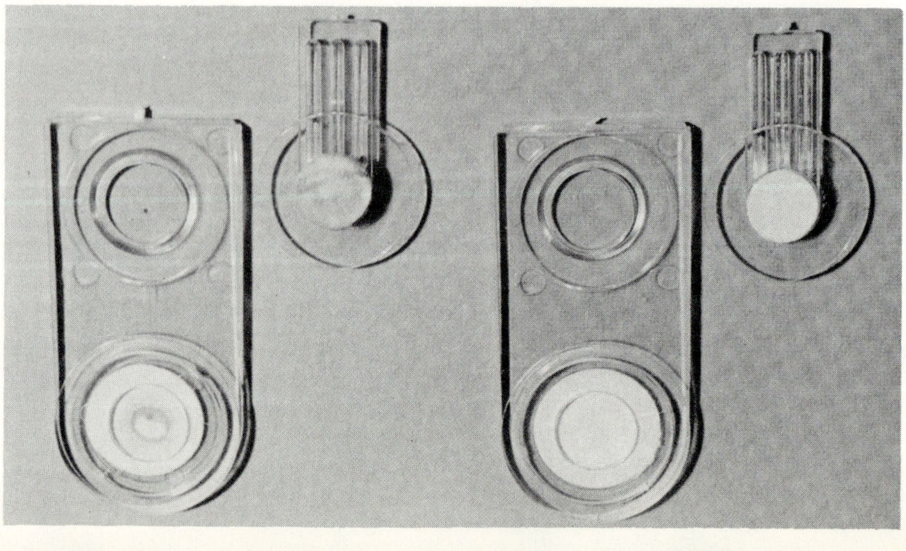

Detector, Chemical Agent, Nerve Vapour

CHINA, PEOPLE'S REPUBLIC

Chemical Warfare Agent Identification Kit, M-75

DESCRIPTION
The M-75 chemical warfare agent indentification kit follows the same general lines as kits such as the Warsaw Pact PKhR series. Contained in a carrying box equipped with a shoulder strap are a small air pump, racks of glass ampoules filled with various reagents and some other items including a set of instructions inside the lid that acts as a working tray when opened. In use, an ampoule is placed into the air pump after one end has been broken off. As air is pumped through the reagent in the ampoule, colour changes will take place if chemical warfare agents are present. The change of colour can be used to identify the agent present. The kit contains ampoules to identify nerve agents (V- and G-agents), mustard (H), lewisite (L), nitrogen mustard (HN), hydrogen cyanide (AC), cyanogen chloride (CK), phosgene (CG), diphosgene (DP), chloroacetonephenone (CN), and adamsite (DM).

The M-75 kit weighs 2.6 kg complete. When packed it is 245 mm long, 105 mm wide and 143 mm high.

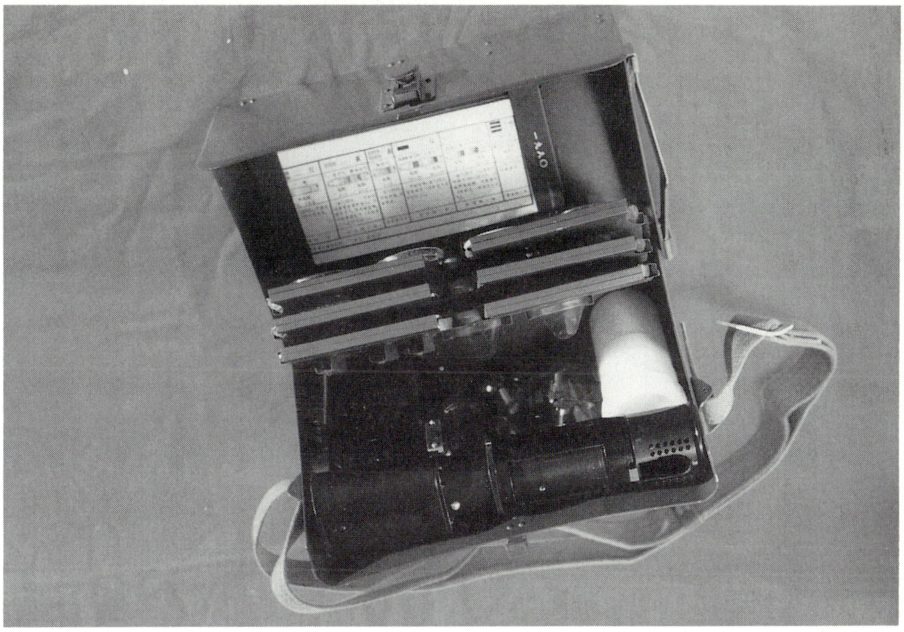
Chemical warfare agent identification kit, M-75

STATUS
In production. In service with the Chinese armed forces.

DEVELOPMENT AGENCY
Research Institute for Chemical Defence, PO Box 1044, Beijing, China.

MANUFACTURER
Great Wall Instrument Factory, PO Box 377, Beijing, China.

Chemical Warfare Agent Vapour Detector Kit, M-86

DESCRIPTION
This kit is used at sub-unit level to detect the presence of chemical warfare agent vapours and determine if it is safe to remove protective respirators and clothing after a chemical warfare attack. The M-86 kit uses the puffer principle in which a small reagent ampoule is placed in the puffer/sampler and the puffer is actuated several times. If the reagent changes to certain colours a chemical warfare agent is present. The exact colour change will provide an indication of the type of agent present.

The kit will detect 0.02 microgramme per litre of nerve agents (VX, GB and GD), 2 microgrammes per litre of mustard agents (H and HD), 40 microgrammes per litre of hydrogen cyanide (AC) and 20 microgrammes per litre of cyanogen chloride (CK).

Each kit is packed into a flat canvas wallet with velcro closures and contains one puffer/sampler, three or four reagent ampoules, chemical agent detector papers and a set of instructions.

The M-86 kit weighs 0.5 kg and measures 160 × 130 × 60 mm when packed.

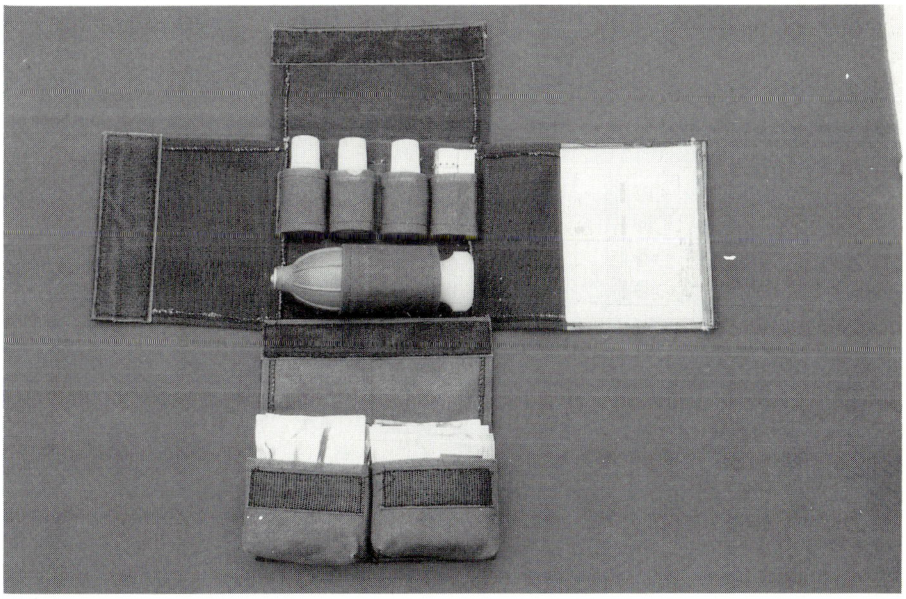
Chemical warfare agent vapour detector kit, M-86

STATUS
In production. In service with the Chinese armed forces.

DEVELOPMENT AGENCY
Research Institute for Chemical Defence, PO Box 1044, Beijing, China.

MANUFACTURER
Great Wall Instrument Factory, PO Box 377, Beijing, China.

Detector Paper Booklet, Chemical Warfare Agent, Liquid

DESCRIPTION
These chemical warfare agent detector papers are issued in booklet form packed into clear plastic envelopes. Each sheet of paper has an adhesive backing. Two types of booklet are produced. In one, the Type X-1, the paper is coloured dark blue and in the other, the Type X-3, the paper is deep red. Exposure of either type of paper to mustard or nerve agents will result in a noticeable colour change.

STATUS
In production. In service with the Chinese armed forces.

DEVELOPMENT AGENCY
Research Institute for Chemical Defence, PO Box 1044, Beijing, China.

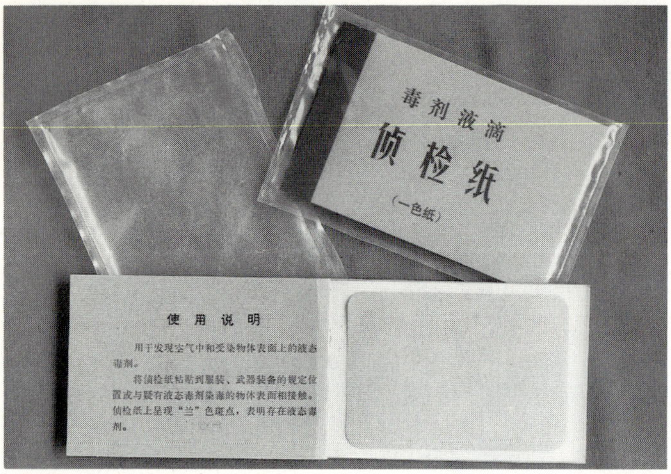

Detector Paper Booklet, Chemical Warfare Agent, Liquid, Type X-1 (dark blue)

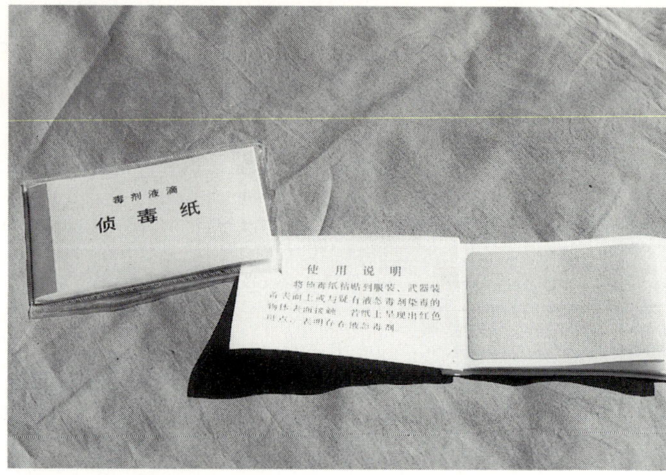

Detector Paper Booklet, Chemical Warfare Agent, Liquid, Type X-3 (deep red)

DENMARK

Brüel & Kjær Toxic-gas Monitor Type 1306

DESCRIPTION
The Toxic-gas Monitor Type 1306 was developed by Brüel & Kjær in close co-operation with the Danish Civil Defence authorities. It is a fixed location, point detector designed for continuous, long term monitoring of nerve agents or mustard agents, depending on the optical filter installed. The measurement principle is based on a physical phenomena known as photo-acoustic spectroscopy, where an acoustic wave is generated by a gas when it absorbs modulated light. The wavelength of the incident light determines which gases are detected.

On the Type 1306 air is drawn into a measuring chamber into which it is sealed by closing the input and outlet valves. On its way into the chamber the air is passed through two filters, the first for coarse particles and the second for fine. Once in the chamber the air is exposed to light from an incandescent lamp seated in a parabolic mirror. Before it reaches the air sample the light is passed through a chopper to pulsate it and an optical filter that passes narrow-band infra-red light. Any nerve or mustard agent present in the air sample will absorb the light (depending on the optical filter installed) and generate heat, thereby increasing the pressure within the chamber. Any pressure increase is periodic at the light-chopping frequency and so acts as a sound wave that can be detected by two sensitive microphones. The microphone signal, which is proportional to the agent concentration, is then processed and the result is transmitted to a central control station. The complete measurement cycle takes 45 seconds.

DENMARK

The monitor can accurately detect down to 0.1 mg per cubic metre for nerve agents and 1 mg per cubic metre for mustard agents. Each monitor is housed in a weather-proof box protected against electromagnetic pulses. The monitor can be located as required. but is usually situated on a mast. Maintenance and calibration are normally required about three times each year. Self-tests are continuously performed.

The Type 1306 monitor can be employed in a network of individual monitors under the control of a central personal computer fitted with a RS 485 interface converter. Monitors can be situated up to 1.2 km apart and connected by balanced, two-wire communication cable to the computer; one computer can control up to 254 monitors and the size of any particular network is dependent on customer requirements. Control of the monitors is automatic and operation is simple, involving user-friendly software. Application Software Type 7619 can control up to 31 monitors whereas Application Software Type 7621 is suited for a smaller system of up to three monitors. Monitoring and alarm triggering systems are fully automated so continuous manning of the system is not required. Under normal conditions each monitor takes an air sample every 10 minutes but if a high concentration of an agent is detected further sampling is automatically performed without an interval.

Each Type 1306 monitor weighs 5.5 kg and measures 400 × 200 × 102 mm. The operating temperature range is from −20 to +50°C. Each monitor is powered by a 12-volt dc supply, either from a vehicle battery or via a power supply unit using a mains voltage. Under stand-by conditions, the power consumption is 2 watts. When operating normally the power consumption rises to 18 watts, with a peak of 34 watts at start-up.

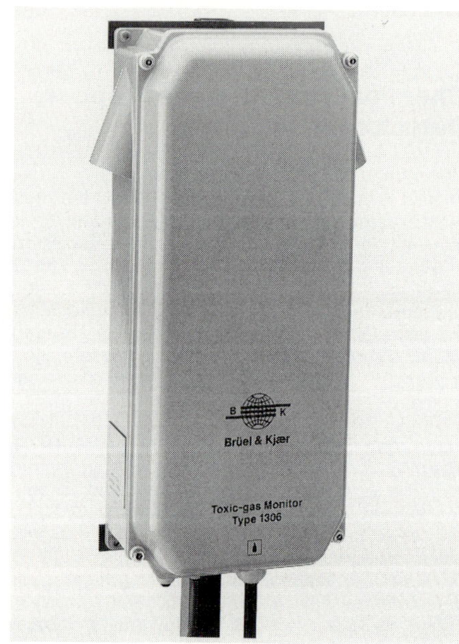

Type 1306 toxic-gas monitor; the small 'ears' either side of the case are the air intake and outlet ports

STATUS
In production.

MANUFACTURER
Brüel & Kjær, 18 Nørum Hovegarde, DK-2850 Nørum, Denmark.

Brüel & Kjær Multi-gas Monitor Type 1302

DESCRIPTION
The Multi-gas Monitor Type 1302 uses the same photo-acoustic spectography principles as the Type 1306 toxic-gas monitor (refer to previous entry for a simple outline of these principles) but by choosing and installing up to five appropriate optical filters on an internal filter carousel arrangement the Type 1302 can selectively measure up to five component gases and water vapour in any air sample. This offers the facility of being able to detect and measure Mustard (H) and nerve (G) agents simultaneously in military and civilian locations, including shelters, command posts and laboratories. The detection threshold of the Type 1302 is gas-dependent but is typically in the 10 parts per million region.

The casing used for the Type 1302 is not weatherproof but it can be mounted in a weather-proof box in external locations. For long-term monitoring air samples are drawn into the instrument through Teflon tubing from points up to 50 metres away. The Type 1302 can also be used as a portable instrument. Used with one or two Multipoint Doser and Sampler Units Type 1303 it can monitor air samples collected from 6 or 12 different locations. The Type 1302 can also be remotely controlled by computer.

The Multi-gas Monitor is controlled by push-buttons on the front panel and is very user-friendly. Each time a button is pressed a short self-explanatory test appears on a flourescent display screen to guide the user through each operating procedure. Thus no special training is required to operate the Type 1302.

Users can 'programme' the Type 1302 to perform almost any type of monitoring task. Measurement results are displayed on the 2 × 40-character fluorescent screen as soon as they are available. Results can be automatically stored in the Type 1302's memory and printed or plotted-out at a later stage.

There are four operation modes: set-up; measurement; memory; and function. In the set-up mode users are able to select items such as parameters, interface and communication codes, and the time over which readings can be averaged. Details of up to ten different monitoring tasks can be stored. In the measurement mode the user selects the monitoring task to be performed and an internal clock will start the task at any pre-allotted time. In the memory mode data is stored in the appropriate memory for later access and in the function mode the user can programme the Type 1302 to automatically perform almost any sequence of operations over any period of time.

Reliability of measurement results is ensured by regular self-tests which the Type 1302 performs to check that it is functioning correctly, as well as by the Type 1302's ability to compensate any measurement for the interference caused by the presence of water vapour and other known 'interferants'. Calibration and changing the filter papers in the air filtration units are usually required no more than twice a year.

The Type 1302 is 155 mm high, 355 mm wide and 300 mm deep. Weight is 9 kg.

STATUS
Due to enter production in June 1989.

MANUFACTURER
Brüel & Kjær, 18 Nørum Hovedgade, DK-2850 Nørum, Denmark.

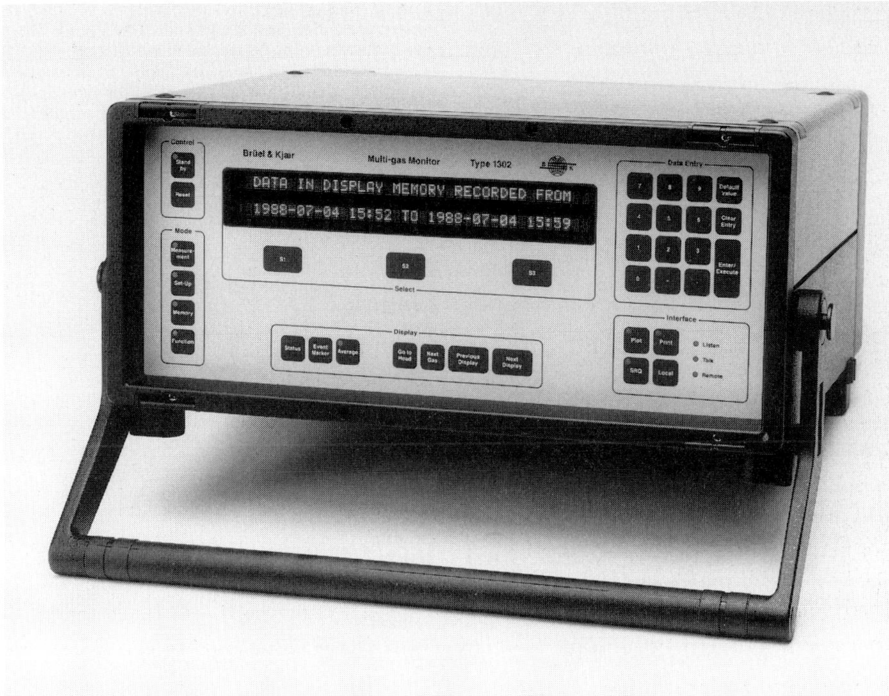

Brüel & Kjær Multi-gas Monitor Type 1302

FINLAND

Chemical Agent Detection System M86/A

DESCRIPTION
The Chemical Agent Detection System M86/A is an automatic system for detecting toxic military chemical agents in the atmosphere. The detection limit of agent concentrations can be set at a value which is not dangerous to personnel.

The systems comprises the following components:

Automatic detector unit M86/A-D
Remote alarm centre M86/A-RAC
Remote horn M86/A-RH.

The system is completed by three alternative power supplies. These are a non-rechargeable lithium battery unit (M86/A-LB), a rechargeable nickel-cadmium battery unit (M86/A-NB), and a power supply unit (M86/A-VMP). The system is completed by a number of accessories including an intake tube and an auxiliary pump.

The basic element of the system is the automatic detector unit (M86/A-D) which exists in several versions. It can be used as an independent portable unit as a personal detector, as a vehicle-carried detector and for monitoring contaminated surfaces using an intake tube. A version for use in civil defence shelters has been developed in co-operation with Temet Oy to form the Electronic Chemical Agent Alarm System Temet ECAS/1. In this application the basic M86/A is provided with a pre-heater, a microfilter, a central unit and an installation rack. The type of power supplied selected will depend on the application.

The M86/A system can be configured to use a remote alarm centre (M86/A-RAC) which can be connected as the central point for up to up to four detector units located up to 2000 metres away from the remote alarm centre. A remote horn to provide an aural alarm (the M86/A-RH) can be connected to either a detector unit or the remote alarm centre. Visual alarms are provided by red LEDs.

In operation the detector unit continuously monitors the ambient atmosphere and when a concentration of agents increases over a set level an alarm is provided. The alarm may be visual and/or aural and can be transmitted by cable to a remote alarm centre (up to 2000 metres away) or to a remote horn (up to 500 metres away. The system can be set for nerve agents or other agents. The alarm threshold for nerve agents (VX, GA, GB, GD) is 0.3 mg/m^3. The alarm threshold for other agents is 10 mg/m^3 for AC and 30 mg/m^3 for H agents. These alarm thresholds can be varied. The reaction time for the nerve agents is less than five seconds. Recovery time following an alarm is ten seconds. Nerve agents

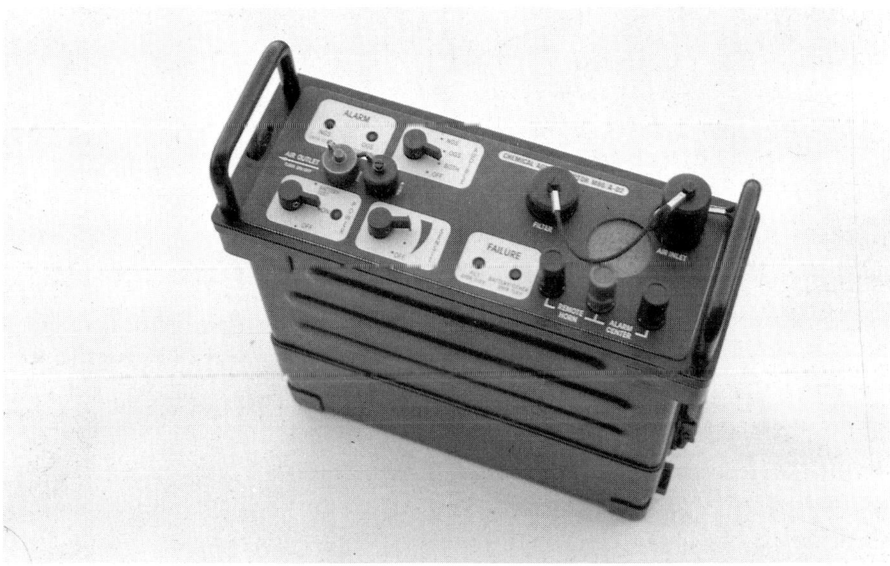

M86/A chemical agent detection system detector unit M86/A-D

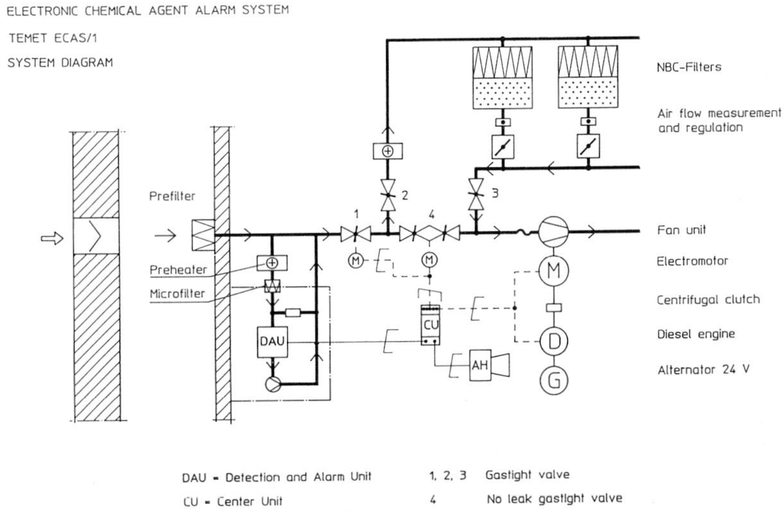

Diagram of the M86/A-D detection and alarm unit in a Temet ECAS/1 shelter installation

Setting up a M86/A-D chemical agent detection system detector unit

produce an intermittent aural alarm; other agents produce a continuous aural alarm.

The microprocessor used with the M86/A system can also be used for self-test and failure diagnostics, including low battery levels. An operational failure produces a different alarm signal from those used for agent alarms.

The M86/A system was designed to operate all year round under Finnish climatic conditions without any protection other than the normal containers. It is also proof against rapid changes in pressure and temperature and transport shocks. It will not cause interference with other electronic equipments and is proof against EMP effects. The estimated service life is from 15 to 20 years.

STATUS
In production. In service with the Finnish armed forces and Finnish civil defence organisations.

MANUFACTURER
Environics Oy, Kotkankallionkatu 11, SF-70600 Kuopio, Finland.

After 1.6.89, PO Box 349, SF-50100 Mikkeli, Finland.

SPECIFICATIONS

Unit	**Detector**	**Remote alarm centre**	**Remote horn**
Designation	M86/A-D	M86/A-RAC	M86/A-RH
Weight*	2.8 kg	2.3 kg	0.6 kg
Length	280 mm	280 mm	39 mm
Width	105 mm	105 mm	63 mm
Height	280 mm	280 mm	125 mm

*without power supply unit

POWER SUPPLIES

Unit	M86/A-LB	M86/A-NB	M86/A-VMP
Type	lithium	NiCd	power supply
Weight	0.65 kg	1.8 kg	2.2 kg
Length	90 mm	92 mm	96 mm
Width	118 mm	241 mm	245 mm
Height	53 mm	55 mm	111 mm
Capacity	12 Ah	4 Ah	–
Voltage	15 V	12 V	220 V*

*input voltage

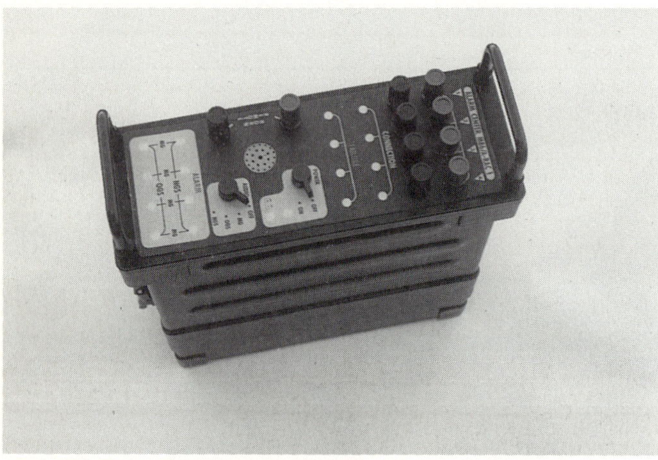

M86/A chemical agent detection system remote alarm centre M86/A-RAC

M86/A chemical agent detection system power supply unit M86/A-VMP

FRANCE

PROENGIN Portable Chemical Contamination Monitor AP2C

DESCRIPTION
The PROENGIN portable chemical contamination monitor AP2C (appareil portable pour la contrôle de la contamination chimique) is a self-contained instrument that operates on the flame photometer principle to detect a wide range of vapours produced by chemical agents, including nerve agents, with a detection response time of less than one second for phosphor-based agents. An ultra-fast return to normal sensitivity is claimed, even after the detection of a high agent concentration.

The instrument is contained within a watertight aluminium housing and can be held and operated by one hand. Power is supplied by lithium batteries or an auxiliary electrical supply. A liquid crystal display panel shows simultaneously the total phosphor- and sulphur-based concentrations of any agents present. Involved in the detection process is a hydrogen supply carried in a rechargeable metal cartridge. When using the internal batteries and hydrogen cartridge the AP2C has an operating autonomy of 12 hours. Two plug-in points at the rear of the instrument body can be used to plug in an optional auxiliary electrical supply and a set of headphones for aural warnings and indications.

SPECIFICATIONS
Weight: 2 kg
Length: 350 mm
Width: 85 mm
Height: 125 mm
Operating temperature: −32 to +50°C

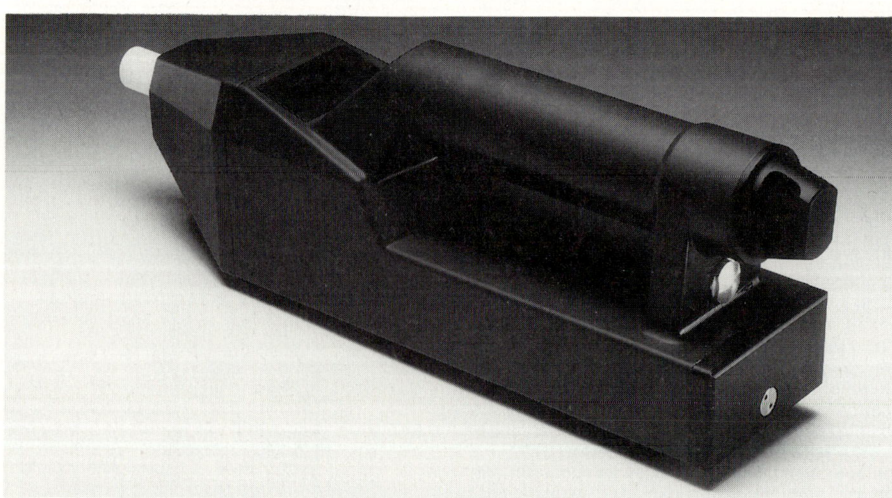

PROENGIN portable chemical contamination monitor AP2C (T J Gander)

Storage temperature: −42 to +75°C
Response time: less than 1 s for 1 microgramme phosphor
Sensitivity:
(total phosphor) 1 microgramme/m^3
(GD) 5 microgrammes/m^3
(VX) 10 microgrammes/m^3
(total sulphur) 25 microgrammes/m^3
(H) 125 microgrammes/m^3
Operating autonomy: 12 h

STATUS
Prototypes under evaluation.

MANUFACTURER
PROENGIN SA, 3 rue de l'Industrie, 78210 Saint Cyr l'Ecole, France. (Under contract DRET-CEB.)

Chemical Detection Kit Model 1 bis

DESCRIPTION
The Model 1 bis kit (Trousse de Detection Chimique de Controle Mle 1 bis – TDCC) is issued in two parts: the kit proper and a kit of spares. The fully equipped kit weighs 1.2 kg and is 270 mm long, 150 mm wide and 65 mm high. It consists mainly of a box containing a hand-operated air pump and various tables and accessory holders. In use air is pumped through a transparent cylinder in which various sensitised papers are placed. As various agents act on the papers the colour changes can be used to identify the agents. The identification range of the kit depends on the sensitised papers used but a normal kit can detect most nerve, blood and choking agents likely to be encountered.

Typical detection capabilities of the kit are as follows:

GA and GB at a concentration of 1 microgramme/m^3

AC at a concentration of 350 microgrammes/m^3

CK at a concentration of 2000 microgrammes/m^3.

To keep the kit fully ready for use a spares kit is issued in a metal case which weighs 7 kg complete and is 330 mm long, 200 mm wide and 205 mm high. The contents can keep the kit operational for a period of ten days intensive use.

A training version of this kit is produced which uses imitation substances in place of the operational chemical reagent papers.

STATUS
In production. In service with the French Army.

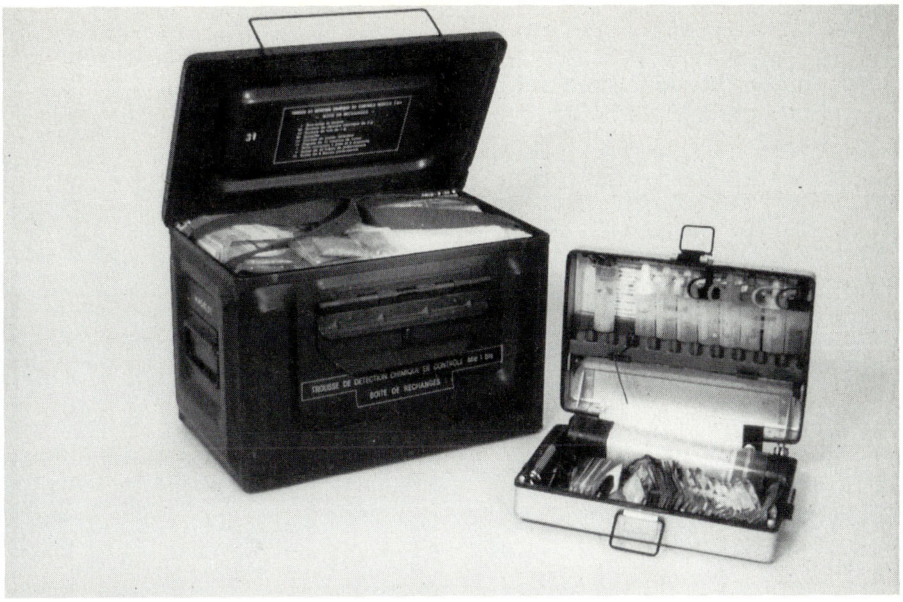

Chemical detection kit Model 1 bis

MANUFACTURER
Groupement Industriel des Armements Terrestres (GIAT), 10 place Georges Clémenceau, 92211 Saint-Cloud, France.

Detalac

DESCRIPTION
Detalac is used for the automatic local detection of chemical agents and operates once the surrounding air contains a concentration of 0.1 mg/m^3 of organophosphorated toxics. It can operate a local or a remote alarm unit which may be either aural or connected to a lamp system. Detalac uses a hydrogen supply for the detection system and a throw-away chemical generator for the hydrogen supply. This generator has to be renewed every 24 hours. Electrical power can be supplied either by a 24 volt external battery or by a cluster of six internal cells (four 3.75-volt and two 1.5-volt). The detector unit is water-proof and when operating weighs approximately 20 kg. Reaction time against toxic agents is less than 2 seconds.

Detalac is produced in two versions. The mle F1 is intended for use at static locations and is supplied with hydrogen by a disposable generator. The mle F2, the portable version, is supplied with hydrogen from a refillable cartridge. Both are supplied with 400 metres of cable on a spool for connection to a remote warning indicator device which has both lamp and aural signal warning devices. Also supplied with each equipment are an external power supply cable, a tool kit and accessories.

A laboratory variant without an alarm is known as Adepte. It can produce a taped record of any agents detected.

STATUS
In production. In service with French Army and Air Force.

MANUFACTURER
Groupement Industriel des Armements Terrestres (GIAT), 10 place Georges Clémencau, 92211 Saint-Cloud, France.

Detalac mle F1 (T J Gander)

Detalac mle F2 (T J Gander)

GERMANY, EAST

Detection and Identification Set, Chemical Agent, Models CHNS and CNS-62

DESCRIPTION
This kit is intended for use by the individual soldier as a 'first-aid' detection kit, but is really meant to be used in conjunction with the PChR-54 equipment for accurate identification and detection of chemical agents. Contained in a plastic wallet measuring 120 × 85 × 20 mm, the kit is carried by every field soldier in a uniform pocket and it contains indicator papers and several reagent ampoules, along with a colour comparison chart. The Model CHNS can detect phosgene, mustard, lewisite and hydrogen cyanide. The Model CNS-62 can detect all the above with the addition of G-type and V-type nerve agents.

STATUS
Probably no longer in production. Model CHNS now held in reserve and for civilian use, and Model CNS-62 in service with the East German forces.

MANUFACTURER
State factories.

Automatic Chemical Agent Indicator GSA-12

DESCRIPTION
The GSA-12 automatic chemical agent indicator determines whether organo-phosphorated vapours (ie nerve agents) are present in the atmosphere. It is an optical-mechanical photo-calirometer which contains the instrument case, an alarm device, two droppers for processing agents, a connecting cable and a power supply pack.

After switching on the instrument a tape transport starts to move and runs through the apparatus for the period of time determined by the working cycle. After a 25 mm section of tape impregnated with reagent has entered the feed system, the air to be analysed is drawn through the tape. If organo-phosphorated agents are present they are absorbed within 120 seconds. A No 1 reagent is then administered to the same tape section and after a certain time a No 2 reagent follows. After a further 30 seconds have elapsed the photometry commences using the comparison between the optical thickness of the tape being analysed and the pre-set expansion/swell value. Should organo-phosphorated agents be present the optical thickness of the tape cannot restrict itself to its expansion/swell quotient and visual and acoustic alarms are initiated. If the air is clean the optical thickness of the tape section under analysis falls rapidly and after 30 seconds goes below the value of the expansion/swell quotient and no alarm is initiated.

The GSA-12 can be installed in UAZ-469-rkh and BRDM-rkh reconnaissance vehicles and may be used for static installations.

STATUS
In service with the East German forces.

MANUFACTURER
State factories.

Mobile NBC Laboratory RChLab-11

DESCRIPTION
Designed for use in rear areas, the RChLab-11 is contained within a special box body carried on a W 50 LA/A (4 × 4) 3000 kg truck. The laboratory is intended to be used for the analysis and identification of NBC agents and is thus equipped with benches, cupboards and drawers for the storage and use of the various types of chemical equipment and instruments necessary for the task. The body roof has partially sloping sides with inset windows to provide good lighting and the door is entered by a ladder at the right rear. A further ladder leads to the roof and a spare wheel is carried on the rear wall. The vehicle carries its own electrical power generator and an air filtration system. It also carries a RWA-72K radiation detection and warning device with an externally-mounted sensor and an internal control unit.

STATUS
In service with the East German Army.

MANUFACTURER
State factories and laboratories.

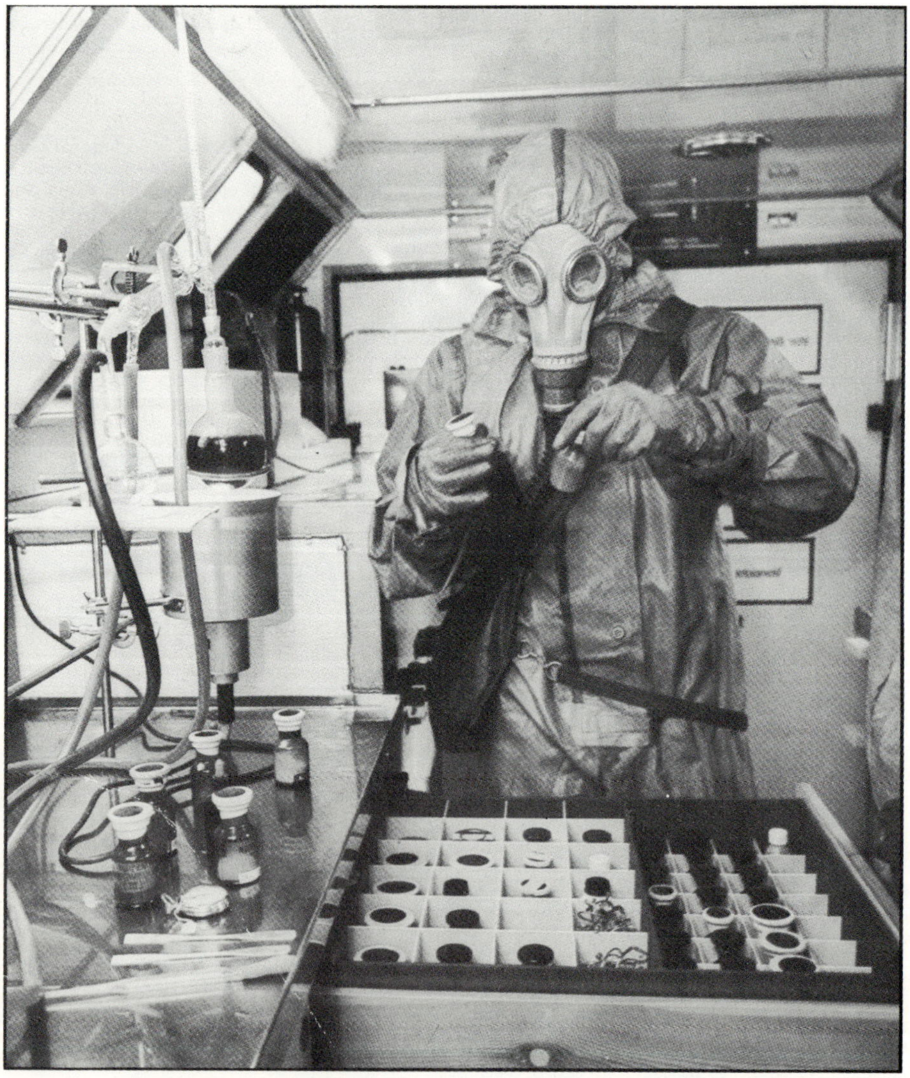

Technician working in interior of RChLab-11

GERMANY, WEST

MM-1 Mobile Mass Spectrometer

DESCRIPTION
The MM-1 mobile mass spectrometer is designed to be capable of detecting chemical agent vapours and surface residual liquids. It is a hardened detector intended for mounting on a vehicle and used for chemical agent reconnaissance. As such it was part of the US Army's now-cancelled XM87 NBC reconnaissance system and is part of the equipment carried by the West German Army's Transportpanzer 1 'Fuchs' NBC reconnaissance vehicle (see entry in *NBC reconnaissance vehicles*) ordered for the US Army. The MM-1 is also part of the equipment carried by the TM 170 NBC reconnaissance vehicle (see entry in *NBC reconnaisance vehicles*). It can be configured with a variety of sampling probes that permit continuous vapour analysis even when the carrier reconnaissance vehicle is in motion.

The MM-1 consists of a sampler, a quadrupole mass spectrometer with an electron impact source and associated power and control electronics. Detection of chemical agents is based on the relative abundance of four unique ions associated with a chemical specimen of interest such as a chemical agent. The MM-1 can be programmed for 22 preselected, different sets of ions, which is equivalent to 22 seperate compounds. The active set of detection criteria can be updated at any time by providing appropriate ion data for a new specimen of interest, such as an entirely new agent.

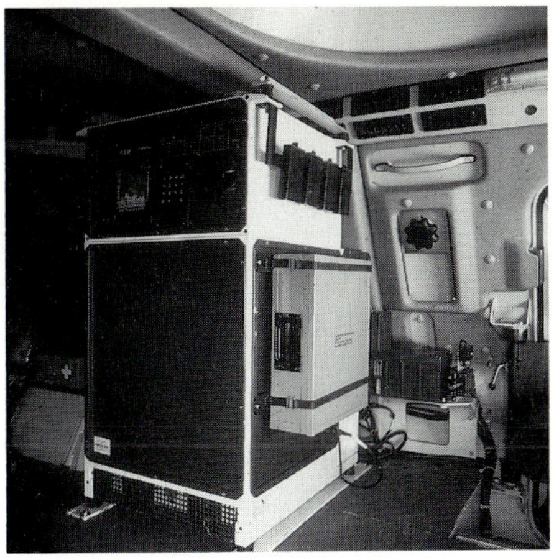

MM-1 mass spectrometer fitted inside a TM 170 NBC reconnaissance vehicle

Sensitivities for surface monitoring are reported to be better than 10 mg/m^2. Figures for vapour monitoring sensitivities have not been released.

The sampler has a sampling head that can be heated to a temperature of 200°C and is fitted with a nickel fabric coated with silicone. An inlet line running from the sampling head to the mass spectrometer is an SE-54 bonded phase fused silica GC capillary column with associated heating elements. This line can be heated by programming the temperature to a range of between 70°C and 190°C, with an optional upper temperature limit of 280°C. Ambient air is used as the inlet gas which is sucked through the sampler and past a silicone membrane by an 80 litres per second ion-getter pump. The membrane provides the pressure drop needed for the device to function and serves to remove certain background constituents by preferentially allowing organic compounds to pass through. This process may improve the overall sensitivity by reducing the background signal.

The mass spectral detection system is a 17-dynode copper beryllium electron multiplier with a dynamic range of 10 million.

STATUS
In production. In service with West Germany and Thailand. Ordered by the US Army.

MANUFACTURER
Bruker-Franzen Analytik GmbH, Kattenturmer, Heerstrasse 12, D-2800 Bremen 61, Federal Republic of Germany.

Marking Set, Contamination: Nuclear, Biological, Chemical (NBC)

DESCRIPTION
Most indication methods for NBC use are cumbersome, heavy and awkward to use. They usually consist of stakes with indicator flags that are often emplaced mechanically. For many units these mechanical methods are not possible, so some form of portable marking equipment is necessary. One form of portable equipment is the Rapp contamination marking set.

The Rapp contamination marking set is contained within a portable slung container that can be easily carried and used by one man. The container may be slung on a canvas strap over one shoulder or slung across the chest. It is a light metal frame holding three marking flag dispensers, 13 rolls of yellow marker tape, and 48 light mounting stakes and crayons.

In the simplest case, the set may be used to dispense the marker flags for attachment to existing objects such as tree branches or fence posts. The flags are simply pulled from one of the three dispensers, a white flag for nuclear contamination, a blue one for biological contamination and a yellow flag for chemical contamination. Using a crayon taken from a clip on the side of the kit container, extra information such as date and extent of contamination can be written onto the flag. The flag can then be fixed to the object by folding the head through a slot in the tail.

When stakes are required, they are removed from the container by opening latches in its side. For most purposes three stakes from the kit are required to make one pole and each stake is provided with a point at one end and a locating collar at the other. Once the first stake has been driven into the ground the others can be fitted into the sockets. The necessary flag can then be fitted onto the top of the pole by impaling it over the top. As before, extra information can be written onto the flag using the crayon from the kit.

At times it will be necessary to isolate an area by fencing it off by poles and tape. The poles are erected in the manner already described and the yellow tape can be taken from the 13 20-metre rolls carried on the kit container. The tape can also be used to form pole stabilising guys with extra stakes as the holding peg anchors. The marker flags can then be hung from the tape lengths connecting the poles.

The kit container is disposed of once used and replaced by a fresh kit. No training in using the kit is required other than a simple handbook providing general guidelines and in use the container is easy to carry and handle. It can be used even when the operators are fully covered by NBC protection clothing, and the only safety point to watch is that the sharp ends of the stakes might cut through protective clothing. The kits are issued ready for use and no form of preparation is required.

SPECIFICATIONS
Container with components
Weight: (approx) 4.5 kg
Length: 345 mm
Width: 235 mm
Height: 90 mm
Mounting stakes
Length: (each) 290 mm
Quantity: 48
Marking ribbon
Length of each roll: 20 m
Quantity: 13
Marking flags
Quantity:
(nuclear) 20 white (marked ATOM)
(biological) 20 blue (marked BIO)
(chemical) 20 yellow (marked GAS)

STATUS
In production. In service with the US Army.

MANUFACTURER
Theodor Rapp KG, Alte Hausacher Strasse 5, Postfach 1320, D-7612 Haslach iK, Federal Republic of Germany.

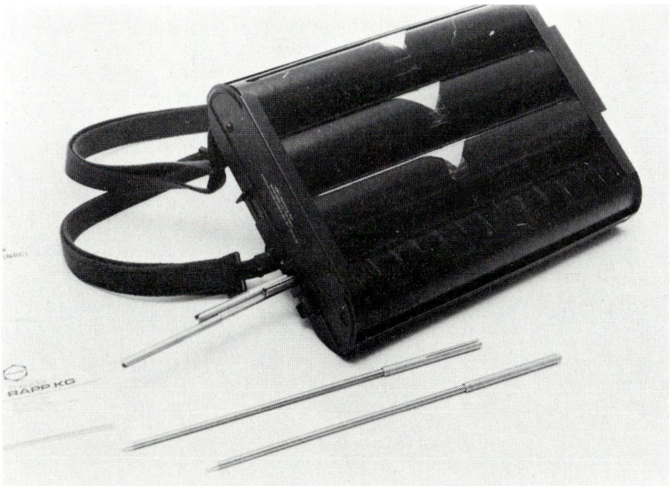

Rapp contamination marking set complete showing stake location

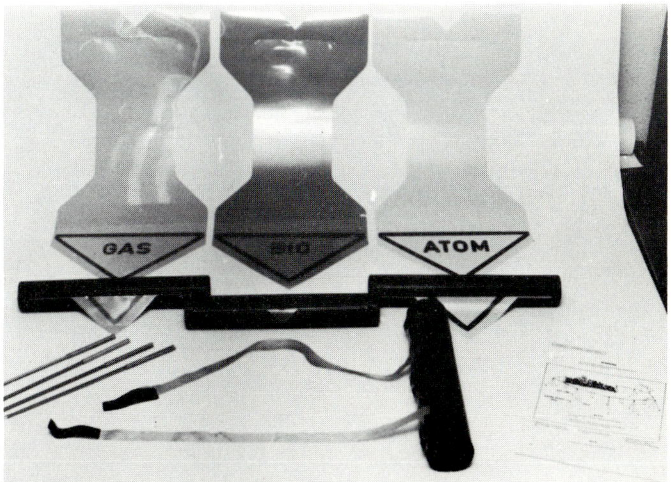

Marker flags and tapes from NBC marking set

HUNGARY

Type 66-M Chemical Reconnaissance Set

DESCRIPTION
The Type 66-M chemical reconnaissance set is the Hungarian version of the Soviet and Warsaw Pact PKhR series of chemical agent and identification kits (see entry under *Warsaw Pact* in this section) and is carried and used in much the same way. While the Type 66-M is similar to the PKhR series in many ways and uses an almost identical hand pump to that used with the PKhR-54, the internal layout and contents of the set differ.

The Type 66-M is used by one man and can be used to detect up to five different chemical agents simultaneously. Detection time for each agent is 1 to 2 minutes. A heater is provided for the detector tubes when the temperature falls below 0°C and a filter disc is used when detecting chemical agents in smoke samples.

The Type 66-M carrying case contains a torch and a silica gel container to protect the contents. Also provided is a carrying strap. A supplementary kit is available for the replenishment of consumable items.

The detector tubes used with the Type 66-M can detect VX, CN, DM, L and BZ. Detector tubes containing simulants are produced for training purposes.

STATUS
In production. In service with the Hungarian armed forces.

Type 66-M chemical reconnaissance set

MANUFACTURER
TECHNIKA Foreign Trade Company, Budapest-Hungary, X., Salgótarjáni.20, 1475 Budapest, POB 125, Hungary.

Type FVJ Chemical Agent Detector

DESCRIPTION
The Type FVJ chemical agent detector is intended for use on vehicles and uses an internal electrically-driven air pump to continually draw air through up to four chemical agent detector tubes. The presence of a chemical agent will be indicated by a change in colour of one or more of the detector tubes. The colour involved will indicate the type of agent involved. At temperatures below +10°C the detector tubes may be heated. A drawer at the bottom of the detector unit can hold spare detector tubes. The detector tubes available can detect HD, CG, AC, CS, VX, CN, DM, L and BZ. Detector tubes containing simulants are available for training purposes.

The Type FVJ requires 10 minutes of preparation before use and thereafter can be left unattended other than viewing the detector tubes through a transparent panel on the unit access door. It is powered from a 24-volt dc battery or a vehicle 24-volt power supply. Maximum power consumption is 2 amps. The detector weighs 6.5 kg and measures 346 × 240 × 135 mm. When not in use the detector is stored in a wooden carrying case.

STATUS
In production. In service with the Hungarian armed forces.

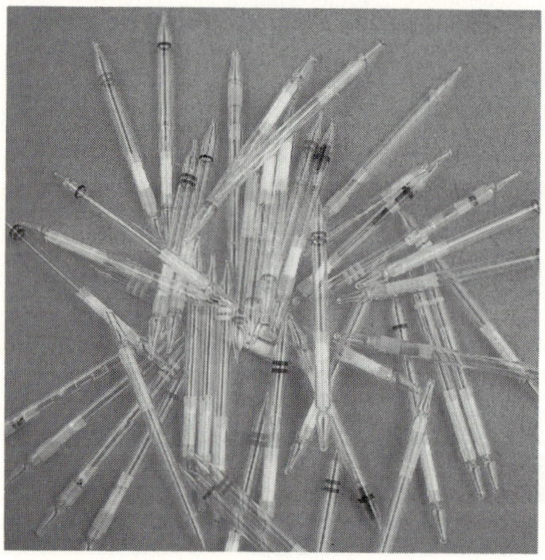

A selection of detector tubes used with the Type FVJ chemical agent detector

Type FVJ chemical agent detector with carrying case

MANUFACTURER
TECHNIKA Foreign Trade Company, Budapest-Hungary, X., Salgótarjáni.20, 1475 Budapest, POB 125, Hungary.

Double Way Chemical Agent Indicator Patch

DESCRIPTION
The Double Way chemical agent indicator patch is intended for use in a number of ways to act as a monitoring device to indicate the presence of chemical agents. Described as a 'stripe', the patch has a self-adhesive backing that allows it to be secured to uniforms, protective clothing, vehicle windscreens or metal and other surfaces. The patch has two separately-treated indicator fields that can indicate the presence of V, G, H and L agents by a colour change. The exact colour change will indicate the nature of the agent which can be indicated by reference to a colour chart provided in the centre of the patch. By using two separately-treated indicator areas spurious indications such as those caused by fuels or explosives can be largely avoided.

For issue each patch is packed into an individual aluminium foil bag.

STATUS
In production.

MANUFACTURER
TECHNIKA Foreign Trade Company, Budapest-Hungary X, Salgótarjáni.20, 1475 Budapest, POB 125, Hungary.

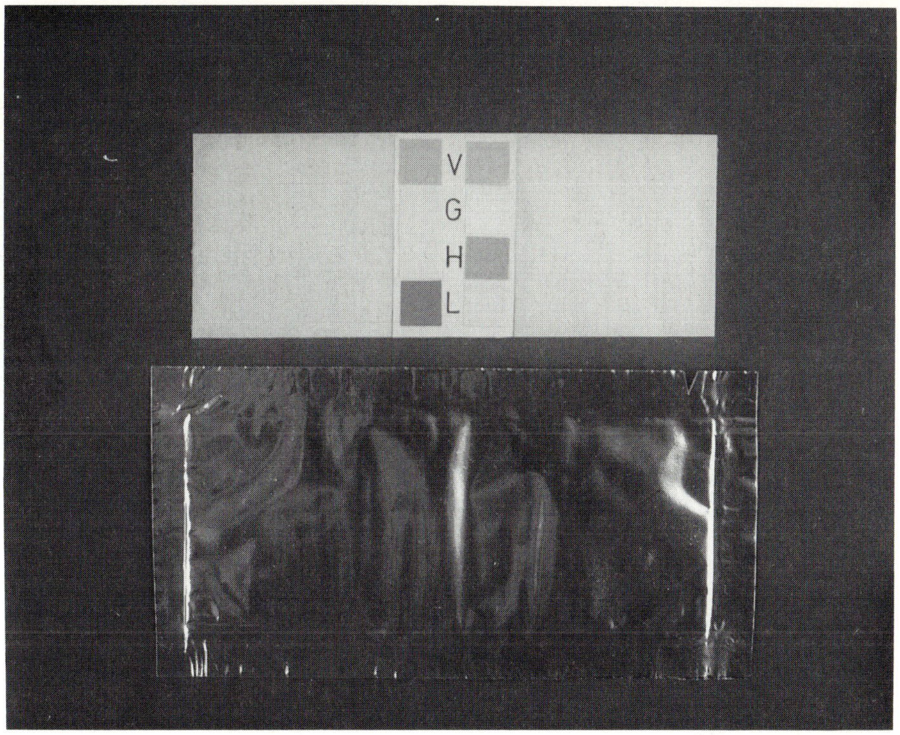

A Double Way chemical agent detector patch with a packed item below

ISRAEL

CHASE Chemical Agents Sensor

DESCRIPTION
CHASE is a continuous chemical agents sensor developed by Elbit Computers Limited for the Israeli Defence Forces. The sensor can be used to provide an effective early warning against a chemical warfare attack, for the identification of a chemical agent type, for reconnaissance and surface monitoring and for decontamination control and protection.

CHASE is a highly sensitive, fast responding and rapidly cleansed differential flame photometer designed for the detection and alarm of the presence of nerve (G and V) and blister (H) agents. Compounds are pumped into the flame photometric detector and are burned to emit light which is then measured. It is possible to measure very low concentration levels (the response threshold is in the range of 10 mg/m^3) with a response of a few seconds.

The equipment is portable and may be carried either on a backpack or on a light vehicle. In both cases it is possible to use a long sniffer probe on an extension hose connected to the main unit. This may be carried in a special mounting rack on a vehicle or in a backpack. A visual concentration gauge indicates the agent level, from safe to high levels of contamination. The device may be controlled remotely by the addition of a repeater unit and may also be used in alarm and monitoring modes to provide remote alarms at ranges of up to 500 metres. Both vapour and aerosol concentrations can be detected and the equipment will respond within seconds to provide various forms of alarm. The equipment is fast-cleansing and will rapidly be ready to detect another concentration and provide another alarm. CHASE is stated to be simple to operate and it has a low false alarm rate.

A field simulator is available. It is claimed that CHASE requires very little maintenance and the instrument has a built-in self-test programme.

CHASE chemical agents sensor

CHASE in use carried on a backpack

STATUS
In production for the Israeli Defence Forces.

MANUFACTURER
Elbit Computers Limited, Advanced Technology Center, POB 5390, Haifa 31053, Israel.

Control panel of CHASE chemical agent sensor

CHASE chemical agent sensor in use with instrument mounted on back of light vehicle and with operator using the sniffer probe

RAFAEL IDENT Semi-automatic Chemical Agent Detector

DESCRIPTION
The IDENT is a semi-automatic chemical agent detector for identifying and detecting a wide range of chemical warfare agents and provides a quick and reliable identification system. The device can be carried and operated by one man.

The IDENT comprises a multi-compartmented carrying side pack holding a set of ten reagent ampoules, 50 testing cartridges, an electronic control unit and a battery unit. The electronic unit is connected to a hand-held sampling and detection unit by a coiled cable. An ampoule crushing device and chemical agent detection paper complete the equipment.

To use the IDENT a reagent ampoule is crushed using the crushing device supplied and some of the ampoule's contents are allowed to drip onto a testing cartridge. The cartridge is inserted into the hand-held unit and pressing a trigger on the unit starts an automatically-controlled air flow and heating sequence for the cartridge. The cartridge will change colour in the presence or absence of chemical agents and any agent detected can be identified by colour reference. The maximum detection and identification sequence takes two minutes.

The sensitivity of the IDENT (in mg/m^3) is as follows:

GB and GD: 0.005
VX: 0.015
HD: 0.75
AC and CK: 1 to 3

Power for the system is provided by a 12-volt lithium battery with enough power for a minimum of 150 tests. Indicators on the electronic unit show a self test/failure state, battery low condition and test in progress.

RAFAEL IDENT semi-automatic chemical agent detector with hand-held sampling and detection unit on left

STATUS
In production.

MANUFACTURER
RAFAEL, PO Box 2082, Haifa 31021, Israel.

NETHERLANDS

Odelft ACAL Chemical Agent Detection and Alarm System

DESCRIPTION
The ACAL chemical agent detection and alarm system may be used as a stand alone unit or as part of an alarm network. It is intended for field use but may also be used in static locations or carried mounted on brackets by a vehicle.

ACAL employs the enzyme principle to detect chemical warfare agents and has a sensitivity of around 0.02 milligrammes per cubic metre with nerve agent gases. The response time is 18 seconds. The detector unit is powered by rechargeable or conventional batteries contained in a battery unit, but may also be powered from a 12 or 24 volt vehicle battery.

ACAL uses a microprocessor-based internal status monitoring system at all times with the exact status shown on a digital display. In the absence of an alarm the proper functioning is indicated as a 'go' signal, audible and/or visible on either the detector or alarm unit, or on both. If applicable the relevant status item is displayed as a warning. An 'end of alarm' situation is displayed automatically.

For maintenance, built-in test equipment (BITE) identifies suspect or defective modules. All modules are easily exchangeable at third echelon maintenance level.

A training mode is a standard feature.

The blast-proofed detector unit measures 200 × 271 × 339 mm, excluding the handgrip and the battery unit. Weight is 10 kg. The operating temperature range is from 25 to +50°C.

STATUS
Prototypes field tested and approved.

MANUFACTURER
Odelft, bv Optische Industrie De Oude Delft, PO Box 72, 2600 MD Delft, Netherlands.

ACAL detector unit (left) with remote alarm unit on right

ACAL remote alarm unit

Duphar 'The Button' Individual Chemical Detector

DESCRIPTION
By using 'The Button' an individual soldier can determine if there is a toxic or non-toxic level of nerve agent in the surrounding atmosphere, thereby permitting him to remove the burden of his respirator if no nerve agent is present. The detector consists of a small, lightweight plastic holder and contains two separated air-permeable filter papers impregnated with reagents, and an air- and liquid-tight container with a wall which can be punctured. The container is filled with a non-contaminated liquid for moistening the reagents to bring about a colour reaction.

After opening the protective gas-tight envelope the detector is placed in the inlet opening of the worn respirator. Air is then sucked through the detector by inhaling 15 times. The detector is then removed and the centre part is pressed inwards causing the filter papers to be pressed together. At the same time this causes the container to be pierced so moistening the reagents after which the colour reaction follows. If no nerve agent is present a blue colour appears. The transparent lid prevents unintentional activation and the whole process can be carried out with gloved hands. A drying agent is contained in the holder to prevent reagent deterioration.

A training version is available.

STATUS
In production. In service with some NATO countries.

MANUFACTURER
Duphar BV, Medical Devices Division, PO Box 7133, 1007 JC, Amsterdam, Netherlands.

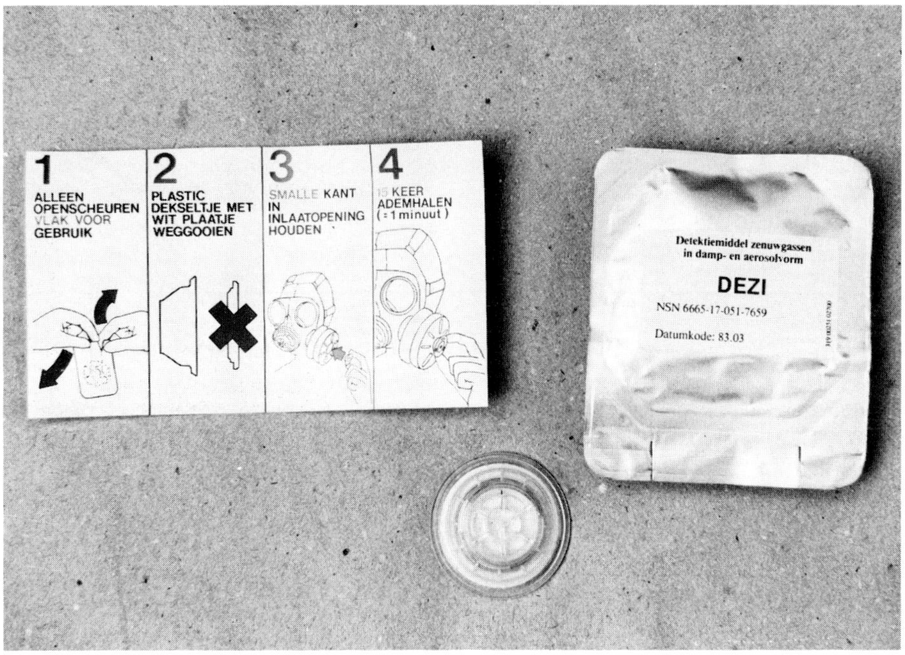

From left; 'The Button' instruction sheet, 'The Button', and a foil-wrapped unit

Duphar Water Testing Kit, Chemical Agents

DESCRIPTION
This kit is designed to determine whether water is contaminated with chemical warfare agents or other toxic substances. The equipment provides a 'go/no-go' indication showing the concentration of the toxic substance above or below a certain detection limit. These requirements are based upon a reduced daily water consumption.

The kit is carried in a hinged metal box that contains enough material for 40 nerve agent tests and 20 other agent tests, along with instructions for the user. The operating time for a trained user is 20 minutes. Some tests are combined.

The equipment will detect the following concentrations in mg per litre:

Sarin	0.02	Arsenics	1–2
Soman	0.02	CN	8
Tabun	0.04	CICN	5
VX	0.02	H	6.5–9
Mustards	2	Cl	5

The above kit is known as the Water Testing Kit I. A more refined Water Testing Kit II is available for use with water purification plants to detect concentrations of contaminating agents under 10 mg per litre thus making it possible to purify water from an available source into potable water.

STATUS
In production. In service with some NATO countries.

MANUFACTURER
Duphar BV, Medical Devices Division, PO Box 7113, 1007 JC, Amsterdam, Netherlands.

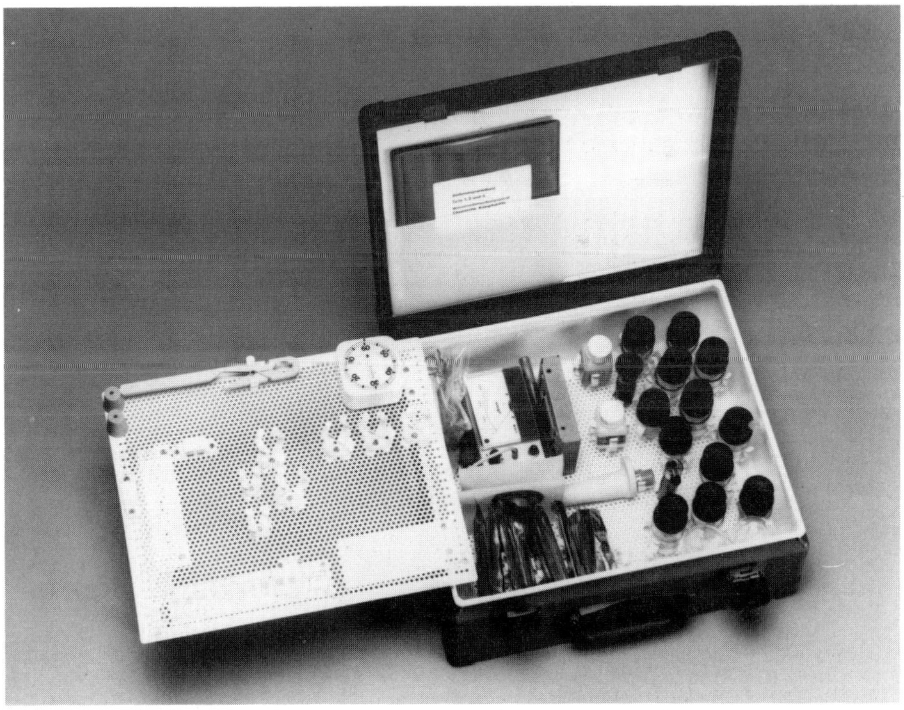

Duphar water testing kit II, chemical agents

SWITZERLAND

Chemical Detection Kit CAD

DESCRIPTION
The CAD chemical detection kit is in service with the Swiss Army and civil defence organisation. The kit is contained in a metal shoulder-slung box with dimensions of 210 × 100 × 150 mm and weighing 1.5 kg. Inside the box are a hand pump, six bottles of reagent and nerve and vesicant gas indicators; the lid interior contains instructions and colour comparison charts. The kit can detect nerve agents, including Sarin, in concentrations of 20 to 100 mg/m^3 of air, and vesicants in concentrations of 2 mg/m^3 of air. An operator can be trained to use the kit in about half an hour.

A special training kit using simulated agents, the CAD T, is available.

STATUS
In production. In service with the Swiss Army and civil defence units.

MANUFACTURER
Louis Schleiffer AG, CH-8714 Feldbach/Zürich, Switzerland.

UNION OF SOVIET SOCIALIST REPUBLICS

Medical-veterinary Chemical Agent Detection and Identification Kit, Model PKhR-MV

DESCRIPTION
This kit is normally used by medical units and is an enlarged version of the standard Model PKhR series kit (see entry under *Warsaw Pact* in this section) to enable it to be used to take samples of water, food and animal fodder. The kit contains all the usual Model PKhR items with the addition of extra test-tubes, tongs, sample jars and bags, and phials. The kit can also be used to analyse water samples for chemical contamination.

STATUS
Production probably complete. In service with Soviet Army medical units.

MANUFACTURER
State factories.

Chemical Agent Detection and Identification Kit, Model UPI

DESCRIPTION
This kit is a simplified version of the Model PKhR series (see entry under *Warsaw Pact* in this section) intended for use by civilian defence agencies, and differs mainly in having no provision for taking or storing samples. It contains the usual cylindrical air pump and attachment, indicator tubes, smoke filters, rubber gloves and instruction sheet. It can detect and identify all the agents that the Model PKhR series can, although early versions had no provision for detecting V-type agents.

STATUS
Production probably complete. In service with Soviet civil defence agencies.

MANUFACTURER
State factories.

Automatic Chemical Agent and Radiation Detector Alarm, Model GSP-1 and GSP-1M

DESCRIPTION
These detectors can be used to detect G-type nerve agents and nuclear radiation, and to provide visible and aural alarms in the presence of both. The GSP-1 and GSP-1M, which are basically similar, provide detection and alarms only and do not determine the nature of the chemical agent or the intensity of the radiation.

The nerve agent detection system consists of a tape transport system, a reagent dropper system, an air system and a photodetector system. The tape transport system and the reagent dropper system are inter-connected, and operate in 5-minute cycles separated by a 5- to 8-second cycle change, both controlled by a clockwork mechanism. At the start of each cycle, reagent is dropped onto the tape which is then moved on so that the fresh drop lies under the photodetector system. An electric pump draws air into the detector system via a screening filter and over the drop on the tape. The reagent changes colour if there is any G-type nerve agent in the air and the photodetector system compares the drop colour with a reference formed from a reference beam. The reference beam is formed in different ways on the two models: on the Model GSP-1 it is derived from splitting the source beam and shining part of it onto an un-moistened part of the tape and on the Model GSP-1M the split light source is passed through a system of filters. In both cases the intensity of the two beams is compared and any change from the norm initiates the alarm systems. Air is drawn into the detector at 1.5 litres a minute and on the Model GSP-1 is then passed directly out of the system whereas on the GSP-1M it is used for interior cooling before being passed out through a grille. The detector-alarm contains enough tape and reagent to run for up to eight hours. Electrical power is provided by two rechargeable batteries, and the system is designed to operate at temperatures from −30 to +40°C.

The nuclear radiation detector system uses a halide tube detection unit. It initiates the alarm system as soon as the radiation level reaches in excess of 0.1 röntgen per hour. The alarm horn and a flashing lamp are both contained inside the case and operate both from nerve agent and radiation detection.

Without the batteries both models weigh 10 kg and with the weight is 18 kg. The dimensions are approximately 450 × 300 × 150 mm.

STATUS
Probably no longer in production. In service with the Soviet and Warsaw Pact armed forces and some Middle Eastern armed forces.

MANUFACTURER
State factories.

Automatic Nerve Agent Detector-alarm, Model GSP-11

DESCRIPTION
The Model GSP-11 differs mainly from the Model GSP-1 in that it can detect G-type and V-type nerve agents; the radiation detection element has been removed. The same basic tape and reagent system as that used in the Model GSP-1 is retained but the reagent uses an enzyme-inhibition reaction using two reagent solutions. This enzyme-inhibition reaction is temperature-sensitive so the interior of the Model GSP-11 is maintained at between +28 and +38°C by a thermostatically-controlled heater operated by external batteries in a battery case.

The GSP-11 has two sensitivity ranges, E1 and E2. The E1 is the most sensitive and operates at intervals of 60 to 80 seconds. This range allows the GSP-11 to remain in operation for two hours on one reagent filling and each operating cycle lasts between 22 to 26 seconds. The airflow through the system using the E1 range is between 0.7 and 1 litre per minute. On the E2 sensitivity range the display interval lasts between 5 to 8 minutes. This allows the unit to remain operational on one reagent filling for 10 to 12 hours. Each operating cycle lasts between 1.5 to 2.5 minutes and the airflow is reduced to 0.5 to 0.7 litre per minute.

Model GSP-11 automatic nerve agent detector-alarm

SPECIFICATIONS
Weights:
(control unit) 12 kg
(alarm unit) 0.5 kg
(battery case) 15 kg
(indicator set) 2 kg
Total weight: 29.5 kg
Warm-up times:
(0°C) 1 h
(–40°C) 3 h
Time taken to refill reagent: 10 min
Operating temperature: – 40 to + 40°C
Refills in indicator set: 3
Voltage: 12 V
Heat output: 120 to 140 W
Number of battery sets: 2
Battery life:
(+ 20°C) 6 h
(–40°C) 1 h

STATUS
Production probably complete. In service with the Soviet and Warsaw Pact armed forces.

MANUFACTURER
State factories.

Field Chemical Laboratory, Model PKhL-54

DESCRIPTION
The Model PKhL-54 field laboratory can be used to detect and identify numerous chemical substances in soil, water and other materials and to analyse water and other similar substances. The equipment in the laboratory can be used for numerous chemical procedures including filtration, distillation, extractions and the like. The laboratory, which has a storage life of five years, is contained in a hinged box which opens into three compartmented sections. The box holds 235 items including many of the items found in the Model PKhR series of kits (but in larger quantities) with the addition of numerous items of laboratory glassware and other hardware including a 50 mg chemical balance and an alcohol burner and writing materials.

The container box is 670 × 160 × 450 mm. It weighs 17.2 kg fully loaded.

STATUS
Probably still in production. In service with the Soviet armed forces.

MANUFACTURER
State factories and laboratories.

Biological Warfare Sampling Kit, Model KPO-1

DESCRIPTION
This sampling kit is used to obtain samples of earth, water, air, vegetation, micro-organisms and insects for analysis. It can also be used to collect unidentified chemical warfare agent specimens. The kit is carried in a case with a shoulder strap and contains, in racks, sample jars, various small sampling tools, an insect net, polyethene bags, a soil sample separator to collect a sample of up to 15 cc, and a water sampler which can collect samples from depths down to 30 metres. Atmospheric samples can be collected on gelatine filters. The kit weighs approximately 2.5 kg complete, and is about 300 mm long, 120 mm wide and 180 mm high.

STATUS
Production probably complete. In service with the Soviet armed forces.

MANUFACTURER
State factories.

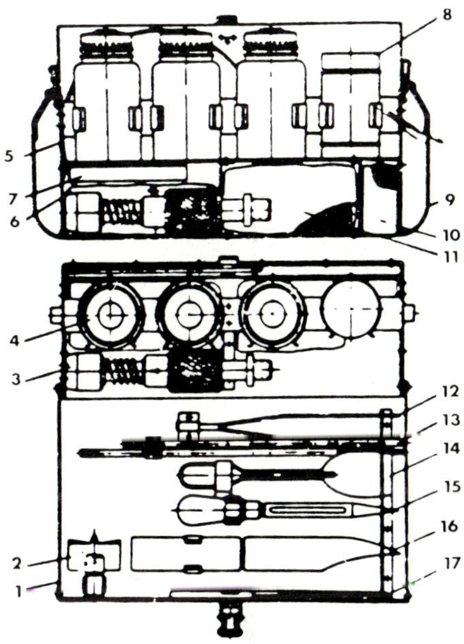

Model KPO-1 biological warfare sampling kit **(1)** *case* **(2)** *soil borer* **(3)** *soil sample separator* **(4)** *sample jars* **(5)** *insect net* **(6)** *cotton wool* **(7)** *gauze* **(8)** *test-tube container* **(9)** *shoulder strap* **(10)** *gelatine filters* **(11)** *polythene bags* **(12)** *forceps* **(13)** *insect net handle* **(14)** *sampling spoon* **(15)** *probe* **(16)** *knife* **(17)** *scissors*

Indicator Powder and Dispenser

DESCRIPTION
This equipment consists of a flattened cylindrical dispenser with a handle at one end, a dispensing nozzle at the other and a shoulder strap. The dispenser contains an inert powder with an added indicator agent which can change colour in the presence of chemical agents. Initially the powder is pale pink but it changes to red or reddish-brown in the presence of mustard gas, reddish-brown in the presence of nitrogen mustard or G-type nerve agents, and lilac in the presence of lewisite. Depending on the amount of agent present the colour change can take place in anything from 2 to 15 minutes. The container can hold from 2 to 4 kg of powder which can cover up to 40 square metres.

STATUS
Probably no longer in production. In service with the Soviet and Warsaw Pact forces.

MANUFACTURER
State factories.

Dispenser for indicator powder

UNITED KINGDOM

Chemical Agent Monitor (CAM)

DEVELOPMENT
The Chemical Agent Monitor (CAM) completed its design and development stage following the award of a United Kingdom Ministry of Defence contract in 1980 to Graseby Ionics Limited, a Cambridge Electronic Industries Company. Following final United Kingdom tri-service approval in July 1984, they were awarded a production contract by the United Kingdom Ministry of Defence in November 1984.

Successful evaluation trials with other overseas government agencies led to the placement of a CAM production contract and an engineering development contract by the US Army in January 1986 and June 1986 respectively, and a CAM production contract by the Canadian Department of Supply and Services in November 1987. CAM passed the US First Article Testing (FAT) and US Army Test and Evaluation Command (TECOM) follow-on evaluation after trials conducted in Panama and Arizona (June 1987). CAM is under evaluation by NATO and other countries.

CAM was used by four United Nations investigative missions to Iran in 1986, 1987 and 1988 where CAM's positive response to mustard agents (H) in the Iran-Iraq Gulf conflict was verified by independent government laboratories.

Following a competitive tender exercise a contract to produce a second CAM batch solely for the United Kingdom Ministry of Defence was awarded to Marconi Command and Control Systems in December 1987. The contract was for an initial 12 159 units and was stated to be worth over £13 million. The contract is managed by Marconi Command and Control Systems from Frimley, Surrey, with production being carried out at the Marconi manufacturing facility at Hillend in Fife, Scotland.

In early 1984 the Environmental Technologies Group Inc (previously Allied Signals Bendix Environmental Systems Division of Baltimore, Maryland, USA) was granted by Graseby Ionics a licence agreement for the marketing and manufacture of CAM for the US Government. In September 1987 Bendix Environmental Systems Division were awarded a contract worth $22 161 053 to supply 3739 CAMs to the US Army.

CAM is a trademark of Graseby Ionics Limited.

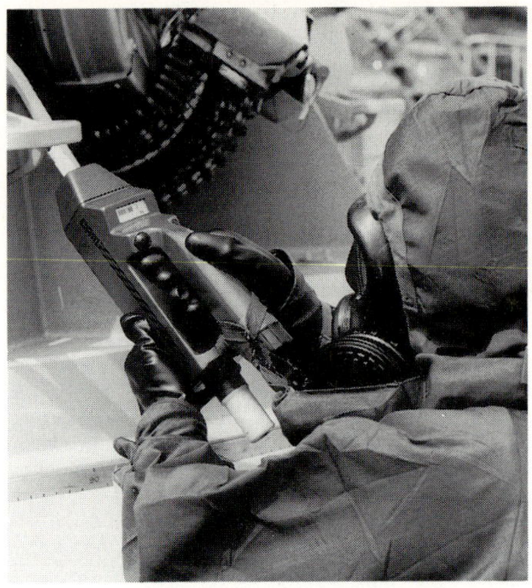

Chemical Agent Monitor (CAM) in use

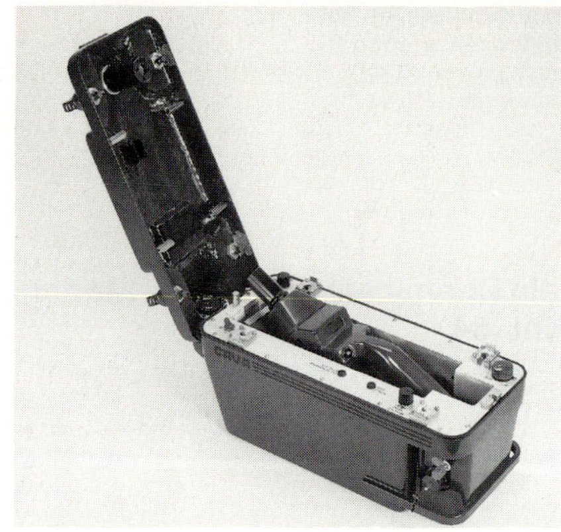

Chemical Agent Vapour Alarm (CAVA) designed and developed by Graseby Ionics for the Canadian Government

DESCRIPTION
The Chemical Agent Monitor (CAM, British Armed Forces designation L1A1) is used to confirm that a chemical attack has taken place; confirm when the threat has passed and enable commanders to order troops to remove their respirators and other protective clothing; delineate areas of ground that have been contaminated; establish any contamination of stores, vehicles, aircraft and personnel; establish that personnel entering collective protection areas are 'clean'; check battlefield casualties for contamination; and confirm the effectiveness of training procedures by using simulated agents. Training in 'real' conditions is possible as the instrument is also programmed to respond to harmless chemical simulants.

CAM makes use of ion mobility spectrometry principles to respond selectively to agent vapours. Air, which is drawn into the unit by a pump, is ionised by a weak radioactive source. As a result of complex charge interchange reactions which then take place the molecules of certain species of trace vapours form ionic clusters of low mobility while others do not. The ions of these species are then further classified according to their relative mobilities. The existence of a given level of a toxic hazard is calculated by a built-in microprocessor and shown on an LCD display. The techniques employed enable specific nerve or blister agents to be monitored at very low level, whether at isolated spots or present in the general atmosphere. Discrimination is also obtained against interfering vapours likely to be found in battlefield situations.

CAM can be carried on a shoulder strap with pouches for a spare battery, a confidence test device and stand-off collars packed in a foam strip. The instrument operates in real time and will recover quickly after exposure to strong concentrations of vapour. A low cost disposable 'stand-off' collar prevents pick-up and retention of agent material on the inlet probe. Warm-up time (at an ambient temperature of 20°C) is about one minute.

After warm-up a built-in test programme is employed to show the user that the equipment is functional and ready for use. The display panel is located in front of the carrying handle which also contains the power supply, a single 6-volt primary battery (a sealed $LiSO_2$ system). On either side of the display are the two push-button controls, on/off and nerve/blister mode. The probe is at the front of the instrument and is normally kept covered by a screw-on cap when not in use. With this cover in place the instrument can be decontaminated by normal methods. If required the instrument can be powered from an external source and the same connector can be used for fault diagnosis. It is possible to identify agents using a remote display/alarm that uses the 19-pin test plug (see under VARIANTS).

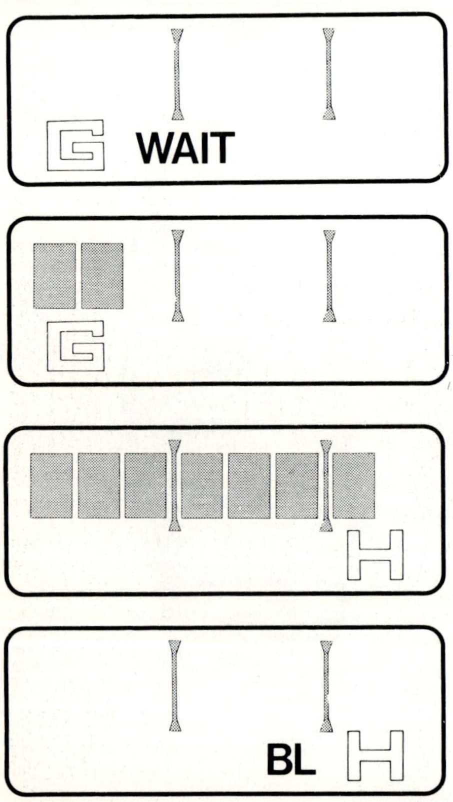

Examples of CAM display from top: nerve agent mode (G) selected, instrument in warm-up phase; nerve agent mode (G) selected with relatively low concentration indicated (two blocks visible); blister agent mode (H) selected with relatively high concentration indicated (seven blocks visible); battery low indication (BL)

Right-hand side of Chemical Agent Monitor (CAM)

Support equipment for the CAM includes a vapour generator to provide accurate and rapid calibration, a diagnostic test set to enable fault diagnosis down the replacement module level and a flexiboard test set to check all aspects of the control and monitoring circuitry within CAM.

A special bag and carrying harness for the CAM has been designed by Remploy Limited for Graseby Ionics. The bag is manufactured using Nomex fabric and can be carried in the most comfortable position to suit any particular circumstance. The carrying harness is fitted with pockets to carry CAM accessories.

VARIANTS

Otto Fuel Monitor (OFM)

The Otto Fuel Monitor (OFM) was developed from CAM by Graseby Ionics for the Royal Navy for use in submarines and with potential for use on surface vessels and armament depots whereOtto fuel is used in torpedos. The OFM is a fixed mode device with one on-off button and is powered by a rechargeable battery. It is in service with the Royal Navy.

Airborne Vapour Monitor (AVM)

The Airborne Vapour Monitor (AVM) is a variant of CAM which is available for commercial users. It can be programmed for the detection of a variety of volatile materials.

Chemical Agent Vapour Alarm (CAVA)

The Chemical Agent Vapour Alarm (CAVA) comprises a weatherproof base unit which houses a standard Canadian CAM. Circuitry within the unit takes over the processing, power supply and mode change functions of the CAM. The resulting detection system is responsive to low levels of both nerve and blister agent vapour. The system is flexible, both in terms of power supply, which can range from internal battery supply to remote input of either ac or dc via a cable link, and in alarm presentation which can be local or remote (up to 200 metres) and visual or audible. Remote visual status information is also provided.

M43

An M43A1 detector product improvement model has been produced by Graseby Ionics using ion mobility technology fitted into an M43A1 case and display system. This development enables the M43 to detect

Special bag and carrying harness developed by Remploy Limited for the CAM

blister agent (H) in addition to nerve agent (G) by switching modes automatically and thus makes the unit less prone to false alarms.

CAM Sprite Installation

Graseby Ionics modified a CAM to be installed in a ML Aviation Sprite remotely piloted mini-helicopter. The Sprite has the ability to hover close to the ground where samples would be taken to establish the extent of chemical agent contaminated areas. If the Sprite is temporarily out of direct radio communication, the information obtained by CAM could be stored and transmitted when contact is re-established.

Remote Display/Alarm for CAM

Graseby Ionics have developed a remote display/alarm (RDA) for use with CAM. This unit increases the operational flexibility and utility of CAM by indicating the agent window code, giving an alarm at a pre-settable level and continuously indicating the increase or decrease in the level of a chemical vapour threat.

The RDA operates by using serial data from CAM and provides: a copy of the CAM display; a decimal numeric display of the analogue value; an audible alarm, with a threshold that is adjustable so that the alarm is triggered by any desired number of 'bars' on the display; and an electrical analogue output corresponding to the internal CAM response strength.

The RDA is housed in a box measuring approximately 150 × 100 × 50 mm and weighs about 0.75 kg. It is powered by a single 6-volt battery connected to the rear of the CAM monitor by a 1.5-metre 19-way cable. The device has a carrying handle with notches to provide lugs for wrapping the cable lead. The operational temperature range for the RDA is from −5 to +50°C.

SPECIFICATIONS

Weight: (hand unit) 1.5 kg
Length: (over longest dimension) 380 mm
Battery life:
(minimum) 6 h continuous
(typical) 10 h
Temperature range:
(operating) −30 to +55°C
(storage) −55 to +70°C
Power supply: single 6 V battery
Sensitivity: to present NATO requirements
Durability: to DEF STAN 07-55

STATUS

In production. In service with Australia, Canada, United Kingdom, the United States (see text) and Spain. Under evaluation by several countries.

MANUFACTURER

Graseby Ionics Limited, 8 Millfield House, Woodshots Meadow, Croxley Centre, Watford, Hertfordshire WD1 8YX, United Kingdom.

NAIAD

DESCRIPTION

NAIAD stands for Nerve Agent Immobilised enzyme Alarm and Detector. It is an automatic alarm system which continually monitors the surrounding atmosphere to provide audible or visual warnings of the presence of nerve agents in either vapour or aerosol forms. The equipment consists of a sampling detector which can provide its own warning plus up to a further three remote alarm units which can be up to 500 metres away.

Its operation depends on the action of an enzyme, cholinesterase, which occurs naturally in the human body and reacts to nerve agents. Inside NAIAD a small sample of this enzyme in a plastic holder is continually irrigated by a supply of an organic ester, butyrylthiochlorine. The two substances combine by hydrolysis to form thiochlorine. When the cholinesterase reacts to a nerve agent the resultant thiocholine shows changes in its electro-chemical

NAIAD detector unit

NAIAD remote alarm unit

reactions which are detected by a graphite measuring electrode which initiates the alarms. NAIAD can be used to detect any nerve agent, even those that are at present unknown, by using natural body chemistry. It can also be used to detect hydrogen cyanide in attack concentrations.

The mechanics of the system depends on electrical power derived from a standard 3.3 Ah Clansman radio battery which supplies the alarm circuits, the reagent liquid pump, the incoming air pump and the air heater. The reagent pack has the same non-stop service life as the battery: both are changed every twelve hours. The reagent pack is sealed, as is the enzyme pad holder, so there is no danger of contaminating the system by handling, and the whole system has a rapid response with a low false alarm rate. The detector is nuclear-hardened and can operate even when tipped on its side or inverted. The alarm units have flashing lights and a clearly audible alarm signal and built-in test equipment automatically tests the cable connections between the alarm and the detector.

The detector is normally carried in a back-pack which also contains spare batteries and reagent packs, but it is possible to fit it into vehicles or fixed installations. A special rack for carrying the detector on the outside of vehicles has been developed. Two test units are available, a base workshop test set and a field test kit.

The Simtec NAIAD is used in a ship chemical alarm system known as Ship Installed Chemical Alarm System (SICAS). This uses the existing hardware and through-bulkhead units of the Royal Navy's Ship Installed Chemical System (SICS – see following entry) and involves the SICAS detector, through-bulkhead units, a remote status panel and a remote alarm unit.

The Airfield Chemical Agent Detection System (ACAS) is a proposed system using NAIAD for airfield use. It involves the placing of a number of detector units around an airfield perimeter with more at selected positions on the airfield. A remote status panel and remote alarm units are involved.

SPECIFICATIONS

Detector unit:
(size) 251 × 209 × 475 mm overall
(weight) approx 12.5 kg with battery

Alarm unit:
(size) 232 × 177 × 99 mm
(weight) approx 2.5 kg with battery

Climatic range: −31 to +52°C, 0–100% relative humidity

Minimum detectable concentration: 0.005 to 0.05 mg/m³ dependent on agent

STATUS

In production. In service with the British Armed Forces, Portugal and Spain.

MANUFACTURER

Jasmin Simtec Limited, Sellers Wood Drive, Bulwell, Nottingham NG6 8UX, United Kingdom.

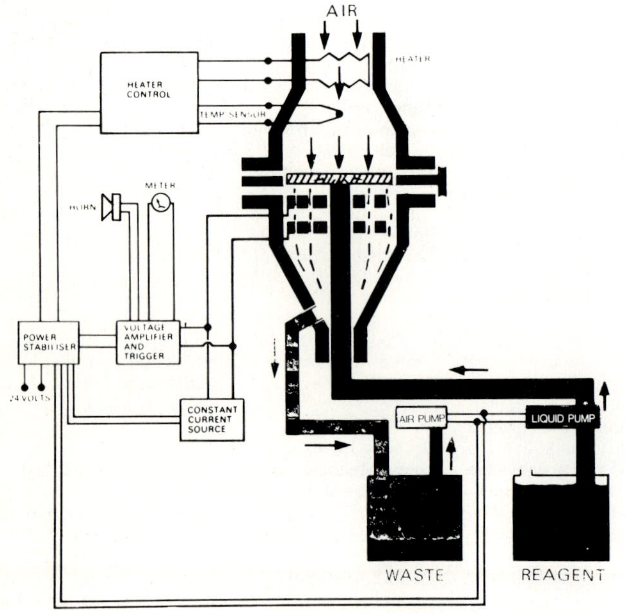

NAIAD detection process

NAIAD interior showing sealed units

Ship Installed Chemical System (SICS)

DESCRIPTION

The Ship Installed Chemical System (SICS) is the outcome of a development programme originally carried out by Thorn Automation Limited in collaboration with the Ministry of Defence. It has been in service since 1974.

SICS is installed in surface ships in order to detect any contamination of the air by chemical agent attack, both within the ship's citadel and around the ship. The system is arranged to provide rapid automatic warning of such an attack to the whole ship's complement so that the necessary defensive equipment and measures can be deployed. Monitoring of the air is continuous and the SICS detector units are sensitive to chemical agents in vapour or aerosol form, with a maximum sensitivity of 0.2 to 0.4 mg per cubic metre. When the presence of contamination is detected both visual and aural alarms are activated automatically. The visual alarm is displayed to the ship's NBC Protection Officer and indicates the detector concerned while the aural alarm is broadcast over the ship's general broadcast system.

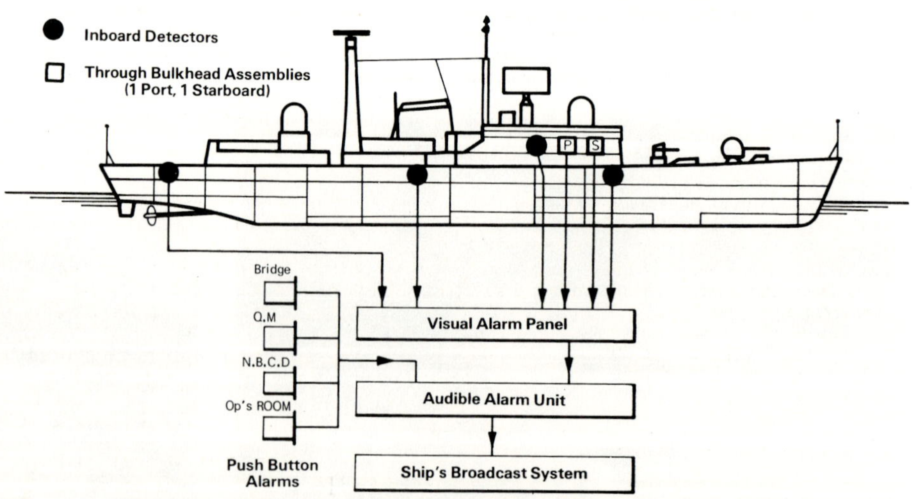

Schematic layout of typical Ship Installed Chemical System (SICS) installation

In operation sampled air is drawn through a SICS detector unit's air inlet, heated to a pre-determined temperature and passed through a filter into a detector cell where it is mixed with a reagent solution. The filter, in addition to removing interfering materials from the incoming air, converts V-agents into corresponding G-type compounds. G-agents, whether from the sampled air or from conversion from V-agents, break down into an oxime present in the reagent. This reaction produces cyanide ions which cause a voltage to be developed between two electrodes and this change in electrical output is detected electronically, activating the alarm relay and stopping the pump motor.

The reagent solution is pumped at a controlled rate into the detector cell from a storage container. After being mixed with sampled air the reagent, now containing air bubbles, is drawn back into the container under partial vacuum provided by an air pump. Inside the container the reagent is separated from the air and recycled.

On detection of a chemical agent only a small amount of reagent solution is contaminated and this is rendered fit for re-use in the normal passage of the reagent through a catalytic scrubber incorporated in the detector cell. Thus the detector unit may be put back into service without delay using normal start-up procedures and obviating the need for cleansing of the system. The chemicals needed for the simple preparation of the reagent are contained in a SICS chemical replenishment kit. This provides for continuous operation of one SICS detector unit for a period of 30 days.

The number of detector units required and their location varies with the class of ship involved. A typical frigate or destroyer installation would consist of six detector units connected to a visual alarm panel. Four of the detector units would be located at selected positions to monitor the atmosphere inside the ship's citadel, while the other two would be sited at either side of the bridge to sample external air via through-bulkhead units. The aural alarm may also be activated by the operation of manual alarm push-buttons strategically positioned throughout the ship.

Replacement of the reagent solution and filter pad in each detector unit is needed once every 24 hours and correct functioning of the unit is checked by a simulant provided in an on-board test kit. A base workshop test set is available for second-line maintenance of detector units and functional modules. Modular design of the units ensures easy and rapid maintenance by replacement when necessary.

A SICS detector unit Mark 7NHA measures 310 × 280 × 230 mm and weighs 13.6 kg. A through bulkhead unit measures 510 × 210 × 300 mm and weighs 9.7 kg.

SICS units can be used to form the basis of the Ship Installed Chemical Alarm System (SICAS) which is based on the use of NAIAD – see previous entry.

STATUS
In production. In service with the Royal Navy.

MANUFACTURER
Jasmin Simtec Limited, Sellers Wood Drive, Bulwell, Nottingham NG6 8UX, United Kingdom.

Chemical Agent Detector MT80 Mark 2

DESCRIPTION
The MT80 Mark 2 chemical agent detector is a complete gas detection outfit designed for use under military conditions. It is able to detect a wide range of war gases and chemical agents in vapourised form. The MT80 Mark 2 is completely self-contained and is powered from an in-built battery which is continuously charged from an integral high efficiency photo-voltaic solar panel. The system was designed to satisfy the requirements of military and law enforcement authorities for a unit that would detect the complete range of chemical agents and gases likely to be encountered in warfare.

The detection unit is contained inside a NATO modular transport case. Inside the case is the alarm and gas concentration indicator plus a removeable cable reel containing 305 metres (1000 feet) of cable and the detection head. In use the detection unit case is set down and the cable reel is removed from the container. The signal cable is anchored to the indicator unit and pays out automatically as the reel is deployed. The reel is then left in the area to be monitored. A control panel on the detector unit has only one switch capable of being operated by a gloved hand. The switch has only three positions; OFF, SENSOR ACTIVE and ALARM MUTED.

SEMA MT80 Mark 2 chemical agent detector with cable reel bearing the sensor removed from the carrying case

The cable reel contains the agent sensor, described as having a unique design. The sensor uses a silicon semiconductor material having a crystalline structure. To obtain a high degree of selectivity to various chemical agents an infinitesimal amount of other elements, each having a different atomic structure, are added to the basic semiconductor. These elements or 'dopes' make the semiconductor alter its conductivity when exposed to chemical agents. To measure the electrical range within the semiconductor a voltage is applied across the silicon chip and the resultant current is measured. This variable current is directly related to agent concentration. The response time is from 1 to 3 seconds and sensitivity is such that a warning will be provided when the atmospheric proportion of a dangerous chemical agent reaches 0.01 part per million.

The sensor housing is manufactured from solid bar stainless steel with the sensor enclosed in a closed end cylinder of stainless steel mesh. The housing body is provided with equi-spaced slots to allow ready access of the agent to be detected, to diffuse through the mesh and thence to the sensor. The sensor is an integral part of the cable reel and is protected by a perforated enclosure.

Included in the detector tranport case is a solid state microprocessor-controlled electronic module incorporating an alpha-numeric display, a dual memory section and sensor control, together with a photo-voltaic panel and an electronic regulator. The supply for the sensor is derived from a switching supply of a buck regulator type and is highly stable. Input signals are presented to a balanced input amplifier which rejects series and common mode noise voltages and ignores transient signals of up to 150 milliseconds duration. The analogue signal from the amplifier is applied to a 12-bit analogue/digital converter with the resulting information fed to a digital processing unit using a 16 bit data highway. The processing unit comprises a hardware stack and a 16 kilobyte data memory. In operation the signal is compared with an algorithm, 12 comparisons resulting in a full converted byte being transferred to the conversion-result store in 40 microseconds. Alarm setting data is supplied to a programmable peripheral interface, this being continuously monitored and compared with the incoming data from the sensor. When a match is found the alarm sounds. Information contained in the memory is continuously supplied to the alphanumeric display which not only indicates the agent concentration detected but also shows the system status.

Operationally the detector will indicate 'ZERO GAS' if the area being monitored is clear. A very small agent concentration will immediately be shown as 'GAS ALARM', accompanied by the audible alarm. The alarm changes to a more insistent sound as agent concentration increases to a higher level and the display shows 'DANGER GAS'.

A diagnostic programme dynamically self-tests the system at regular intervals giving an audible alarm for any faults found, such as sensor disconnected or sensor shorted. The programme also monitors the battery condition and state.

In-built nickel cadmium batteries are fitted as standard and have capacity to power the system non-stop for 24 hours without recharging. Automatic charging is by means of a fitted photo-voltaic panel that continuously trickle-charges the batteries in most light conditions short of total darkness. Recharging when the detector is not in use is by means of a mains operated charger unit that plugs in in place of the photo-voltaic panel. The charger unit can be left connected indefinitely until the detector is required for service. The battery charging state is shown on the detector display panel.

Dimensions of the MT80 Mark 2 carrying case are 600 × 415 × 262 mm and the all-up weight is 33 kg. The operating temperature range is from −30 to +80°C.

STATUS
In production and in service.

MARKETING AGENCY
J & S Franklin Limited, Franklin House, 151 Strand, London WC2 1HL, United Kingdom.

Chemical Agent Detector MT90

DESCRIPTION
The MT90 chemical agent detector was designed for chemical agent reconnaissance in areas where it is physically impossible for personnel or vehicles to penetrate. Chemical agent detection capabilities are similar to that of the MT80 Mark 2 (see previous entry). The system employs gas detection projectiles, launched using compressed air, that can transmit chemical agent detection data to a radio receiver.

A complete MT90 detection outfit comprises 10 gas detection projectiles, a receiver/transmitter module, an air-powered projectile launcher and a compressed air cylinder. All components are enclosed in a metal carrying case. The gas detection projectiles can be supplied in cases of 50.

In operation the receiver/transmitter module is switched on and connection is made into a control radio net. The air-powered projectile launcher is then connected to the compressed air cylinder by a light pneumatic hose and the safety pin is removed from a gas detection projectile. The projectile status is immediately transmitted and displayed on the receiver/transmitter module to verify that the projectile is operational and that an identification can be made. Transmission then ceases. The projectile is muzzle loaded into the launcher and fired into a designated area. On landing the gas detection projectile again transmits status information and shuts down. Transmission recommences only on detection of a chemical warfare agent. Information is then transmitted to the receiver/transmitter module in 1 millisecond pulses spaced 10, 20 and 30 seconds apart.

Each gas detection projectile is 270 mm long and 50 mm in diameter. Their appearance is similar to that of a smoke mortar bomb. They can be deployed up to 20 km from the receiver/transmitter module. Each projectile contains a battery-powered transmitter and will operate continuously for 24 hours in the transmit mode and 200 hours in the monitor mode. The receiver/transmitter module is powered by replaceable batteries which operate continuously for at least seven days. Each gas detection projectile can be checked by the module before firing to check signal efficiency. On receiving a gas signal from a projectile the module will indicate the projectile number, the chemical agent detected and its concentration. If two or more projectiles are transmitting the receiver will automatically scan each in turn and relay changes in information as they occur. The receiver/transmitter module can operate as a net radio and re-transmit information as data over the 30 to 88 mHz frequency range with a frequency hopping capability. The module can be supplied to operate in Band 1, 2 or 3 with an encryption facility.

The air-powered projectile launcher is a compact, lightweight, shoulder-fired unit developed in accordance with DEF STAN 07-55 in conjunction with the British Army. The launcher is constructed from black anodised aluminium and steel, can be safely immersed in salt water and has a calibre of 50 mm. Safety controls include a mechanical safe fire interlock and a projectile interlock. The muzzle energy is 2 kJ and the launcher is fired using a trigger. A compressed air cylinder with a charge pressure of 215 bar giving over 100 firings is supplied. Air pressure controls include a launcher charging valve, a pressure limiter and a dump valve to allow the launcher to be discharged at any time.

The system is carried in a stackable aluminium alloy weatherproof case measuring 800 × 600 × 400 mm. Each component in the case is housed in a rigid wooden frame with a support frame mounted inside the lid for further security. The compressed air cylinder is strapped into place. Also included is a pneumatic hose approximately 5 metres long to couple-up the compressed air supply to recharge the launcher.

The system is designed to be carried and operated by two men.

STATUS
In production.

MARKETING AGENCY
J & S Franklin Limited, Franklin House, 151 Strand, London WC2R 1HR, United Kingdom.

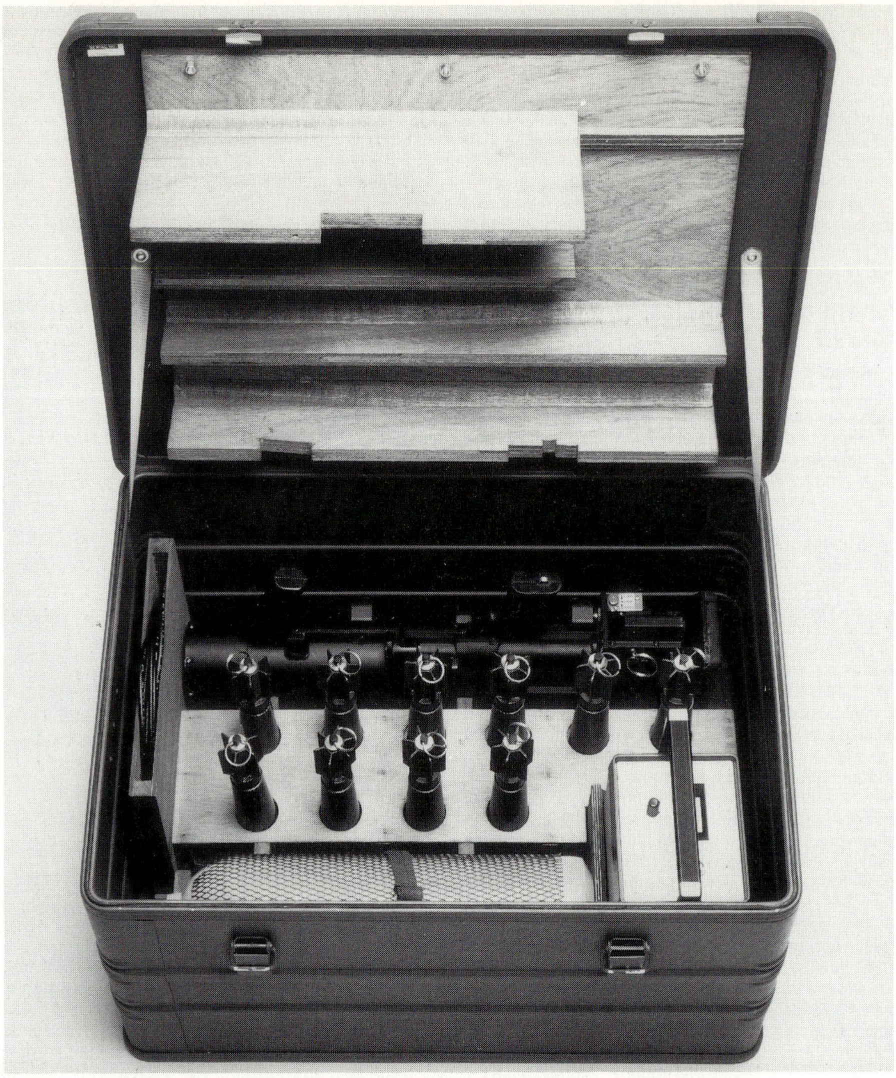

A complete MT90 chemical agent detector in its carrying case, showing the gas detection projectiles

Individual Chemical Agent Detector MT100

DESCRIPTION
The MT100 individual chemical agent detector is a small lightweight gas detection unit designed as a personal protection alarm against a wide range of chemical agents. It has high sensitivity with multiple agent detection and a fast response activating an audible alarm to provide warnings of hazardous situations.

The detector element used with the MT100 is protected from the environment by a hydrophobic membrane. The detector itself is a unique design and was specially developed for the detection of war chemicals and gases. The sensor is a carefully prepared silicon semiconductor material having a crystalline structure. To obtain a high degree of selectivity to various gases, an infintesimal amount of other elements having a different atomic structure are added to the basic semiconductor. These elements, or 'dopes', make the semiconductor alter its conductivity when exposed to gases, so to measure the electrical range within the semiconductor a voltage is applied across the silicon chip and the resultant current is measured. This variable current can be directly related to gas concentrations.

Power to operate the MT100 is derived from disposable lithium batteries which are accessible from the top of the instrument. Weight of the MT100 is 0.5 kg and it measures 155 × 92 × 57 mm.

STATUS
In production.

MARKETING AGENCY
J & S Franklin Limited, Franklin House, 151 Strand, London WC2R 1HR, United Kingdom.

Detector Paper Chemical Agent No 2 Liquid

DESCRIPTION
This detector paper is issued in booklets in a sealed clear plastic pack. Normally the paper is a neutral grey/green but exposure to liquid or aerosol chemical agents produces distinctive navy blue spots. The individual sheets of detector paper (there are 12 to a booklet) measure 115 × 60 mm and have an adhesive backing to secure them to clothing, equipment or structures.

In addition to booklets this detector paper can also be supplied in 60 mm wide strip form or in rolls measuring 126 mm × 20 metres, both with the same adhesive backing as the booklets.

STATUS
In production. In service with the British Armed Forces.

MANUFACTURER
Field Packaging, Limberline Road, Hilsea Industrial Estate, Portsmouth, Hampshire PO3 5JF, United Kingdom.

Detector Kit Chemical Agent Residual Vapour No 1 Mark 1

DESCRIPTION
The Detector Kit No 1 Mark 1 is a personal piece of equipment normally issued to squad leaders to detect the presence of mustard or nerve agents after an attack. The kit is carried in a folding stiff canvas pack which contains instructions, reagent capsules and a small hand-pump operated sampler. In use a reagent capsule is placed in the sampler and the hand pump is actuated. If any agent is present the reagent changes colour. The kit has a shelf life of about four years imposed by the active life of the reagents employed.

STATUS
Production complete. In service with the British Armed Forces.

SUPPLIER
Richmond Packaging, New Road, Winsford, Cheshire, United Kingdom.

No 1 Mark 1 detector kit

UNITED STATES OF AMERICA

ICAD Miniature Chemical Agent Detector

DEVELOPMENT/DESCRIPTION
The Bendix Corporation's Environmental Systems Division carried out the initial development of a miniature chemical agent detector, the ICAD, which is intended to address the chemical agent threat of the 1990s. The hardware underwent laboratory and field evaluation in Europe and is now in production. Final development was carried out by Environmental Technologies Group Inc.

The ICAD is a low-cost, miniature detector which can also be used as a personal detector badge. Based on electrochemical detection phenomenon, the ICAD is compact and lightweight, measuring 122 × 71 × 36 mm and weighing 227 grammes.

The system consists of an electronics module and a replaceable sensor module. The electronics module, consisting of the detector processor and an audible alarm and warning light, is reusable. The sensor module, which contains the battery power source and sensor cells, is disposable and designed for three months of continuous use. The sensor contains three independent interference-rejection techniques and provides an alarm response to nerve agents, mustard, lewisite, phosgene, hydrogen cyanide and cyanogen chloride. After an initial alarm the sensor will automatically reset and be capable of accepting subsequent chemical agent challenges.

A training system using a hand held transceiver and electronic pagers has been developed. Potential applications for the ICAD include a remote monitor for fixed and mobile installations and a filter monitoring device for collective protection equipment. The ICAD has been dropped from a remotely-piloted vehicle in a remote detection role.

STATUS
In production. In service with NATO forces.

MANUFACTURER
Environmental Technologies Group Inc, 1400 Taylor Avenue, Baltimore, Maryland 21284, USA.

ICAD miniature chemical agent detector

ICAD miniature chemical agent detector showing sensor module on left and electronics module on right

XM85/XM86 Automatic Liquid Agent Detector System

DESCRIPTION
This system is under development to detect thickened nerve agents, mustard gas and lewisite down to droplets 200 microns in diameter. The system consists of two units. One is the XM85 Central Alarm Unit (CAU) which continuously monitors a network of individual XM86 Detector Units (DU). The XM86, also known as ALAD (Automatic Liquid Agent Detector) can detect a single drop (200 microns and larger) of threat chemical agents. The detection mechanism is based on the physical-chemical integration of the agent with a special paint resin in which are suspended very fine elemental silver flakes. This silver-bearing paint acts as an electrical conductor which swells when attacked by an agent, causing physical separation of the conductive silver flakes and a resulting change in the electrical resistance of the detector grid. This change activates an alarm function which is transmitted automatically by radio to a CAU. The sensor element can be inserted into the detector unit.

The two units operate together to warn of a liquid chemical rain attack by allocating each detector unit a unique identification code. Using this coding the CAU can identify which detector unit is providing an alarm. By using the known location of each detector the aerosol cloud characteristics causing the alarm can be plotted and assessed, eg course, velocity and dimensions.

STATUS
Advanced development.

DEVELOPMENT AGENCY
Chemical Research, Development and Engineering Center, Aberdeen Proving Ground, Maryland, USA (now closed).

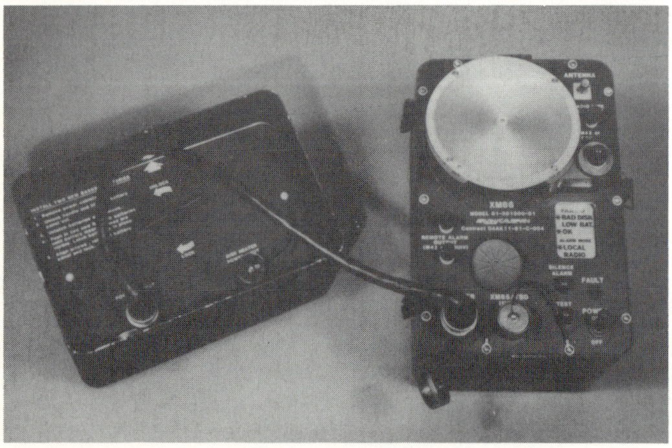

XM86 Detector Unit showing sensor element and unit cover, removed before use but containing the unit battery power supply (US Army)

XM85 Central Alarm Unit (CAU) (T J Gander)

Alarm, Chemical Agent, Automatic: Portable, M8 and M10 to M18

DESCRIPTION
This alarm system has ten variations all based on the use of two basic components, the M43 detector and the M42 alarm. Each detector unit can activate up to five alarm units. By the use of various components and associated equipment these two basic units can form the basis for the following alarm systems:
Alarm, chemical agent, automatic: portable, manpack, M8 or M8A1
Alarm, chemical agent, automatic: portable, fixed emplacement, M10
Alarm, chemical agent, automatic: portable, for Truck, Utility,-ton, M11
Alarm, chemical agent, automatic: portable, for Truck, -ton, M12 (no longer in production)
Alarm, chemical agent, automatic: portable, for Truck, 2-ton, M13
Alarm, chemical agent, automatic: portable, for full-tracked APCs and ARVs, M14
Alarm, chemical agent, automatic: portable, for Carrier, Command and Recce, M15
Alarm, chemical agent, automatic: portable, w/power supply for Truck, Utility, 1-ton, M16
Alarm, chemical agent, automatic: portable, w/power supply for Truck, Utility, -ton, M17 (no longer in production or service)
Alarm, chemical agent, automatic: portable, w/power supply for Truck, 2-ton, M18

These systems differ mainly in the type of mounting used, but common to all is the M253 winterisation kit which has two special batteries and a cable assembly. This kit has not been purchased by the US Army.

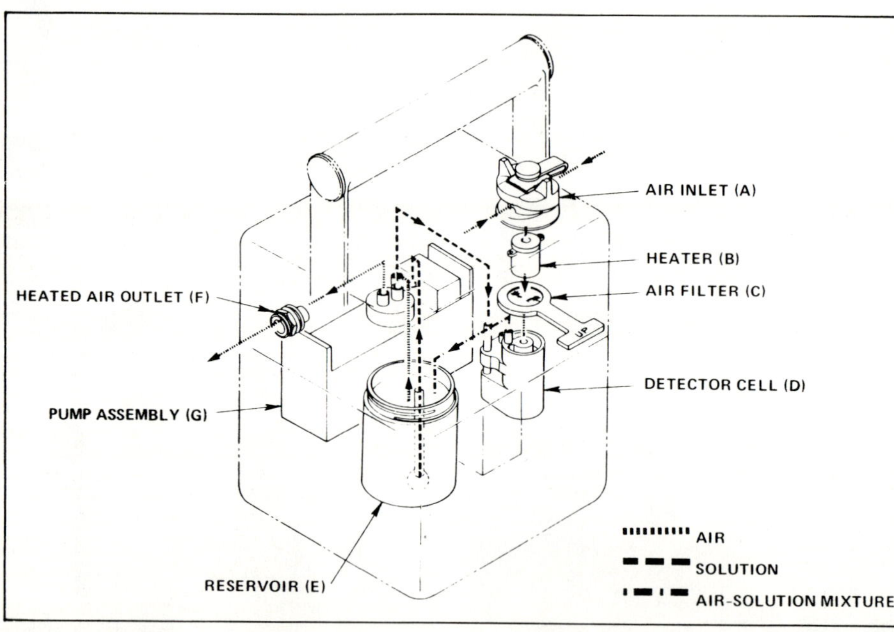

M43 detector unit air-solution flow diagram

The M43 detector operates by passing air through an oxime solution surrounding a silver analytical electrode and a platinum reference electrode. If a chemical agent is present a reaction occurs in the solution which increases the potential between the two electrodes. Any such potential change is amplified and the alarm signal is triggered.

The M43A1 Chemical Agent Detector, developed at the US Army Armament Research and Development Command's Chemical Systems Laboratory, CB Detection and Alarms Division, serves as a direct replacement for the M43. The M43A1 directly replaces the M43 and is designed to detect low concentrations of nerve agents in less than one minute. The M43A1 is structurally and functionally similar to the M43 but it differs mainly in its internal sensor characteristics and the way in which it operates. For operation, air is drawn continuously through the internal sensor by a pump. The air consists of its normal constituents plus any chemical agents which may be present. As the molecules are drawn past the radioactive source, a small percentage are ionized by alpha-rays. Through multiple collisions some of the ions cluster with agent molecules which then become very stable ions. As the air and agent ions are drawn through the baffle sections of the cell, the lighter and less stable air ions diffuse to the walls and are neutralised more quickly than the heavier and more stable agent ions. As a result, the collector senses a greater current when nerve agents are present compared to the current when only clean air is sampled. An electronic module monitors the current produced by the sensor and triggers the alarm when a critical threshold is reached. The alarm has a variable pitch with a maximum range of 70 decibels.

The M43A1 HD Chemical Agent Detector (manufacturer's designation, not type classified) is a modification of the M43A1G (manufacturer's designation, nottype classified) and can be field retrofitted. The modification allows the M43A1 HD to detect not only nerve agents but mustard (HD) agents as well. The M43A1 HD detector is created when the M43A1G is retrofitted with a modified cell module (containing a corona tube and Hopcalite filter) and a high voltage power supply. These modifications add approximately 0.9 kg to the weight and 57 mm to the height of the unit. The operation of the detector is basically the same as the standard M43A1 except that prior to reaching the radioactive source the air being sampled passes through the corona tube which creates ions allowing the unit to sense HD. The operation of the corona tube inhibits the detection of nerve agents so a timing circuit turns the corona on and off at intervals to allow detection of both HD and nerve agents. The electronic module functions exactly like that of the M43A1 and triggers the alarm when a critical threshold current is reached. M43A1 HD detectors have been tested by the US Marine Corps.

The Detector Unit, Simulator Automatic Chemical

Agent Alarm, XM81, has been developed for use with the M8 system and when fitted is attached to the M43 detector unit (see *Chemical training and maintenance* section).

STATUS
In service with the US Army and West German forces (M8A1). 26 000 portable manpack M8 sets were ordered for the US Army from 1975 to 1982.

MANUFACTURERS
M8 and M43: Environmental Technologies Group Inc, 1400 Taylor Avenue, Baltimore, Maryland 21284, USA.
M8A1, M43A1G and M43A1 HD: Brunswick Corporation, Defense Division, One Brunswick Plaza, Skokie, Illinois 60077, USA.
M43A1, M43A1G, M43A1 HD: Honeywell Chemical Defense Center, 13350 US Highway 19 South, Clearwater, Florida 33546, USA.
M43A1: Inframetrics Corporation, 16 Esquire Road, North Billerica, Minnesota 01862-2597, USA.

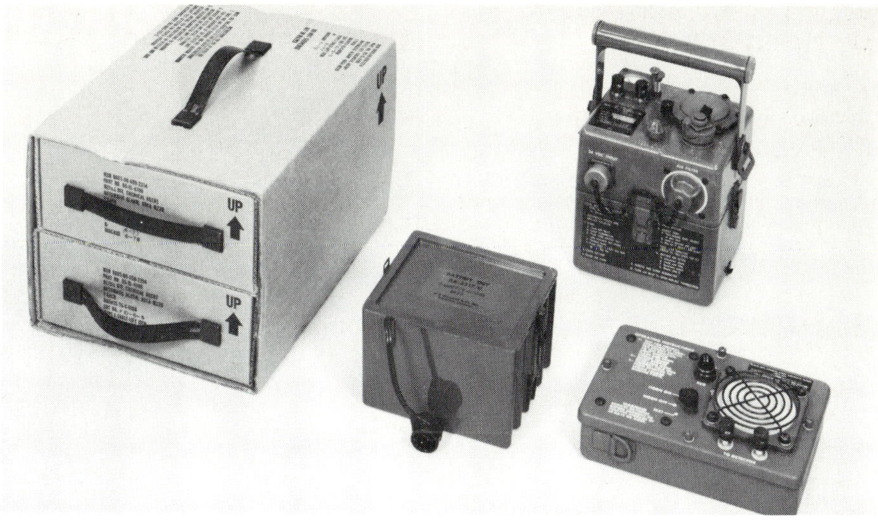

Basic alarm, chemical agent, automatic, portable, M8. From left: M229 refill kit, BA 3517 battery, M43 detector unit (at rear) M42 alarm unit

SPECIFICATIONS

Component	M43 and M43A1 Detector	M42 Alarm Kit	M229 Refill Supply	BA3517 Battery Kit	M10 Power Kit	M228 Mounting	M182 Mounting
Length	160 mm	224 mm	330 mm	160 mm	15 mm	191 mm	191 mm
Width	196 mm	152 mm	264 mm	196 mm	163 mm	254 mm	254 mm
Height	279 mm	89 mm	229 mm	127 mm	188 mm	318 mm	203 mm
Weight	3 kg	1.7 kg	6.3 kg	3.4 kg	8.2 kg	7.3 kg	6.8 kg
Maximum distance of detector from alarm	400 m	400 m	400 m	400 m	400 m	400 m	400 m
Life on one refill (continuous use)	15 days	n/a	15 days	n/a	n/a	n/a	n/a

Detector Kit, Chemical Agent: ABC-M18A2

DESCRIPTION
The ABC-M18A2 detector kit is normally issued to specialist chemical warfare teams for the detection of a range of blister, blood and nerve agents, and can also be used to collect samples for further analysis. The kit is contained in an olive-green carrier slung from the shoulder and it holds an aspirator bulb assembly, detector paper, sampling tubes, packets of reagents and substrate solution, ABC-M8 detector paper, M7A1 detector crayons and a few other accessories. When packed the kit measures 203 × 152 × 76 mm.

STATUS
In service with the US Army, Navy and Air Force.

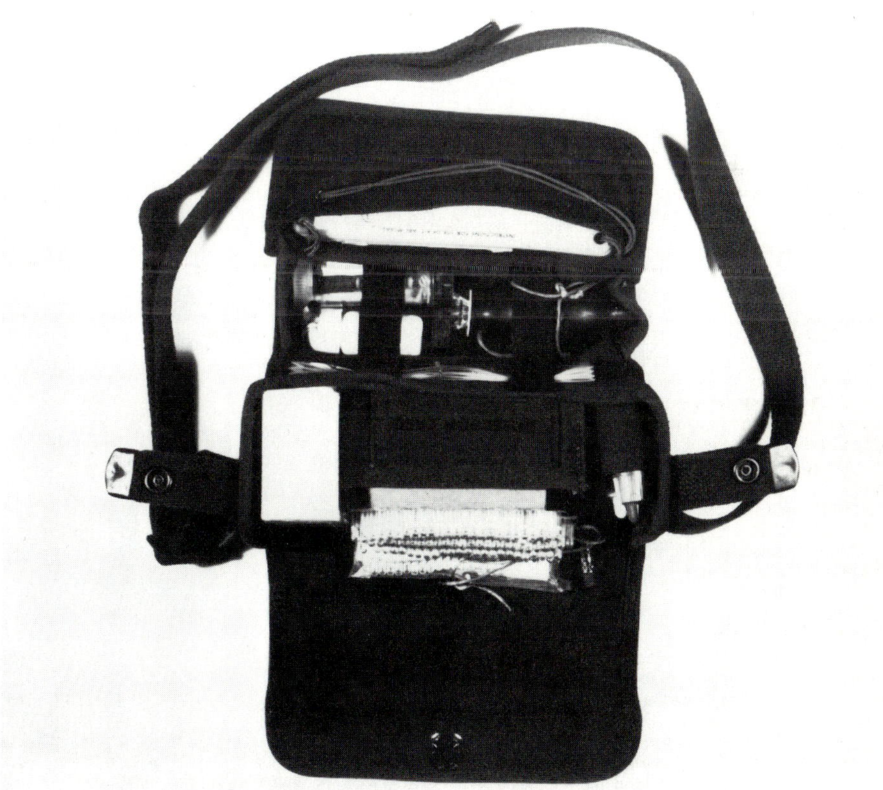

Detector Kit, Chemical Agent: ABC-M18A2

Detector Kit, Chemical Agent, M256

DESCRIPTION
The M256 chemical agent detector kit is replacing the earlier ABC-M15A2 and ABC-M18A2 kits for most field use applications and is produced so that troops in the field can determine within ten minutes whether it is safe to remove protective masks and clothing. The kit contains 12 pocket-sized sampler-detectors, a book of M8 chemical detection paper and instruction cards, all carried in a rugged case.

Each sampler-detector contains an impregnated test spot for blister agents, a circular test spot for blood agents, a star test spot for nerve agents and a lewisite pellet and marking pad. There are eight glass ampoules, six containing reagents for testing and two in an attached chemical heater. In use the reagent ampoules are crushed by hand and preformed channels in the plastic sheet of the sampler-detector direct the flow from the ampoules to the appropriate test spot. Simple safe/danger indicators are printed on the back of each sampler-detector and show the colour that each test spot develops if a chemical agent is present. The foil packet that encloses each sampler-detector also contains instructions printed on its outer surface.

The M256 kit measures 175 × 125 × 75 mm and

weighs 0.54 kg. The kit requires about four hours training to use it with skill. A Training Kit (Simulator) for the M256 is available.

STATUS
In production. In service with the US Army.

MANUFACTURERS
Chemical Compounding, New York, New York, USA.
Anachemia Canada, 500 Second Avenue, PO Box 147, Lachine (Montreal), Québec H8S 4A7, Canada.

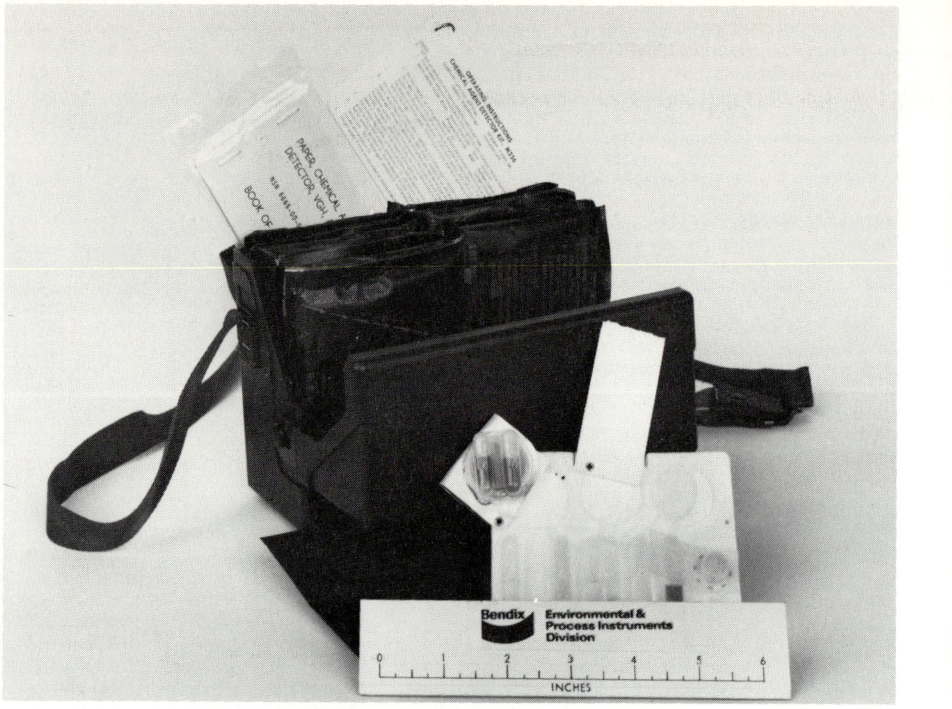

Chemical agent detector kit M256

Sampling and Analysing Kit, CB Agent, ABC-M19

DESCRIPTION
The contents of this kit are used by specialist personnel for detecting and identifying chemical warfare agents. By using the kit some preliminary processing of unidentified agents is possible. The kit is carried in a storage case measuring 356 × 432 × 178 mm which is internally compartmented to accommodate the various components. The actual components themselves are not standard but may be selected from three other kits, the M32 CB agent sampling and analysing parts kit, the M33 CB agent analysing components refill kit, and the M34 agent sampling components refill kit.

This kit is a Standard A item.

STATUS
In service with the US Army.

Sampling Kit, CB Agent, M34

DESCRIPTION
The M34 kit can be used to sample soil, water and other items for the presence of various chemical and biological warfare agents. The kit is issued in a fibreboard box which weighs just over 2 kg, and consists of two soil sampling kits, a phial container, and two pairs of gloves. The kit can be used for some preliminary soil testing.

STATUS
In service with the US Army, Navy and Air Force.

XM21 Remote Sensing Chemical Agent Alarm

DESCRIPTION
The XM21 remote sensing chemical alarm is an automatic scanning passive infra-red sensor which detects nerve agent clouds by detecting changes in the infra-red spectrum from remote objects or the sky. It is effective up to a range of approximately 5000 metres and can operate unattended for up to 24 hours in all weathers. The detector is mounted on a tripod and is capable of scanning a horizontal arc of 60 degrees in seven steps. A complete scan lasts less than 60 seconds.

The detector design is based on a Michelson interferometer and the use of a MC68000 microprocessor. The complete equipment consists of the detector which weighs about 20.4 kg, a tripod weighing 6.8 kg, an M42 remote alarm (used with the existing M8 alarm system), a power cable and a power source. There is also a transit case. Two men can carry the complete equipment and it takes ten minutes to prepare for unattended use.

The power source for the XM21 is to be a thermoelectric generator (TEG) under development by Teledyne Energy Systems who are under contract to the Electronic Devices and Technology Laboratory, ERADCOM. The XM21 has a power consumption of less than 80 watts.

The XM21 detector is constructed from nine major hardware components. These are the base assembly, the spatial scanner, the interferometer, an infra-red detector, analogue signal electronics, a signal processor, a power supply, a control panel and a sighting device. Using the sighting device on top of the detector the XM21 is normally placed looking into the wind to view a scene that consists of

XM21 Remote Sensing Chemical Agent Alarm together with its Thermo Electric Generator (TEG) (US Army)

the background and the air path along its line-of-sight to the background. The XM21 measures and stores a background spectrum that consists of the ambient energy contained in the scene. When an agent cloud enters the line-of-sight of the XM21, the scene is spectrally altered by the absorption/emission characteristics of the cloud. The XM21 software detects the changes and provides an alarm. The M42 alarm may be more than 400 metres from the detector.

It has been proposed that detection devices based on the XM21 could be mounted on reconnaissance vehicles, helicopters and remotely-piloted drones.

STATUS
Engineering development. First production deliveries scheduled for 1988.

MANUFACTURER
Brunswick Defense, 2000 Brunswick Lane, DeLand, Florida 32724 USA.

XM22 Automatic Chemical Agent Detector Alarm (ACADA)

DEVELOPMENT/DESCRIPTION
The XM22 Automatic Chemical Agent Detector Alarm (ACADA) is a man-portable equipment that is being developed to replace the existing M8 and M8A1 series of alarm systems. A point sampling alarm system, it is used to detect and identify all nerve agents, mustard gas and lewisite. Using a surface sampling probe with an integral heater the XM22 detects agents on the ground or other battlefield surfaces. The detector uses ion mobility spectrometry, a 'dry' system providing improved sensitivity compared with the M8/M8A1 systems.

The research and development programme commenced in 1982 with the award of a $10 million contract to the Bendix Baltimore plant. Development was overseen by the US Army Chemical Research, Development and Engineering Center at Aberdeen Proving Ground, then part of the US Army Armament, Munitions and Chemical Command (AMCCOM). Continued development was assured during 1984 by the award of a new development contract worth $27 163 733, again to the Bendix Baltimore Plant.

Development has reached the point where the detection unit is known as the XM22. This is powered by a battery pack secured to the base of the detector unit and the probe unit, complete with its flexible probe pipe, is connected to the top. The complete unit is carried slung from a webbing strap although the unit can be used on a fixed site. The detector can have a five-second response to provide a five-minute or continuous horn alarm. A discrete LED display is also provided on command. Built-in field diagnostics are provided along with a simple maintenance kit.

The surface sampler is an optional item as is a remote alarm and a mains power supply.

A redesign of the XM22 was completed in May 1988 and engineering development will begin during FY90.

Early prototype of XM22 Automatic Chemical Agent Detector Alarm (US Army)

SPECIFICATIONS
Weight: 5.44 kg
Length: 165 mm
Width: 165 mm
Height: 330 mm

STATUS
Advanced development.

DEVELOPMENT AGENCY
Chemical Research, Development and Engineering Center (CRDEC), AMCCOM, Aberdeen Proving Ground, Maryland 21010, USA.

MANUFACTURER
Environmental Technologies Group Inc, 1400 Taylor Avenue, Baltimore, Maryland 21204, USA.

Hughes Remote Active Spectrometer

DEVELOPMENT/DESCRIPTION
Following the award of a $3 million contract from the US Army Night Vision Electro-Optical Center, the Hughes' Electro-Optical and Data Systems Group produced an exploratory development model of a Remote Active Spectrometer (RAS) that was successfully field-tested at the US Army's Dugway Proving Grounds in Utah during 1986. During 1987 design changes for an advanced Remote Active Spectrometer were being evaluated in expectation of a production contract for 4000 units to be issued in 1994.

The Remote Active Spectrometer employs four carbon dioxide laser beams that detect the infra-red light absorbtion patterns inherent in toxic chemical agents and assists troops to avoid areas contaminated by persistent and non-persistent agents. By analysing the return energy reflected by chemical agents from up to 4.8 kilometres away the Remote Active Spectrometer can determine the type, quantity and location of contaminants. In addition to the lasers the Remote Active Spectrometer employs a precision 60-element optical system and an advanced signal processing computer system. In the later stages of development the instrument will be smaller and lighter than the early models to enable the instrument to be carried more readily by reconnaissance vehicles and helicopters.

STATUS
Development.

MANUFACTURER
Hughes Aircraft Company, Electro-Optical and Data Systems Group, PO Box 902, El Segundo, California 90245, USA.

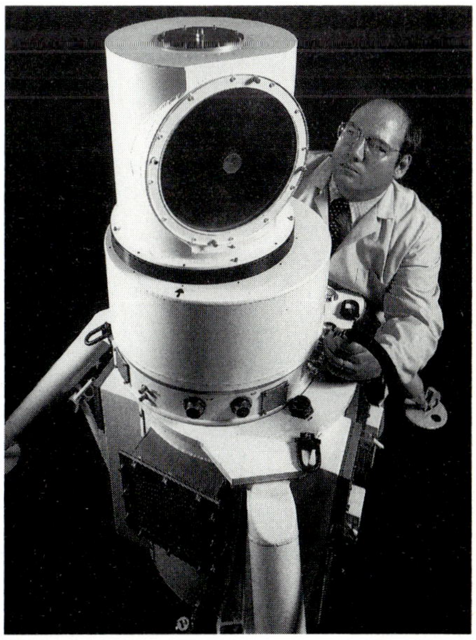

Development model of Hughes Remote Active Spectrometer (RAS)

Alarm, G-Agent, Automatic, Fixed Installation, M5

DESCRIPTION
Designed for use in manufacturing installations and other such premises, the M5 automatic alarm operates on the continuous trace principle. Detector chemicals are fed onto a continuous paper trace which is constantly scanned for any alteration of pigment caused by the presence of G-type nerve agents. If any G-type agent is present, a photometer unit will initiate the circuits which control the alarm horn and a flashing red light. The equipment is powered by a 110-volt ac source and when installed it weighs 328.8 kg. It is 2 metres high, 0.66 metre wide and 0.61 metre deep. When packed the full equipment weighs 840 kg, and is contained in four packing cases, one of which contains repair parts.

STATUS
In service with the US Army in base installations and manufacturing facilities.

MANUFACTURER
Leeds and Northrop, Philadelphia, Pennsylvania, USA.

Paper, Chemical Agent Detector: AN-M6A1

DESCRIPTION
This type of detector paper can be used only for detecting G-type nerve agents. The papers are issued in booklets of 25 with each paper measuring 63.5 × 101.6 mm. Each sheet of paper is coated on one side with M5 liquid vesicant detector paint which when exposed to a liquid G-type nerve agent turns from its original olive-green colour to red or orange.

STATUS
In service with the US Army, Navy and Air Force.

WARSAW PACT

PKhR Series of Chemical Agent Detection and Identification Kits

DESCRIPTION
This series of chemical agent identification and detection kits has been under development since the late 1940s and has undergone gradual and slight changes over the years. The kits are now in service in several forms, all using the same basic operating processes, and some have been produced in East Germany as well. Some of the earlier kits are now obsolete but are still in use for training and by some of the smaller Warsaw Pact nations. The following list summarises the main types:

Soviet designation	East German designation	Comments
PKhR-46	n/app	Obsolete
PKhR-50	n/app	Obsolete
	Used by Bulgaria	
PKhR-51	n/app	Obsolete
	Used by Bulgaria	
PKhR-54	KA-54, PChR-54	
PKhR-63	PChR-54 (improved)	
None ,02version of PKhR-63	PChR-54U	East German
PKhR	WPChR-64	

If the Model PKhR-54 is taken to be typical of the above series the kit is contained in a metal case fitted with waist and shoulder straps. The case lid opens forward to form a working tray. The kit box contains a hand-operated cylindrical air pump, nine or ten different types of indicator tubes (packaged in tens), a pump attachment, smoke filters, perforated plastic caps, a spatula, two sample jars, tape for marking affected areas, a torch, gloves, report forms and an instruction sheet. The kit is capable of detecting and identifying mustard, nitrogen mustard, lewisite, hydrogen cyanide, cyanogen chloride, phosgene, disphosgene, chloropicrin, adamsite, chloroacetophenone and G-type nerve agents. V-type nerve agents can also be dealt with by the addition of suitable indicator tubes. Each type of tube can normally identify and detect one type of agent but some of the more recent indicator tubes can detect up to four types of agent. The tubes are 7 mm in diameter and 110 mm long. They are packed in tens and are colour-coded as follows:

Tube marking	Agent detected
1 yellow ring	mustard (H)
2 yellow rings	nitrogen mustard (HN)
3 yellow rings	lewisite (L)
1 black ring	hydrogen cyanide (HC)
1 green ring	chloropicrin
2 green rings	cyanogen chloride (CK)
3 green rings	hydrogen cyanide (HC) cyanogen chloride (CK) phosgene (CG) diphosgene (DP)
1 white ring	chloroacetophenone (CN)
2 white rings	adamsite (DM)
1 red ring	G-type nerve agents
1 red ring with 1 red dot	G-type and V-type nerve agents

Each tube contains a measured quantity of reagent in a sealed ampoule. The end of the tube is broken off by inserting it in a hole on the end of the pump handle and snapping it. The ampoule is then pierced by a stiff wire contained in the pump handle and the tube placed in a socket in the head of the air pump. Hand pumping draws air through the tube and the ampoule will change colour according to the type of agent present. The pump can hold several tubes at once and the degree of agent concentration can be determined by reference to a calibrated colour scale. Fitting the air pump attachment enables sampling of soil, fabric or surfaces and a smoke filter can be used to screen smoke fumes that might affect detection. The sample jars can be used to take samples for further laboratory analysis. The kits from the Model PKhR-54U onwards contain an indicator tube heater as an extra in place of one of

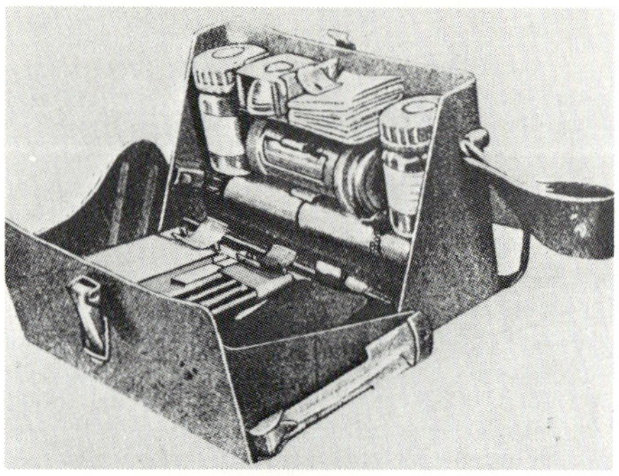

PKhR-63 chemical agent detection and identification kit

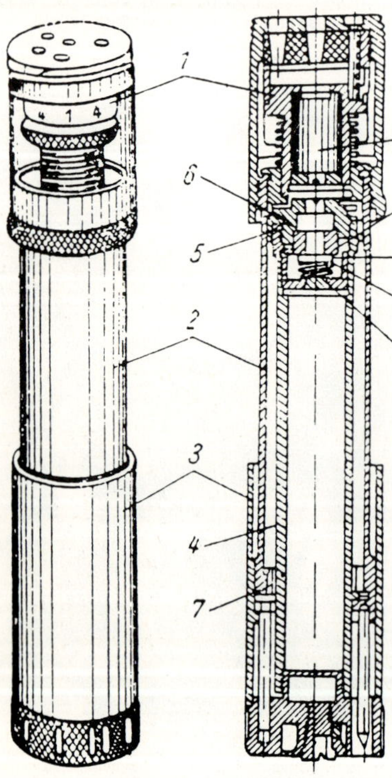

Piston-type air pump PKhR-54 **(1)** *collector* **(2)** *pump housing* **(3)** *handle with ampoule opener* **(4)** *piston rod* **(5)** *sleeve* **(6)** *seat* **(7)** *spacer ring* **(8)** *protective cartridge* **(9)** *valve* **(10)** *valve seat* **(11)** *spring* **(12)** *spring stop*

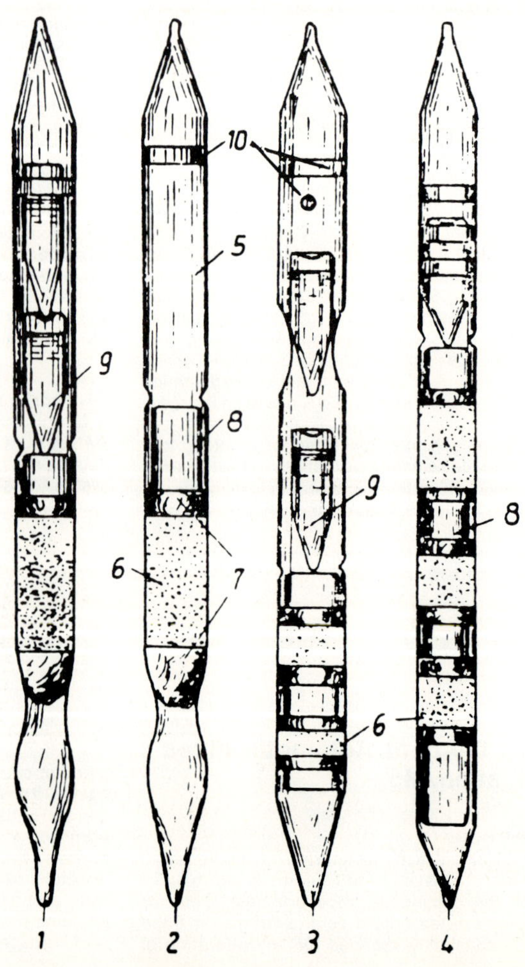

Indicator tubes **(1)** *red ring tube* **(2)** *1 yellow ring tube* **(3)** *1 red ring with 1 red dot tube* **(4)** *3 green ring tube* **(5)** *transparent glass tube* **(6)** *inert filter* **(7)** *cotton plug* **(8)** *flow-through canal* **(9)** *reagent ampoule* **(10)** *colour-code marking*

the sample jars. This heater is a hollow plastic cylinder closed at both ends and filled with cotton wadding. The top has four wells to hold three indicator tubes and a heating ampoule. When the ampoule is pierced using one of the wires in the pump handle, an exothermic reaction heats the tubes, which are placed in the heater for one minute. The heater is used only when the temperature is below 15°C.

The latest and current Soviet kit is the Model VPKhR which is smaller and lighter than the previous models with a more limited detection capability. Only three types of indicator tube are carried, but by using multi-detection tubes the kit can identify and detect mustard, phosgene, diphosgene, hydrogen cyanide, cyanogen chloride and G-type and V-type nerve agents. The sample jars have been replaced by plastic bags and a tube holder is also included. The East German Model PChR-54U kit also contains the CHNS indicator paper kit.

SPECIFICATIONS

Model	**PKhR-54**	**VPKhR**
Length	240 mm	205 mm
Width	100 mm	100 mm
Height	140 mm	140 mm
Weight	2.7 kg	2.3 kg

STATUS
Production and development continuing. Models PKhR-50 and PKhR-51 now obsolete. The rest are in service with the Soviet and Warsaw Pact forces, with the Model VPKhR being the current version.

MANUFACTURER
State factories.

Semi-automatic Chemical Agent Detection and Identification Kit, Models PPKhR and PPChR

DESCRIPTION
This kit is carried on various types of radiological and chemical reconnaissance vehicles such as the BRDM-1 RKh. Basically it is a Model PKhR series kit with an electrically-driven rotary air pump in place of the cylindrical hand pump, and an electric heater replacing the indicator tube heater. Both these components are driven by the vehicle battery. The East German version is the Model PPChR which is a modified Model PChR-54U kit.

STATUS
Probably still in production. In service with the Soviet and Warsaw Pact armed forces.

MANUFACTURER
State factories.

YUGOSLAVIA

DHM-11B Chemical Detection Kit

DESCRIPTION
The DHM-11B chemical detection kit is used for taking and identifying detection samples from chemical agents in the air, on the ground, on equipment, in loose material and in granular forms of food. It uses the pump and indicator tube method of sampling involving colour changes of chemicals. Agents that can be detected include the nerve gases, hydrocyanic acid, cyanogen chloride, phosgene, diphosgene, mustard agents and the nitrogen mustards.

The DHM-11B consists of a shaped sheet steel carrying case with a shoulder or belt strap, a manual piston pump with an adapter, an indicator tubes holder, indicator tubes, an accessories holder and various accessories and spare parts. In use various indicator tubes are placed in the end of the pump and air is hand pumped through them for about five to ten pump actions. The indicator tubes are carried in a double-ended and tubular aluminium holder with provision for 20 indicator tubes at each end. Before use each indicator tube has both ends cut off by using a cutting device on the pump. The tubes are then inserted into the pump for the sampling process. Various colour codings on the tubes denote their detection capabilities, as follows:

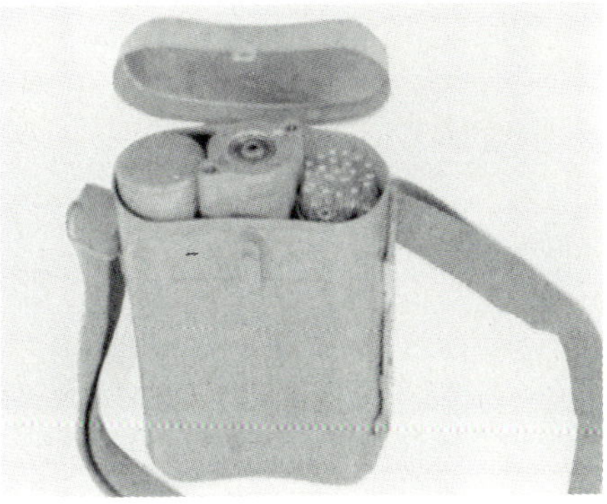

DHM-11B chemical detection kit

Tube with yellow and green ring with yellow point: phosgene, diphosgene, mustard, nitrogen mustard
Tube with blue ring and blue point: cyanogen chloride, hydrocyanic acid, tabun, CN group
Tube with red ring: nerve agents

The normal operating procedure is to use the tube with the red ring first, then the tube with the blue ring and blue point, and finally the tube with the yellow and green rings and yellow point. Preparation time is between 1.5 to 3 minutes and each sampling takes between 1.5 to 5 minutes.

There is also a tube filled with silica gel for the sampling of aerosols and another filled with cotton wool for sampling unknown fumes. After the pumping action has been completed these tubes are resealed at both ends with plastic plugs and sent to a mobile chemical laboratory for further analysis. The DHM-11B kit also contains various colour-change detection papers and a small sampling spoon is carried on the outside of the case for taking samples to be placed into jars or sample tubes. A torch is normally carried with the kit. The pump adapter is used to take agent samples from equipment.

Extra indicator tubes are delivered in sealed packs, each pack holding five tubes.

STATUS
In production. In service with the Yugoslav armed forces.

MANUFACTURER
Federal Directorate of Supply and Procurement (SDPR), 9 Nemanjina Street, 11105 Belgrade, Yugoslavia.

DRHT Radiation and Chemical Detection Equipment

DESCRIPTION
The DRHT radiation and chemical detection equipment is intended for use in armoured fighting vehicles such as tanks and is used for the detection of nuclear radiation, for the measuring of nuclear radiation dose rates from the surrounding terrain, for warning the vehicle crew when a pre-selected dose rate is exceeded either from the local terrain or within the vehicle, and detecting nerve agents and other chemical agents in the surrounding atmosphere and warning the vehicle crew. The DRHT may also be used to automatically transmit alarm signals to other locations via a radio link.

The complete equipment weighs less than 30 kg and has a power consumption of less than 7 amps when operating off the vehicle's 24-volt dc power supply. There are five main units. They are the radiation detector, an automatic chemical agent detector, a measuring and control unit, a compressor unit and a cyclotron. There is also a consumables kit and various tools and accessories including the inter-connecting cables.

The DRHT operates over a temperature range of −15°C to +55°C. The radiation dose rate measuring range is from 0.05 cGy/h and the radiation dose from 0.1 to 999.9 cGy. The vehicle internal dose rate sensitivity is such that an indication is given when primary gamma radiation is detected in excess of 5 cGy/h. Agents such as nerve gases can be detected

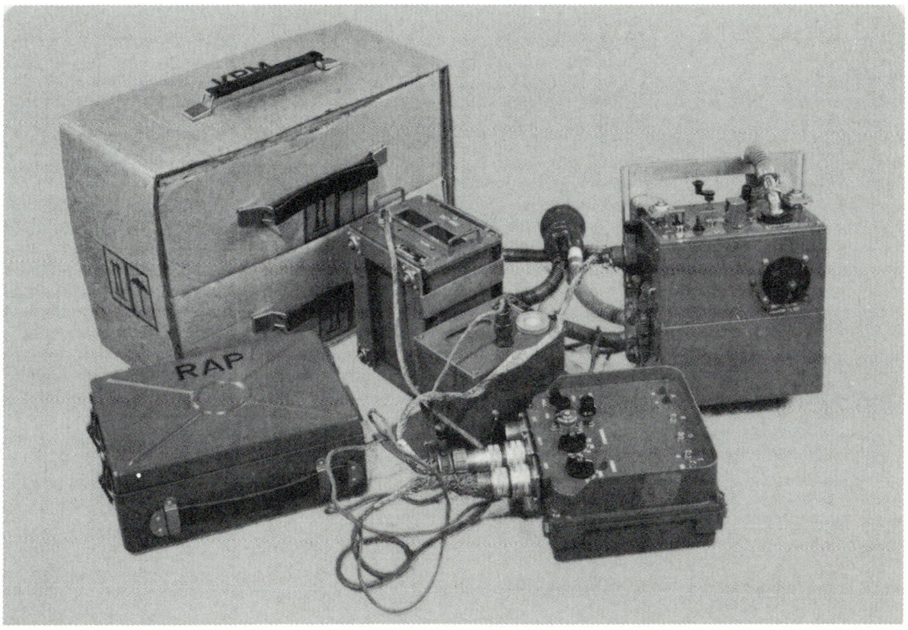

DRHT radiation and chemical detection equipment complete with consumables kit in cardboard case

in concentrations of less than 6×10^6 mg/dm^3. Nuclear dose rate indications can be provided after a period of 8 seconds while chemical agent indications are provided after 90 seconds.

STATUS
In production. In service with the Yugoslav armed forces.

MANUFACTURER
Federal Directorate of Supply and Procurement (SDPR), 9 Nemanjina Street, 11105 Belgrade, Yugoslavia.

PHD Semi-automatic Chemical Agent Detector

DESCRIPTION
The PHD semi-automatic chemical agent detector is a portable unit connected by a cable to a power supply contained in a separate case provided with a shoulder strap. A second case is required to carry consumables, spare parts and accessories. It is intended both for continuous and occasional monitoring for the presence of chemical agents in gaseous and aerosol form. The detection response time for agents such as mustard (H) and phosgene (CG) does not exceed 10 to 20 seconds for battlefield agent concentrations but may be of the order of 3 to 5 minutes for small agent concentrations. Alarms are provided by flashing lights on the detector unit.

The detector is powered by two 6-volt batteries or from a vehicle's 12-volt supply; power consumption is less than 700 mA. Weight of the complete equipment is 5 kg.

STATUS
In production. In service with the Yugoslav armed forces.

MANUFACTURER
Federal Directorate of Supply and Procurement (SDPR), 9 Nemanjina Street, 11105 Belgrade, Yugoslavia.

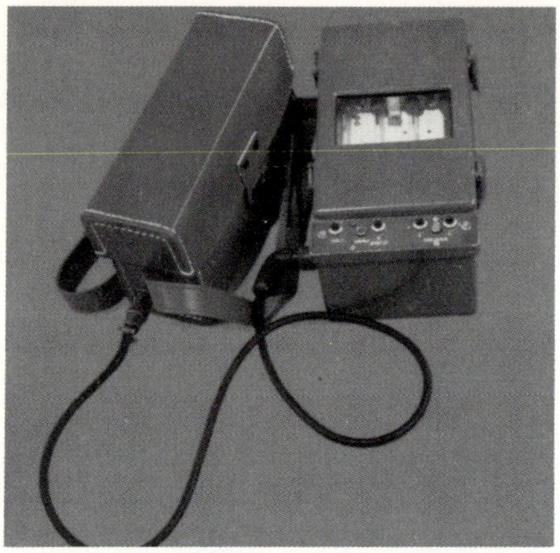

PHD semi-automatic chemical agent detector

Contaminated Area Marking Kit KOZ

DESCRIPTION
The Contaminated Area Marking Kit KOZ can be used for the day or night marking of NBC contaminated areas and also for providing an indication of the direction of travel to avoid such areas.

The complete KOZ kit comprises a set of 20 flags to mark the boundaries of contaminated areas, a set of two signboards for directions, a lighting kit for night use, writing instruments and two bags to carry the kit. The carrying bags can be used to hold contaminated signs and flags ready for decontamination.

STATUS
In production. In service with the Yugoslav armed forces.

MANUFACTURER
Federal Directorate of Supply and Procurement (SDPR), 9 Nemanjina Street, 11105 Belgrade, Yugoslavia.

Contaminated area marking kit KOZ in use

RADIATION SURVEY METERS AND DETECTION EQUIPMENT, DOSIMETERS AND CHARGING UNITS

CANADA

Radiac Set, Remote Monitoring and Alarm, AN/FDR-502(V)

DESCRIPTION
This multi-unit equipment is intended to continuously monitor and detect gamma radiation from mixed energy spectra, and is designed to be installed at fixed installations, although semi-fixed installations such as mobile containers and ships can accommodate the equipment. The basic unit of the equipment consists of a control group mounted in a 482 mm rack which may be connected to up to ten sensors, each of which is connected to a sensor mount and junction box. The junction box may be connected to the control group by up to two interconnecting boxes, so that the individual sensors (which are connected to their sensor mounts by a one metre long length of pipe) may be up to 3048 metres away from the control group.

Each sensor is nuclear hardened and the control group is protected against electromagnetic pulses. An indicator on the control group provides a visual display in digital form of the current dose rate. The control group also incorporates an audio-visual alarm that can be preset to any level between 0.1 and 100 rad/h. The maximum reading possible is 5000 rad/h.

The power supply for the equipment may be either 115 volts, 60-hertz single phase, 220 volts 50- or 60-hertz single phase or 24 volts dc from vehicle or other batteries.

A simulator for training purposes is available.

SPECIFICATIONS

Unit	Control group	Sensor and junction box	Sensor mount	Interconnecting box
Weight	13.6 kg	1.4 kg	8.2 kg	11.4 kg
Width	440 mm (diameter)	92 mm	400 mm	400 mm
Height	131 mm	158 mm	101 mm	101 mm
Depth	437 mm	–	464 mm	464 mm

STATUS
In production. In service with the Canadian Armed Forces.

MANUFACTURER
R-Metrics Limited, 295 Alliance Road, Unit 10, Milton, Ontario, Canada L9T 4W8.

Radiacmeter IM-5016/PD

DESCRIPTION
This is a hand-portable instrument for the measurement of low and medium-range gamma radiation of mixed energy spectra. It is of the Geiger-Müller type powered by two 9-volt alkaline batteries. Radiation readings are given in röntgens per hour on two scales, the lowest from 0 to 100 mr/h and the highest from 0.1 to 10 r/h. The readings are taken from a single meter which is raised in its housing from the die-cast aluminium case. A handle below the meter is so arranged that the single operator's switch can be moved by the user's thumb. This switch has four positions; off, check, high and low. Instructions for use are printed on the body in both French and English and the instrument can be carried on a web carrying strap with a hook at each end.

There is a non-military version of this instrument known as the RD-5016.

SPECIFICATIONS
Weight: (with battery) 1.6 kg
Length: 230 mm
Height: 120 mm
Width: 120 mm

Radiacmeter IM-5016/PD

STATUS
In production. In service with the Canadian Armed Forces.

MANUFACTURER
R-Metrics Limited, 295 Alliance Road, Unit 10, Milton, Ontario, Canada L9T 4W8.

Radiacmeter IM-108S/PD

DESCRIPTION
The Radiacmeter IM-108S/PD is a portable instrument used for the measurement of gamma radiation of mixed energy spectra. It is of the ion chamber type and is powered by a single 9-volt battery with a minimum operating life of 300 hours. Radiation readings are displayed on a single logarithmic scale of 2.5 decades with range 0 to 500 r/h. Calibration in Greys is also available.

The die-cast aluminium case is waterproof and the instrument operates from −40 to +52°C. Operating instructions in both English and French are printed on the identification plate.

Operation is fully automatic with the only operator control being an on-off switch. The instrument is self-testing. When switched on the sensor is disabled while the electronics are tested for 5 to 8 seconds. The operational condition is displayed by a meter pointer in the 'check' band and when the meter needle returns to zero the sensor is enabled and the survey begins. A battery monitor warns of impending battery exhaustion up to several hours in advance and is continuously active while the equipment is on. Exhaustion is indicated by the needle swinging back and forth to prevent a reading.

A non-military version of this instrument is known as the RD-108.

SPECIFICATIONS
Weight: (with batteries) 1.3 kg
Height: 120 mm
Width: 106.9 mm
Length: 171.4 mm

STATUS
In production. Previous model IM-108C/PD is in service with the Canadian and US armed forces.

MANUFACTURER
R-Metrics Limited, 295 Alliance Road, Unit 10, Milton, Ontario, Canada L9T 4W8.

Radiacmeter High Range IM-108S/PD

CHINA, PEOPLE'S REPUBLIC

Radiac Meter M-78

DESCRIPTION
The radiac meter M-78 is a simple portable meter with a radiation detection probe carried on a telescopic rod. The meter can be used to detect either gamma radiation or, with a cover over the end of the sampling probe open, alpha and beta radiation. Readings are given in digital form on a display located on top of the main meter unit. Also on the unit is a single control knob.

The measuring range for alpha and beta radiation is quoted as 5 to 5 × 10^4 Bq/cm, and 1.29 × 10^{-7} to 0.129 C/kg per hour for gamma radiation.

The M78 weighs 2.5 kg and the main meter unit measures 195 × 100 × 160 mm. The telescopic rod can vary in length from 205 to 545 mm.

STATUS
In service with the Chinese armed forces.

DEVELOPMENT AGENCY
Research Institute for Chemical Defence, PO Box 1044, Beijing, China.

MANUFACTURER
Jianan Instrument Factory, PO Box 2509, Chongqing, Sichuan, China.

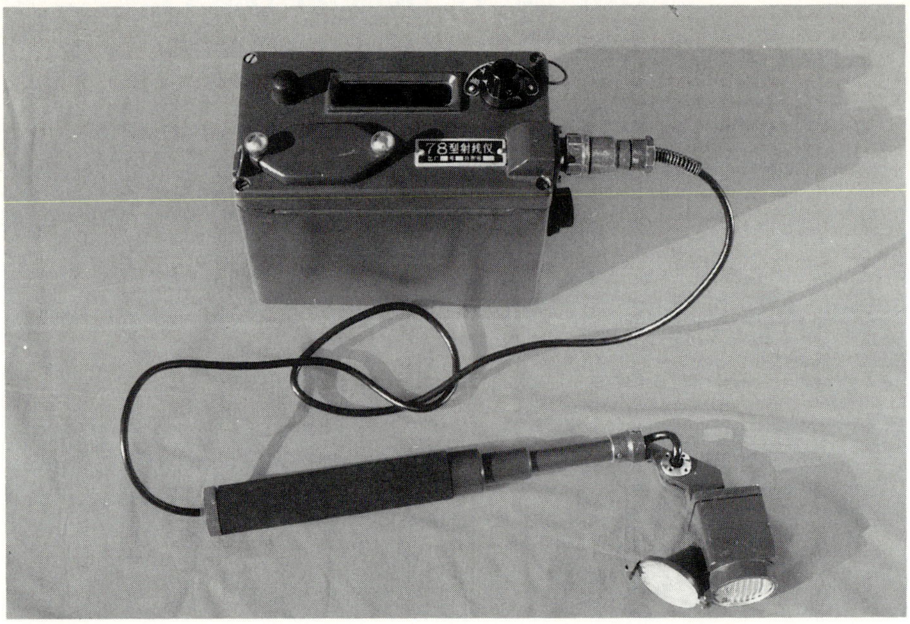

Radiac meter M-78

Gamma Ray Monitor M-84

DESCRIPTION
The gamma ray monitor M-84 appears to be a light and portable monitoring device that indicates the presence of low levels of gamma radiation by a lamp on the monitor flashing when radiation is present. The measuring range of the instrument is given as 2.58 × 10^{-5} to 5.16 × 10^{-2} C/kg. The instrument weighs 0.5 kg and measures 140 × 70 × 40 mm.

STATUS
In service with the Chinese armed forces.

DEVELOPMENT AGENCY
Research Institute for Chemical Defence, PO Box 1044, Beijing, China.

MANUFACTURER
Jianan Instrument Factory, PO Box 2509, Chongqing, Sichuan, China.

Gamma ray monitor M-84

Dosimeter System M-83

DESCRIPTION
The M-83 dosimeter system consists of a shoulder-slung plastic box containing ten individual dosimeters, a dosimeter charger/reader unit and carrying case, and a set of instructions. The measuring range is quoted as 0 to 0.2 C/kg.

Each fountain pen-type individual dosimeter weighs 50 grammes and is 120 mm long with a diameter of 16 mm. The complete boxed system weighs 2 kg and the carrying case measures 225 × 145 × 110 mm.

STATUS
In service with the Chinese armed forces.

DEVELOPMENT AGENCY
Research Institute for Chemical Defence, PO Box 1044, Beijing, China.

Dosimeter system M-83

CZECHOSLOVAKIA

Tactical Dosimeter, Model DK-62

DESCRIPTION
The Model DK-62 dosimeter operates on the thermoluminescence principle using a single-crystal phosphor glass as the energy-storing component. The dosimeter is contained in a round black plastic case 40 mm in diameter and 20 mm thick, and the energy component is 24 × 18 × 11.5 mm and weighs 30 grammes. The case can be worn around the neck or in a uniform pocket with the dosimeter itself held inside the case on a piece of white plastic foam. For reading the case is opened and the dosimeter is placed inside a VDK-62 dosimeter reader, where it is heated. Heating reduces the energy level of the dosimeter to zero and the energy produced by the passage of radiation is released in the form of light which is optically matched to a colour scale to determine the level of radiation. The Model VDK-62 dosimeter reader is powered by a 12-volt power source such as a vehicle battery and measures 200 x 150 x 120 mm.

STATUS
Production probably complete. In service with the Czechoslovakian armed forces.

MANUFACTURER
State factories and laboratories.

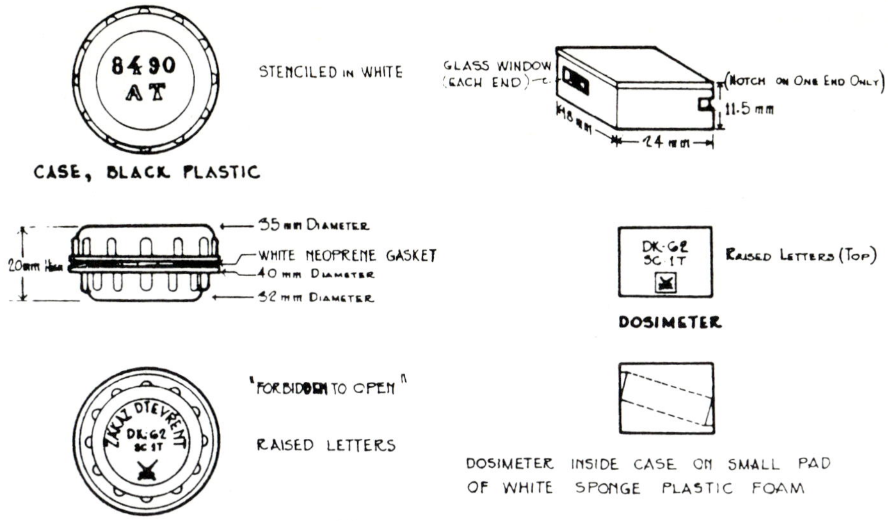

DK-62 tactical dosimeter

FINLAND

Alnor RDS-100 Radiation Survey Meter

DESCRIPTION
The Alnor RDS-100 radiation survey meter is a hand-held instrument designed to meet the requirements of civil defence and para-military personnel. The energy range of the instrument is from 50 keV to 1.25 MeV over a dose range of 0.1 to 999μSv/h and a dose value range of 0.001 to 999.9 mSv. One halogen quenched and energy compensated Geiger-Müller tube provides the detection method; gamma and X-rays can be detected.

A microprocessor inside the RDS-100 enables a number of 'user-friendly' functions to be incorporated. These include checking all segments of the four-digit LCD display when first switching on, a non-volatile RAM memory, an automatic indication of overload (a blinking display), and continuous function status checking, including a defect display, if the detector tube provides too few pulses in a specified period. Other features include a two-stage alarm for low battery condition and a 'two keys' philosophy to reset to zero and avoid accidental reset. A buzzer can be switched on to provide audible information of each pulse provided by the detector tube. A scale illumination feature allows the instrument to be read in low light conditions.

The RDS-100 has a robust splash-proof aluminium case and is powered by three AA cells. The instrument dimensions are 88 × 196 × 42 mm and it weighs 650 grammes with the batteries inserted (570 grammes without).

STATUS
In production.

MANUFACTURER
Alnor Oy, Ruissalontie 11, PO Box 506, SF-20101 Turku, Finland.

MARKETING AGENCY
Wallac (Newbury) Limited, Crown House, Kings Road West, Newbury, Berkshire, United Kingdom.

Alnor RDS-100 radiation survey meter

Alnor RD-10 Universal Survey Meter

DESCRIPTION
The RD-10 survey meter was designed for operation under military and other demanding conditions to the extent that the usual moving coil meter display has been replaced by a line of 40 light-emitting diodes that are easy to read under a wide range of user environments. The RD-10 uses solid state circuitry housed in a rugged polycarbonate spray-tight case and is able to withstand hard knocks, including falls onto a hard floor.

There are two gamma and X-ray measurement ranges; from 0.03 to 300 mR/h and from 0.03 to 300 R/h. The ranges are selected using a two-way switch; the only other controls are an on/off switch and a scale illumination switch. An overload display is provided for readings up to 5000 R/h. The radiation detectors are two halogen quenched, energy compensated Geiger-Müller tubes. Power is produced by two standard 1.5-volt dry cells located in their own sealed compartment.

The RD-10 can be supplied with a carrying strap and the instrument may be connected to a loudspeaker to produce an audible pulse rate signal. A variant, the RD-10 B, can be supplied together with an external detector for the detection of the full range of alpha, beta, gamma and X-ray detection.

The RD-10 measures 210 × 155 × 56 mm and weighs 1.1 kg complete with batteries.

STATUS
In production.

MANUFACTURER
Alnor Oy, Ruissalontie 11, PO Box 506, SF-20101 Turku, Finland.

MARKETING AGENCY
Wallac (Newbury) Limited, Crown House, Kings Road West, Newbury, Berkshire, United Kingdom.

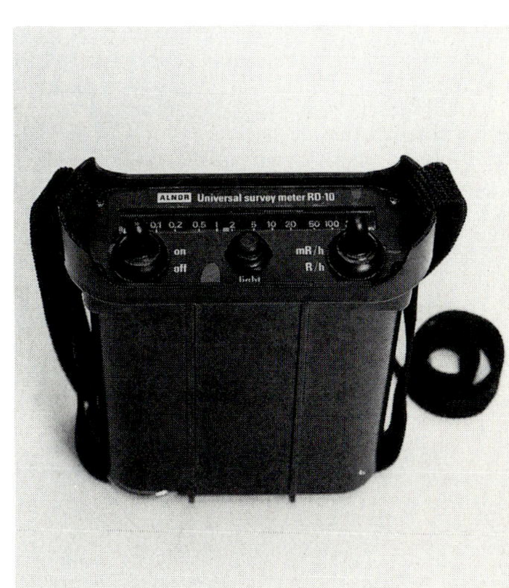

Alnor RD-10 universal survey meter

FRANCE

Radiation Meter Model DOK 803 and DUK 807 Collective Unit

DESCRIPTION
The basic Radiation Meter DOK 803 is an instrument contained in a lightweight rectangular moulded plastic case that can be easily carried and used in one hand. Using a Geiger-Müller tube as a detector it can detect gamma radiation. It has two indicators. One is a small lamp that flashes when radiation present is less than 0.01 r/h. At rates from 0.01 r/h up to 120 r/h the dose rate is indicated in digital form on a display panel. The only control is a single on/off switch and power is produced by two internally-housed 1.5-volt BA30 cells. These cells provide power for the instrument to be used continuously for 20 hours and intermittently for 40 hours.

The DOK 803 may be used by itself and may be carried and used in a stout fabric case with clear panels for observation of the panel and the lamp. It may also be removed from this case and used in what is described as a 'Collective Unit' which converts the DOK 803 from a portable detection unit into a monitoring and alarm unit for static use at various locations. In this arrangement the unit forms part of the DUK 807, which, apart from the DOK 803, has two radiation probes and an electronics unit into which the DOK 803 is plugged. One of the probes is a low level unit used from 10 mr/h to 120 r/h. The other is for use at higher dose rate levels from 10 r/h to 1200 r/h. These probes are plugged via cables into the side of the electronics unit as appropriate. When the DOK 803 is plugged into the electronics unit it then provides the read-out for whatever dose rate the probe in use detects.

The electronics unit may be used to provide an aural warning whenever the dose rate level reaches 10 mr/h when the low level probe is in use or 10 r/h when the high level probe is in use. The alarm may be switched off using a switch on the electronics unit. The electronics unit has its own internal battery supply and when the voltage level drops too low for continued use the lamp on the DOK 803 will flash. The DUK 807 may be powered by a mains voltage.

Each of the probe units is 150 mm high and has a diameter of 65 mm. They may be connected to the electronics unit by cables up to 50 metres long.

SPECIFICATIONS (DOK 803 only)
Weight: 450 g
Length: 160 mm
Width: 75 mm
Height: 41 mm
Power supply: 2 × 1.5 V BA30 cells

STATUS
In production. In use with the French civil defence authorities.

MANUFACTURER
SAPHYMO-PHYSIOTECHNIE, 18 rue de Villeneuve, Silic 551, 94643 Rungis Cedex, France.

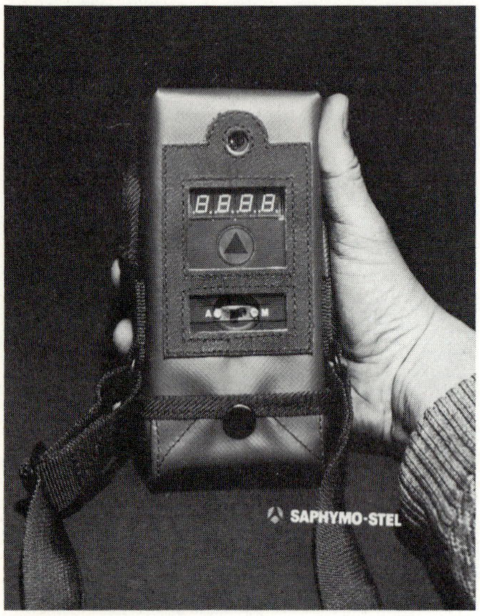

Radiation Meter Model DOK 803

DUK 807 collective unit with DOK 803 inserted and showing probes

Radiation Meter Model DOK 810 (TOTAL 860)

DESCRIPTION
The Radiation Meter Model DOK 810 has the manufacturer's model designation of TOTAL 860 and is a small, light and portable digital survey meter that is completely automatic. The only control is a single press to start, press to stop push button. It detects gamma and X-rays by means of two GM tubes connected to an autoranging microprocessor-based electronic control circuit. The range covered is from 0.1 mrad/h to 100 rad/h with plus or minus 15 per cent accuracy. The value is displayed on a lightened liquid crystal display with four large figures and arrow indicators for reading trends. Using an optional model S 860-100 probe it is possible to detect beta contamination. There is permanent auto-monitoring of the electronic circuitry and of the battery level.

The instrument can be used for up to 40 hours of continuous operation powered by four 1.5-volt alkaline batteries. It is waterproof at a depth of one metre and is proof against electro-magnetic pulses. Normally the instrument is carried slung from the shoulder on a sling. The case is ruggedised.

Weight complete with batteries is 520 grammes and the dimensions are 160 × 30 × 90 mm.

STATUS
In production for the French civil defence.

MANUFACTURER
SAPHYMO-PHYSIOTECHNIE, 18 rue de Villeneuve, Silic 551, 94643 Rungis Cedex, France.

Radiation Meter Model DOK 810 (TOTAL 860) with Beta radiation probe

Radiation Meter Model DOM 410 and Accessories

DESCRIPTION
The basic unit of this equipment is the Model DOM 410 survey meter to which other parts can be fitted to undertake a variety of tasks. The basic Model DOM 410 is a survey meter containing a Geiger-Müller counter tube. The unit is portable and waterproof and may be fitted to vehicles or fixed installations. The meter has six scales: 0 to 10, 100 and 1000 mr/h and 0 to 10, 100 and 1000 r/h.

Power comes from two 1.5-volt type BA30 dry cells in the meter carrying handle but if required an external 110/220-volt ac supply can be connected. The 1.5-volt cells can keep the meter in use for about 16 hours. The meter weighs 2.5 kg and is 230 mm high, 100 mm wide and 190 mm deep. Issued with the meter are a carrying strap, a hermetically sealed carrying case, an extension cable for a probe, a set of instructions and a small radiation test source.

Two other units can be fitted to the basic Model DOM 410 meter, a visual alarm device and a decontamination probe which can be used to detect beta radiation. If required both devices can be fitted to the basic meter at the same time.

For training purposes the Model ROK 410 radiation simulator has been designed consisting of a transmitter, various cables and batteries, and ten receivers. The receivers are all identical to the normal Model DOM 410 meters but their interiors have been altered to pick up the variable signals emitted by the equipment's transmitter. The transmitter has a range of up to 8 km and can be fitted with either directional or omni-directional aerials. Operating on a frequency of between 3250 and 3400 kc/s the transmitter can emit signals that simulate radiation readings on all the receivers in range. The transmitter weighs 33 kg and each receiver weighs 2.6 kg.

STATUS
In production. In service with the French armed forces and several other overseas armed forces.

MANUFACTURER
SAPHYMO-PHYSIOTECHNIE, 18 rue de Villeneuve, Silic 551, 94643 Rungis Cedex, France.

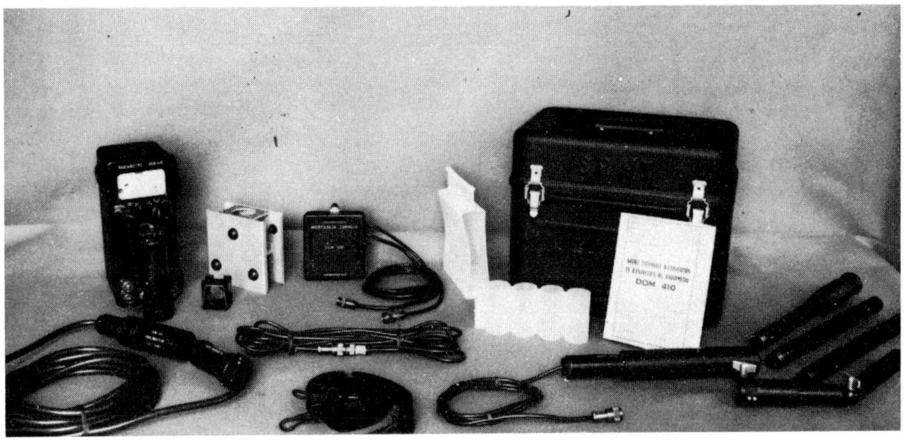

Radiation meter Model DOM 410 and accessories

Radiation Meter Model DOK 420

DESCRIPTION
Designed as a portable radiation meter, the Model DOK 420 can detect and indicate both gamma and X-ray radiation. It is a single unit with a single scale indicating from 0 to 500 r/h. The power is supplied by two 1.5-volt type BA 58 cells which can power the meter for up to 150 hours of use. Accuracy is stated to be ±30 per cent. The unit measures 120 × 75 × 35 mm and weighs 0.35 kg.

STATUS
In production. In service with the French armed forces.

MANUFACTURER
Nardeux SA, ZA de Courteboeuf, 91940 Les Ulis, France.

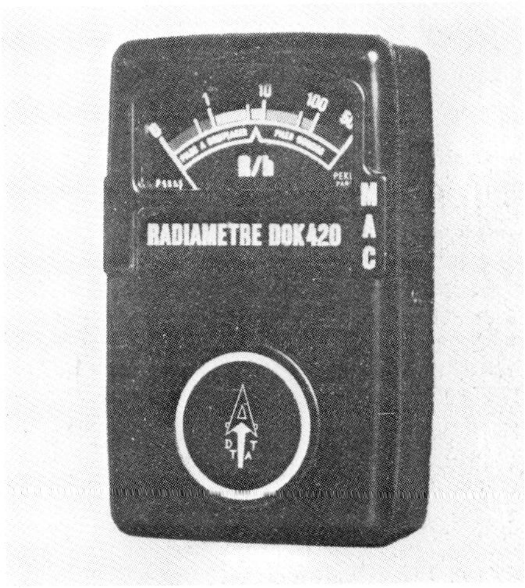

DOK 420 radiation meter

Control and Alarm Radiation Meter Model RA 73

DESCRIPTION
The Model RA 73 is a small radiation meter designed to be carried on the user's belt and capable of emitting an audible alarm once a preset level of radiation exposure has been reached. The alarm threshold is set at the manufacturing stage and may be from 10 to 100 mr/h. Apart from the alarm, the meter can measure from 0 to 1000 r/h on three logarithmic scales. The power comes from a single 4.5-volt battery which can supply the meter for up to 100 hours. The meter measures 165 × 75 × 50 mm and weighs 0.7 kg.

The RA 73 was licence-produced by the Swiss firm of Autophon which produced 25 000 examples for the Swiss armed forces and a similar number for the Swiss civil defence organisations. Autophon, in conjunction with SAPHYMO-PHYSIOTECHNIE produced a radiation simulator for use with the RS 73 known as the SIMA-80. This is described in an entry under *Switzerland* in this section.

STATUS
Production complete. In service with the Swiss armed forces and civil defence organisations.

MANUFACTURERS
SAPHYMO-PHYSIOTECHNIE, 18 rue de Villeneuve, Silic 551, 94643 Rungis Cedex, France.
Licence production by Autophon AG, Military Products Division, Ziegelmattstrasse 1–15, CH-4500 Solothurn, Switzerland.

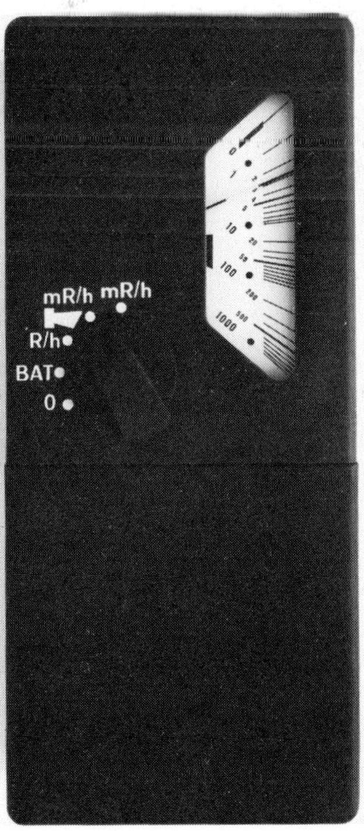

Control and alarm radiation meter Model RA 73

Airborne Radiation Reconnaissance System JUK 450

DESCRIPTION
The JUK 450 airborne radiation reconnaissance system is designed for installation on slow- and low-flying aircraft such as helicopters. It is used to survey areas exposed to radiation and to determine safe routes through such areas. The JUK 450 is used by the armed forces of France, West Germany and Italy.

The system includes an ionisation chamber to be mounted externally on the carrier aircraft, a data processing unit and a control/display unit. The ionisation chamber can measure a dose rate of from 0.02 to 250 r/h. The data processing unit uses microprocessor techniques under the control of the control unit which also has a display panel and controls for a paper printer which uses paper tape to record the required outputs. These outputs are delivered in a series of figure codes as follows: two figures for position indication, three figures or symbols for indication of printing rate or time in minutes, three figures for altitude indication, six figures for indication of dose rate, plus two symbols for overload (one for measurement out of range and the other for hot-spot marking), and one letter to indicate the selected mode. Print-outs can be made automatically every 5, 10 or 20 seconds or under operator control. The total recording time is three hours.

The JUK 450 system can be used to indicate the dose rate at 1 metre above ground level, the dose rate at H + 1 or the dose rate at 1 metre above the ground level and at H + 1.

The total weight of the JUK 450 is approximately 15 kg. The main processor unit is normally contained in a case measuring 380 x 200 × 200 mm. Power is derived from the aircraft's normal 19 to 32 volt dc supply and the operational temperature range is from −30 to +60°C.

STATUS
In production. In service with the French, West German and Italian armies. A few low range detector units are in service with the French civil defence.

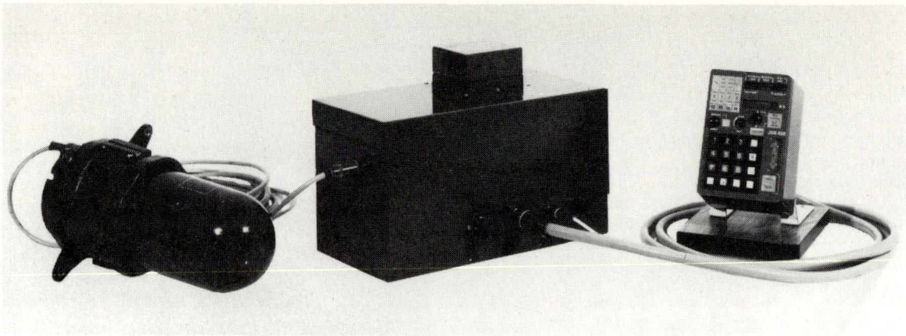

Pre-production version of Airborne Radiation Reconnaissance System JUK 450 with ionisation chamber (left), data processing unit (centre) and control/display unit (right)

MANUFACTURER
BEFIC, 18 rue de Villeneuve, Silic 551, 94643 Rungis Cedex, France.

Airborne Radiation Meter/ Dosimeter Model DUK-DUR 430

DESCRIPTION
Although this equipment is intended primarily for use in aircraft it can also be fitted to land vehicles and maritime craft. It consists of a single unit which can be mounted onto an instrument panel and incorporates a radiation meter capable of indicating up to 500 r/h and a dosimeter which automatically indicates the dose absorbed up to 999.9 r. The unit requires a power source of 27.5 volts ac, and it measures 80 × 80 × 188 mm. The weight of the complete unit is 1.2 kg.

STATUS
In production. In service with the French Army.

MANUFACTURER
SAPHYMO-PHYSIOTECHNIE, 18 rue de Villeneuve, Silic 551, 94643 Rungis Cedex, France.

Airborne radiation meter/dosimeter Model DUK-DUR 430

Vehicle-mounted Radiation Meter/Dosimeter Model DUK-DUR 440

DESCRIPTION
Designed to be mounted in a range of vehicle types, the Model DUK-DUR 440 is made up from two units. One is the main component case that contains the system electronics and mounts the two main indication panels and controls, and the other is the external sensor. The external sensor is a sealed metal case containing a Geiger-Müller counter and is connected to the electronics unit by a cable. The electronics unit is intended to be mounted under the vehicle dash-board or some similar location, with the indicators facing towards the user. One of the meters indicates the dose rate in digital form between 0.1 and 999 r/h. The other digital meter indicates the accumulated dose between 0.1 and 999 r. The response time is less than 3 seconds and the accuracy ±20 per cent up to 999 r/h. Power is provided from the vehicle's 19- to 33-volt dc source. Both units are shockproof and can withstand immersion in water to depths of up to 4 metres. The electronics unit may also be used to provide an aural and visual alarm when a preset threshold is reached.

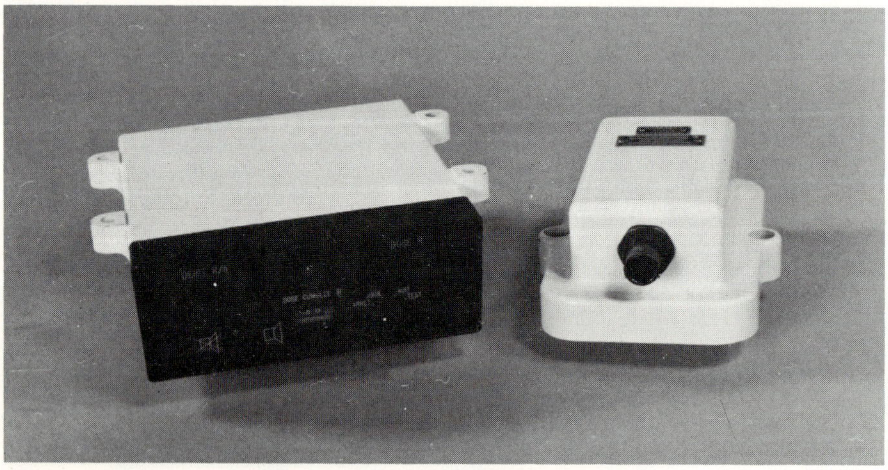

DUK-DUR 440 electronics unit (left), *sensor* (right)

SPECIFICATIONS

Unit	Sensor	Electronics unit
Depth	170 mm	140 mm
Height	60 mm	60 mm
Width	140 mm	70 mm

STATUS
In production. In service with the French Army.

MANUFACTURER
Groupement Industriel des Armements Terrestres (GIAT), 10 place Georges Clémenceau, 92211 Saint-Cloud, France.

Individual Dosimeters and Charging Unit

DESCRIPTION
The normal individual dosimeters used by the French armed forces are the Type JER. There are seven different versions of this dosimeter with ranges of 0 to 0.2, 0 to 0.5, 0 to 5, 0 to 50, 0 to 200, 0 to 300 and 0 to 500 r.

They are all conventional pen-type dosimeters, but they require a charger-reader. The normal type in French service is the Type XOP 402 which is a small unit powered by a single 1.5-volt BA30 battery which is capable of charging several hundred dosimeters. The unit measures 93 × 93 × 53 mm and weighs approximately 0.5 kg ready for use.

STATUS
In production. In service with the French armed forces.

MANUFACTURERS
Dosimeters: La Physiotechnie, 34 avenue Aristide Briand, 94110 Arcueil, France.
Charging unit: Groupement Industriel des Armements Terrestres (GIAT), 10 place Georges Clémenceau, 92211 Saint-Cloud, France.

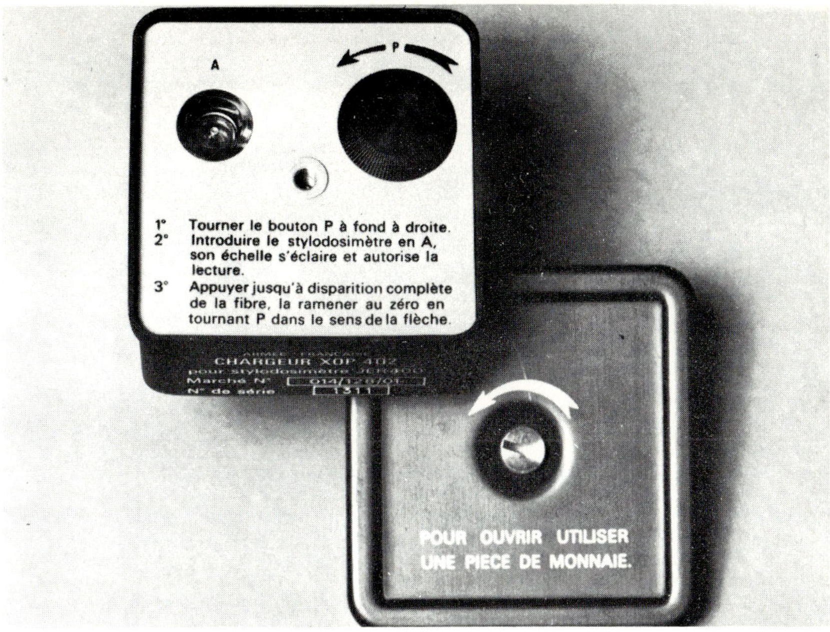

Type XOP 402 charger/reader for dosimeters

Dosimeter Model LIR 305 and Reader

DESCRIPTION
The Model LIR 305 dosimeter relies on the photoluminescent principle. It is normally worn slung around the neck on a beaded chain with a plastic case containing the tube support and the sensitive element in a tube 6 mm long and 3.7 mm in diameter. The complete unit weighs 35 grammes and measures 62 × 29 × 14 mm. The dosimeter can indicate doses of up to 1000 r of gamma radiation.

The dosimeter reader UCPL Model XOR 305 is used for reading the Model LIR 305. The reader is issued in a metal carrying case. The case also contains cables, tools for dismantling the Model LIR 305 dosimeter, spares and charging and decharging units. The reader is powered by either four 1.5-volt batteries or a 110/220-volt supply. It can give an immediate reading from a Model LIR 305 of up to 1000 r. If required the dosimeter can then be recharged or left with the accumulated dose still present. Weight of the reader (with batteries) is 5.6 kg and dimensions are 255 × 160 × 160 mm.

STATUS
Production complete. In service with the French Air Force.

MANUFACTURERS
Dosimeter: Société de Carbonisation Entreprise et Céramique CEC, 8 place des Etats-Unis, 92120 Montrouge, France.
Reader: Groupement Industriel des Armements Terrestres (GIAT), 10 place Georges Clémenceau, 92211 Saint-Cloud, France.

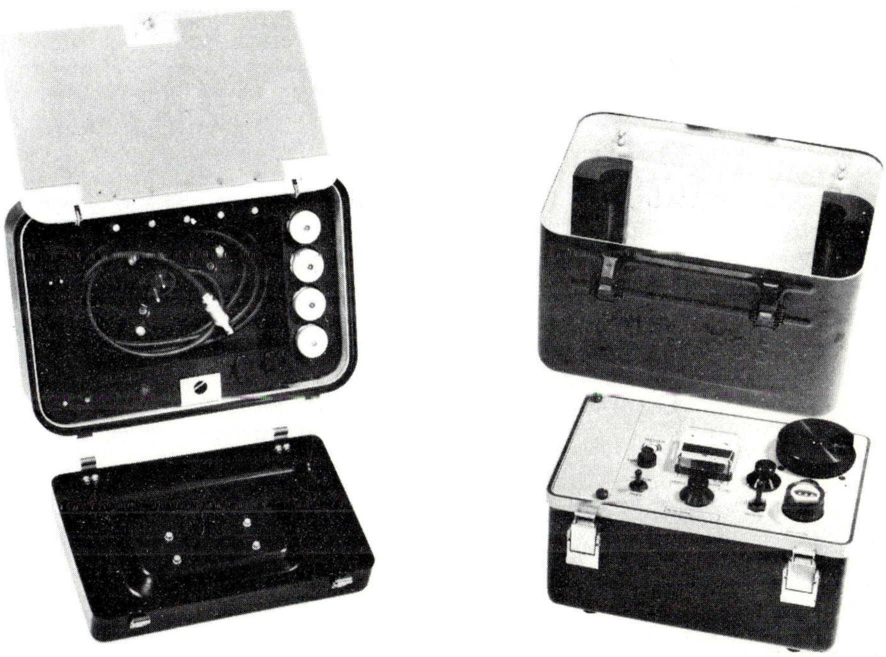

Dosimeter reader Model XOR 305

BEFIC Individual Dosimeters

DESCRIPTION
The BEFIC individual dosimeter is termed a *Stylodosimetre* as it resembles a fountain pen in size and construction, being 122 mm long and 13 mm in diameter. Weight is 45 grammes. The dosimeters measure the irradiation by photons or electrons with energies higher than 60 KeV; a version for levels of 10 KeV is also available. The dosimeters are read by looking through the instrument towards a source of light.

Six versions are available measuring radiation doses of up to 0.1 rad, 0.2 rad, 0.5 rad, 1 rad, 5 rad, 100 rad and 200 rad (other ranges are available on request). The dosimeters are normally issued in multiples in canvas carrying bags that also contain the dosimeter charging unit.

STATUS
In production. In service with the French armed forces and civil defence and some other European nations.

MANUFACTURER
BEFIC, 18 rue de Villeneuve, Silic 551, 94643 Rungis Cedex, France.

GERMANY, EAST

Radiation Meter, Model RAM-60A

DESCRIPTION
The Model RAM-60A is a highly sensitive instrument for monitoring radiation levels deposited on personnel, equipment, clothing and food. It is a fully-transistorised equipment contained in a carrying case and consisting of a meter, a probe unit, headphones and several accessories. The probe contains a Geiger-Müller tube and has a shield to eliminate beta radiation from gamma radiation readings. Beta radiation readings can be made with the shield open. The meter has three ranges, × 1, × 10 and × 100, and thus the gamma radiation scales are from 0.01 to 0.4, 0.1 to 4 and 1 to 40 mr/h. The Model RAM-60A is powered by six rechargeable nickel-cadmium batteries, and the weight of the meter and probe is approximately 4 kg. There is a training version in the form of the Model RAM-60A Üb in which the probe unit is inert and readings are fed into the meter electrically via an instructor's control unit.

STATUS
Production complete. In service with the East German forces but being replaced by the Model RR-66.

MANUFACTURER
VEB Vakutronik, now VEB Robotron-Messelektronik, Dresden, German Democratic Republic.

Radiation Measuring Instrument, Model RAM-63

DESCRIPTION
The Model RAM-63 is a scintillation-type measuring instrument intended to measure both alpha and beta-gamma radiation. It is only partially transistorised and power is provided by five batteries which can be either conventional or rechargeable. The complete equipment includes the basic instrument, two scintillation heads (one for beta-gamma radiation and the other for alpha), a photo-multiplier tube, a tube stand, a carrying case and straps, connecting cables, various tools, check sources for calibration and testing, and an extra set of batteries. The meter unit weighs 3.5 kg and measures 245 × 118 × 176 mm.

This instrument is in use with the Soviet Armed Forces as the WPChR.

STATUS
Production probably complete. In service with the East German and Soviet armed forces.

MANUFACTURERS
State factories in East Germany and USSR.

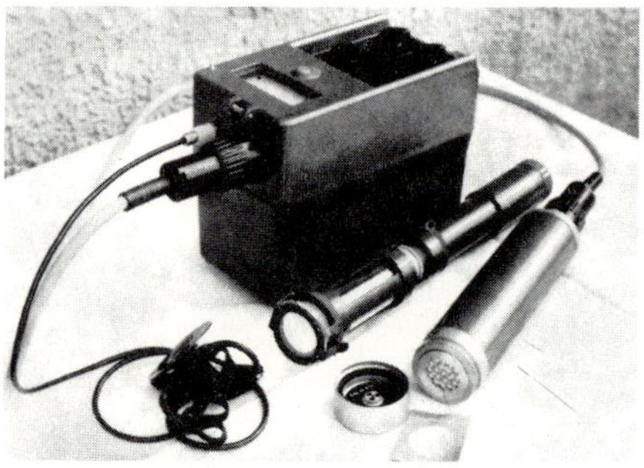

Radiation measuring instrument Model RAM-63

Aerial Radiation Survey Meter, Model RAM-63L

DESCRIPTION
The Model RAM-63L is an adaptation of the Model RAM-63 for use on aircraft or helicopters, but in its airborne form it has two main components. One is an aluminium case containing the control and measuring circuitry and the other component is a highly sensitive photo-multiplier tube mounted on the outside of the aircraft or helicopter. The two units are connected by cable and both units have special shock-absorbing mountings.

STATUS
Production probably complete. In service with the East German armed forces.

MANUFACTURER
State factories.

Radiation Detection Meter, Model RSA-64D

DESCRIPTION
This instrument is used to detect and measure beta and gamma radiation in the field and as such its level of accuracy is reported to be of the order of ±30 per cent. It is hand-held and has a hand generator, but it also has the facility for a jack-socket to give aural indications on a set of headphones. The Model RSA-64D has three sub-scales which can measure from 0.0025 to 50 r/h. The instrument measures 100 × 200 × 140 mm. There is a training version of this instrument, the Model RSA-64 Üb, which is identical to the Model RSA-64D except that the internal circuitry has a sensitivity factor of 1000 so that a source of 50 mr/h will give a scale indication of 50 r/h.

STATUS
Probably no longer in production. In service with the East German armed forces.

MANUFACTURER
State factories.

Radiation Warning System, Models RW-64 and RW-64S

DESCRIPTION
The Model RW-64 and RW-64S are both gamma radiation warning instruments intended for use in radiological storage and handling areas. They have two main components connected by cables, the largest of which is the main instrument which has an external detection probe. The other component is an audible warning alarm, but the main instrument also has a red flashing light that operates in conjunction with the sound alarm. The operating sequence will initiate when the general radiation levels in the vicinity of the detection probe rise above one of five possible preset levels which vary from 0.05 to 2 r/h. The power supply is a 12-volt battery.

The RW-64 is fully contained inside a suitcase-type carrier and has a probe unit on the end of a long co-axial cable. The Model RW-64S is a version used by the East German Navy for use on ships. There is also a training version, the Model RW-64 Üb, which is externally identical to the Model RW-64. However, its circuitry has been altered so that its sensitivity is increased by a factor of 1000 to enable small radiation sources to simulate radiation alarms and initiate drills.

STATUS
Probably no longer in production. In service with the East German armed forces.

MANUFACTURER
State factories.

Radiation Monitoring and Survey Meter, Model RR-66

DESCRIPTION
This equipment combines the functions of both area surveying and personnel monitoring of radiation. For general area radiation survey the Model RR-66 is used as issued, but for radiation monitoring of personnel, water, food, equipment, etc, a three-section probe is used, connected to the instrument by a flexible cable. For general survey the Model RR-66 has a scale range of from 25 mr/h up to 250 r/h. For monitoring the range is from 0.025 to 25 mr/h. Normally the probe head and the three-section probe handle are stowed on the side of the carrying case, and the weight of the entire equipment is approximately 3.9 kg. Headphones are supplied for optional aural indication and the instrument and extras are all contained in a carrying case suspended from the shoulder.

STATUS
In production. In service with the East German armed forces. In time this equipment will replace the earlier Models IH-2, RAM-60A and DP-12 in East German service.

MANUFACTURER
State factories.

Dosimeter Set, Model RDC 2

DESCRIPTION
There are two types of Model RDC 2 dosimeter set in service, one with 50 Model RDC 2 dosimeters and the other with 150. In both cases half the supplied dosimeters are of the lower range type and the other half have a higher range. Both are conventional fountain-pen type dosimeters which differ as follows:

	Range	Cap Colour	Grooves
Low range	0–10 r	yellow	one
High range	0–100 r	red	two

Both types are charged and read by the supplied charger-reader unit which can read dosimeter charges of up to 100 r. The unit, which is powered by a 4.5-volt battery, has an optional scale light and is contained in an aluminium case.

STATUS
No longer in production. In service with the East German armed forces.

MANUFACTURER
VEB Vakutronik, now VEB Robotron-Messelektronik, Dresden, German Democratic Republic.

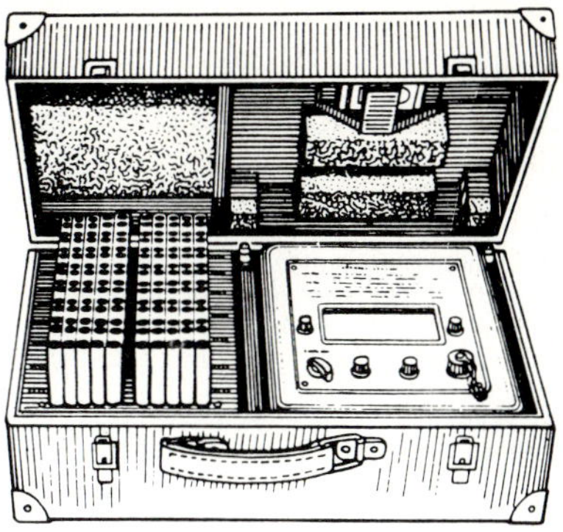

Model RDC 2 dosimeter set

Individual Dosimeter, Model RDA-64

DESCRIPTION
This type of dosimeter requires no external equipment as it relies on the ion chamber principle with a built-in electrostatic generator. The initial charge is produced by turning a knob on the instrument and resetting the dial needle to zero. It is possible for radiation doses of up to 50 r to be recorded, and the dosimeter is normally carried in a uniform pocket. Instructions are given on the back of the instrument. The Model RDA-64 measures approximately 90 × 120 × 40 mm. There is a training version in the form of the Model RDA-64 Üb in which the sensitivity is reduced by a factor of 1000 so that a scale reading of 2.5 r can be produced by a 2.5 mr source. The two models are otherwise identical.

STATUS
Probably no longer in production. In service with the East German armed forces and some civilian organisations.

MANUFACTURER
State factories.

Dosimeter System, Model RDC-64

DESCRIPTION
This system depends on the thermoluminescence principle in which the passage of radiation through calcium fluoride powder contained in a glass tube produces an excited energy state proportional to the amount of radiation passed. The dosimeter system based on this principle consists of two components, the RDC-64D thermoluminescence dosimeter and the RDC-64A dosimeter reader. The dosimeter, which can absorb doses of from 3 to 1000 r, consists of two plastic cases, one orange and the other black. Both are worn around the neck with the wearing cord fitted to the lighter half. The black half is intended for fairly frequent readings and the orange half for long-term measurements. Each half has a metal cap connected to the glass tube containing the calcium fluoride powder. For reading the glass tube is removed from the dosimeter and placed into the Model RDC-64A dosimeter reader. The tube is rotated into position and placed under a photo-electric cell. The glass tube is then heated for 20 seconds which releases the energy in the powder in the form of light and also reduces the powder to its normal energy level for re-use. The energy released in the form of light is measured by the photo-electric cell and the radiation dosage absorbed can then be read off on a scale.

This dosimeter system is expected to replace the earlier Model RDC 2 dosimeter system.

SPECIFICATIONS
Dose reading range: 0.03 to 10 Gy
Rate of readings: max of 100 readings per hour
Power supply: 220 V or 127 V, 50 Hz or 12 or 24 V dc
RDC-64A dosimeter reader
Weight: 24 kg
Length: 330 mm
Width: 465 mm
Height: 290 mm
RDC-64D thermoluminescent dosimeter
Weight: 15 kg
Length: 47 mm
Width: 33 mm
Height: 14 mm
RDC-64Z 24 V adaptor
Weight: 1 kg
Length: 25 mm
Width: 56 mm
Height: 62 mm

Dosimeter System, Model RDC-64

STATUS
In production. In service with the East German armed forces.

MANUFACTURER
State factories and laboratories.

Dosimeter System RDC 111

DESCRIPTION
The Dosimeter System RDC 111 is designed to measure gamma and neutron levels in a contaminated area after the detonation of a nuclear device. The complete system consists of the RDC 111D dosimeter, the RDC 111A dosimeter measuring instrument and the power supply unit RDC 111N. Operating on the thermoluminescence principle the system can measure from 0.03 to 15 Gy (3 to 1500 r) and uses luminous material placed in a container inside a light-proof and water-tight cassette. Each soldier carries two of these dosimeter cassettes carried in dosimeter containers. The containers are carried inside combat clothing between the armpit and waist belt in such a way that the flat side of the container is next to the body. For reading each dosimeter is placed inside the RDC 111A measuring instrument and readings are taken. Each reading takes from 20.4 to 28 seconds and each dosimeter can be used up to 250 times. The factor of error for gamma radiation is 20 per cent and in neutron or combined gamma-neutron radiation up to 25 per cent.

The RDC 111A and its RDC 111N power unit may be carried in a vehicle and powered by the vehicle's 12- or 24-volt power supply. It may also be used in static installations operating off a 220 or 127 volt, 50 Hertz, mains supply. Normally the measuring instrument is placed on top of the power supply unit and connected to it by a cable but the power supply unit may be located remotely.

SPECIFICATIONS
Dosimeter RDC 111D
Weight: 20 g
Length: 56.5 mm
Width: 25 mm
Depth: 17 mm
Dosimeter measuring instrument RDC 111A
Weight: 11 kg
Length: 205 mm
Width: 278 mm
Depth: 214 mm
Measuring instrument power supply RDC 111N
Weight: 5 kg
Length: 185 mm
Width: 245 mm
Depth: 94 mm

STATUS
In service with the East German armed forces.

Dosimeter system RDC 111 showing measuring instrument and power supply with inter-connecting cable

MANUFACTURER
State factories.

GERMANY, WEST

Radiation Detection System MVM 1000

DESCRIPTION

The Radiation Detection System MVM 1000 provides continuous measuring and recording of gamma radiation in a radiation environment. The equipment may be used to provide an alarm and can simulate radiation environments.

The MVM 1000 is used to measure and display gamma radiation levels, to sound acoustic and visual alarms when the radiation level exceeds a set threshold, to record measured values and to simulate radiation environments for training and servicing purposes.

The system consists of a central main unit containing evaluation circuitry, a display, controls, monitoring and alarm components and a simulation system; up to five sensors, each of which can be up to 150 metres away (10 km if a repeater device is used); and a recorder. The complete equipment may be installed in a vehicle or a static location. Normally the recorder is located close to the main unit but a second recorder can be located at a distance with the aid of a special cable and can serve as a remote display.

SPECIFICATIONS

Main Unit
Weight: 35 kg
Width: 330 mm
Height: 430 mm
Depth: 390 mm
Sensor
Weight: 1 kg
Diameter: 110 mm
Height: 160 mm
Recorder
Weight: 10 kg
Width: 220 mm
Height: 230 mm
Depth: 360 mm
Supply voltage: 115 V, 60 Hz
Power consumption: 100 VA

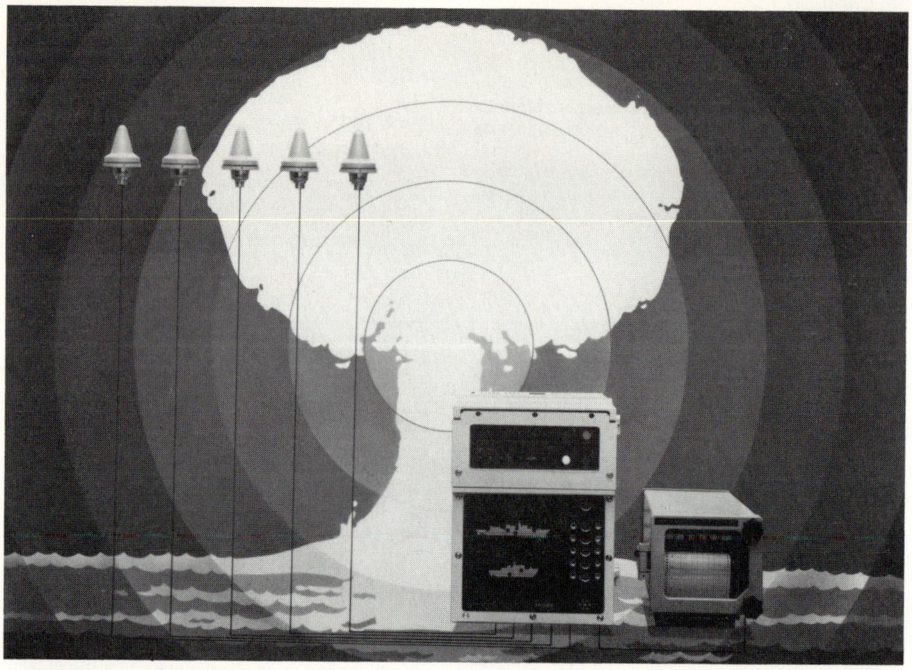

Diagrammatic illustration of Radiation Detection System MVM 1000 with five sensors and recorder

Measurement range: 0.1 mrad/h to 1000 rad/h or 0.0001 to 1000 cGy/h
Simulation: microprocessor controlled, 10 to 90 min in 10 steps
Recorder: two-channel compensation recorder
Sensor: watertight to a pressure of 70 kp/cm^2

STATUS

In production. In service with the West German Navy and other navies.

MANUFACTURER

Philips GmbH, Unternehmensbereich Systeme und Sondertechnik, Postfach 44 87 40, Hans-Bredow-Strasse 20, D-2800 Bremen 44, Federal Republic of Germany.

HUNGARY

Radiation Detection Meter, Model IH-2

DESCRIPTION

The Model IH-2 is used by Hungarian and East German field units for detecting beta radiation and measuring gamma radiation. It is a small lightweight meter carried in a case suspended from the user's neck, and it operates on the conventional ion chamber principle. A beta particle window is situated on the lower side of the instrument. This equipment is used by mobile Hungarian radiological and chemical reconnaissance teams, and some have been issued to the East German forces. Total weight of the equipment is approximately 2.2 kg. It is a very simple instrument with three sub-scales measuring up to 200 r/h.

STATUS

Production complete. In service with the Hungarian armed forces and in partial service with the East German armed forces.

MANUFACTURER

State factories.

Radiation Measuring Instrument, Model IH-12

DESCRIPTION

The Model IH-12 is used for the low-level monitoring of personnel, weapons and other equipment as well as the monitoring of food and water supplies. It can detect and measure alpha, beta and gamma radiation on four sub-ranges, with the actual type of radiation being selected by a source selector switch – the gamma range is from 0.02 to 500 mr/h. The complete equipment consists of the instrument proper, a probe head, an aluminium telescopic rod, a leather carrying case, control samples for calibration and checking, a screwdriver, and an instruction manual. The instrument is powered by a battery giving an operational life of over 40 hours. The probe is connected to the end of the telescopic rod by a polyethylene-covered cable and the probe head is prismatic. The complete equipment weighs approximately 2.65 kg and the case measures 105 × 180 × 210 mm.

STATUS

Production probably complete. In service with the Hungarian armed forces.

MANUFACTURER

State factories.

Radiation Meter IH-5

DESCRIPTION
The IH-5 radiation meter is used to detect gamma and beta radiation in a field environment and may be either man-carried or mounted in a vehicle. It can be used to detect radiation on equipment, clothing, water and soil.

The IH-5 consists of the meter carried inside a carrying case with a window in the top flap to allow viewing of the instrument measurement scales. The meter is powered by an external 1.2-volt 3 Ah battery and is connected by a cable to a detector unit carried on the end of a telescopic probe arm that can be extended to a length of 840 mm. When not in use the telescopic arm is stowed inside the carrying case. Weight of the complete instrument is 3 kg and packed it measures 220 × 110 × 210 mm. Battery life is 24 hours of continuous use.

The IH-5 can measure radiation levels over a range of 5 millröntgens to 200 röntgens per hour using seven measurement scale ranges. The instrument may be calibrated in Greys. Accuracy of measurement is ±30 per cent over a temperature range of from −40 to +50°C.

STATUS
In production.

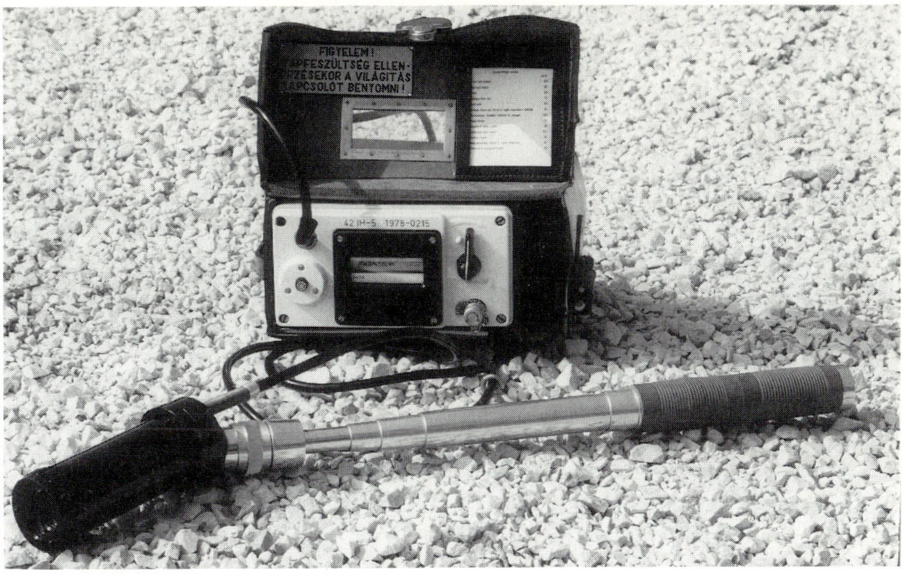

Radiation meter IH-5 showing detector probe arm in retracted position

MANUFACTURER
TECHNIKA Foreign Trade Company, Budapest-Hungary X, Salgótarjáni.20, 1475 Budapest, POB 125, Hungary.

Radiation Hazard Monitor Type SVJ-1 AND SVJ-2

DESCRIPTION
The SVJ-1 and SVJ-2 radiation hazard monitors are intended for use in field or other sheltered areas and consist of an indicator unit and a probe unit. The two units are connected by a co-axial cable that can be either 15 or 200 metres long. Once installed the units continuously monitor the surrounding atmosphere for the presence of radioactivity. Once radiation is detected above a threshold level the units provides visual and audible warnings.

The visual warnings are displayed on a bank of lamps on the front of the indicator unit. Lamp 1 is on all the time the unit is in use to provide a confidence check when switching on and general background indication. Lamp 2 comes on when radiation levels are between 100 and 500 mr/h. Lamp 3 comes on when levels reach 25 r. By the time Lamp 4 comes on the radiation levels will have reached 50 r. Lamp 5 denotes that levels have reached 100 r and the sheltered area involved will have to be prepared for evacuation. By the time Lamp 6 comes on the levels will be above 200 r and the area will have to be evacuated within 24 hours. Lamp 7 denotes levels of 400 r, limiting the evacuation time to 8 hours. By the time Lamp 8 has lit the radiation levels outside the sheltered area will have reached 800 r and evacuation time will be only 30 minutes.

The Type SVJ-2 differs from the Type SVJ-1 in having a digital display on the front of the indicator unit.

The Type SVJ-1 and SVJ-2 can powered by five internal 1.2-volt, 3 Ah storage batteries and can also be powered by a 12-volt vehicle battery. A 220-volt 50 Hertz ac mains adaptor is supplied with each unit.

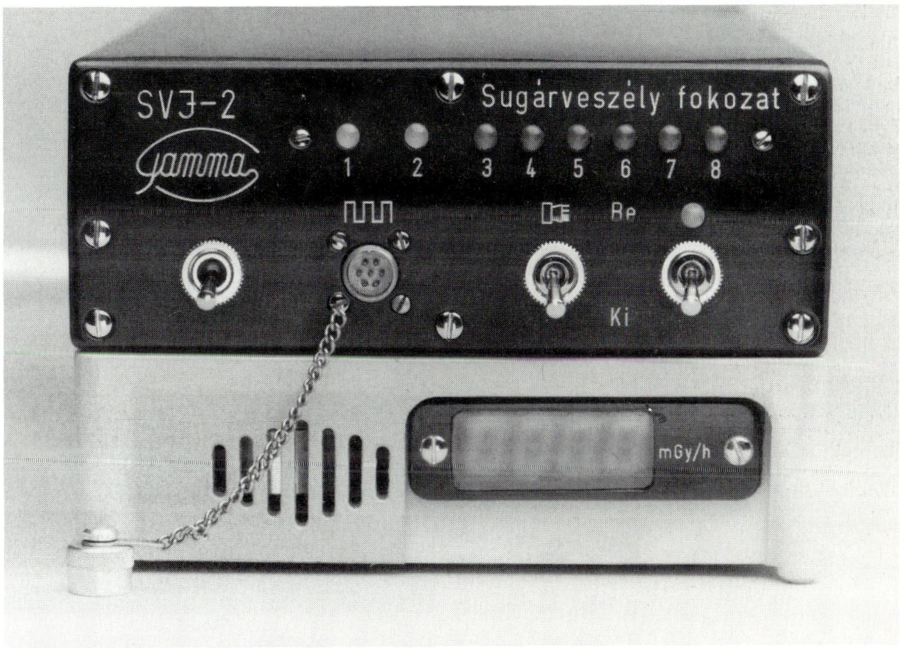

Indicator unit of radiation hazard monitor Type SVJ-2

Both models weigh 5.6 kg complete with mains adaptor and storage batteries.

STATUS
In production.

MANUFACTURER
TECHNIKA Foreign Trade Company, Budapest-Hungary, X., Salgótarjáni.20, 1475 Budapest, POB 125, Hungary.

Radiation Detection Meter Type SZÉM-1

DESCRIPTION
The Type SZÉM-1 radiation detection meter is intended for fixed or portable use to detect gamma radiation contamination of food, soil, material, and personnel. The detector consists of an indicator unit and a detection probe connected to the indicator unit by a cable. The unit can detect gamma radiation in the range from 0 to 2000 mr/h with an accuracy of ± 10 per cent. Readings are provided on a digital meter.

The Type SZÉM-1 is intended for use by one operator and weighs 3.2 kg complete with its 220-volt 50 Hertz mains adaptor unit. When the mains adaptor is not in use the detector can be powered for up to 8 hours by five internal 1.2-volt 3 Ah storage batteries.

STATUS
In production.

MANUFACTURER
TECHNIKA Foreign Trade Company, Budapest-Hungary, X., Salgótarjáni.20, 1475 Budapest, POB 125, Hungary.

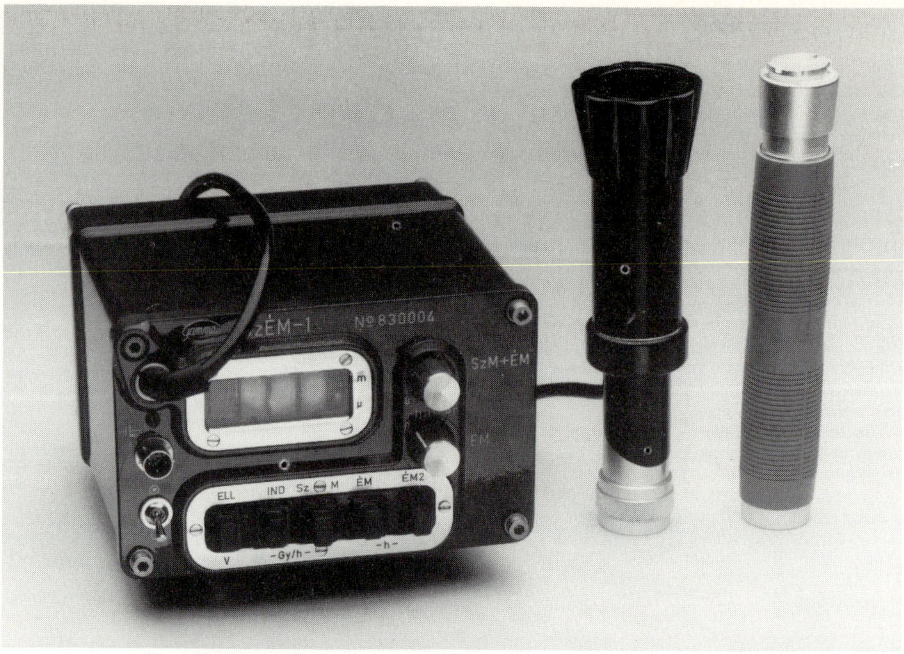

Radiation detector Type SZÉM-1

Radiation Dosimeters Type DKP-50

DESCRIPTION
The designation DKP-50 covers a range of pocket radiation dosimeters of conventional design that are charged before field use so that the charge leaks away when the instrument is exposed to a source of radiation, thereby recording the dose received by the user. The instrument is read by holding the dosimeter towards a light source and viewing the scale through one end of the dosimeter body. The Type DKM-50 range can be used to record gamma doses of up to 0.5 Gy/50 r.

STATUS
In production.

MANUFACTURER
TECHNIKA Foreign Trade Company, Budapest-Hungary X, Salgótarjáni.20, 1475 Budapest, POB 125, Hungary.

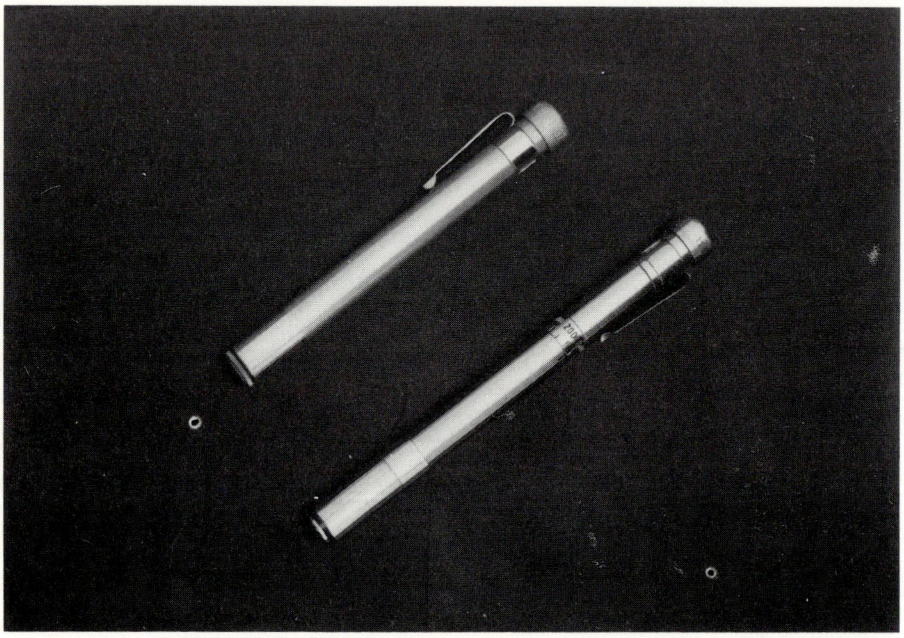

Type DKP-50 radiation dosimeters

ITALY

Tirrena Multi-Function Radiation Meter EC 2000

DESCRIPTION
The EC 2000 is a portable multi-function radiation meter designed for the field monitoring of gamma radiation and beta contamination, and to provide the adsorbed dose as well as integral count values. The EC 2000 is based on the use of an advanced microprocessor and has two compensated Geiger-Müller tubes for an extended gamma operating range (5×10^{-5} to 10^3 cGy/h). There is an external probe for beta (and alpha) contamination monitoring.

The EC 2000 may be used as a portable or vehicle-mounted detector and may also be used as a remote location unit using a serial data output point. The operator's functions are usually limited to switching on the instrument and reading measurements on the large six-digit LED display; the display is refreshed every 1, 2 or 4 seconds, depending on the radiation intensity. The operator may make some adjustments such as inserting an alarm threshold reading or switching to a dose rate meter function. Calibration is required only once a year.

Power for the EC 2000 is supplied by alkaline batteries with an operating life of more than 24 hours; there is provision for providing power from an external source. A multi-compartment carrying bag is provided.

The EC 2000 weighs 1.85 kg (the full production version may well be lighter) and measures 180 × 180 × 160 mm.

STATUS
Ready for production. Under evaluation by the Italian Army.

MANUFACTURER
Tirrena SpA, Via del Quirinale 22, 00187 Rome, Italy.

Tirrena multi-function radiation meter EC 2000

POLAND

Radiation Detection Meter, Model D-08

DESCRIPTION
The Model D-08 is an entirely conventional radiation detection meter utilising the ion chamber principle. It is battery-operated, operates over a range of from 0.1 up to 300 r/h, and weighs about 3 kg.

STATUS
Production probably complete. In limited service with the Polish armed forces and some Polish civil defence organisations.

MANUFACTURER
State factories.

Radiation Alarm Device, Model RIK-59

DESCRIPTION
This small alarm device is about the size of a cigarette packet and is worn in the breast pocket by medical and laboratory personnel working in potential radiation hazard areas. The device detects beta and gamma radiation and once the radiation intensity exceeds 7.5 mr/h the device starts to emit a buzzing tone that increases in pitch and intensity as the level rises. A neon bulb on the side of the device glows when the device is switched on and goes out only when the alarm signal reaches a continuous state.

STATUS
Production probably complete. In limited service with the Polish armed forces but probably still used by various civilian organisations.

MANUFACTURER
State factories.

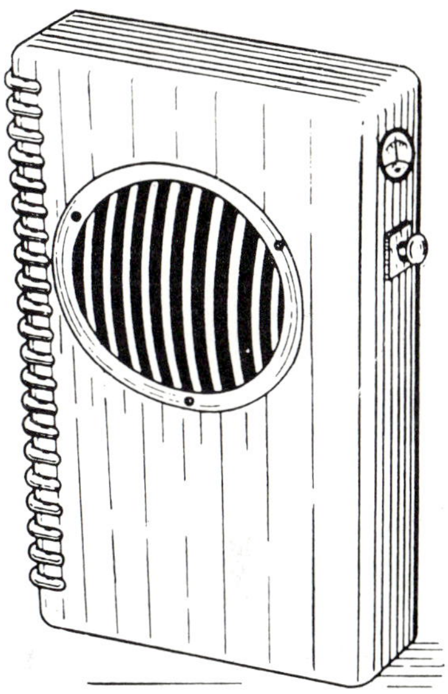

Model RIK-59 radiation alarm device

Radiation Detection Meter, Model DP-66

DESCRIPTION
The Polish Model DP-66 radiation detection meter can be used either as an area survey meter or a low-level radiation detection meter. Normally contained in a fitted case, the equipment consists of the measuring instrument, a probe, a probe extension handle, a small earphone, internal or external power supplies, a carrying strap and instructions. The Model DP-66 can be used to measure gamma radiation at intensity levels of from 0.05 mr/h up to 200 r/h but the general accuracy is only ±25 per cent. The measuring instrument and probe together weigh about 3 kg, and the equipment can operate for some 70 hours on one set of batteries.

STATUS
Production complete. In service with the Polish armed forces.

MANUFACTURER
State factories.

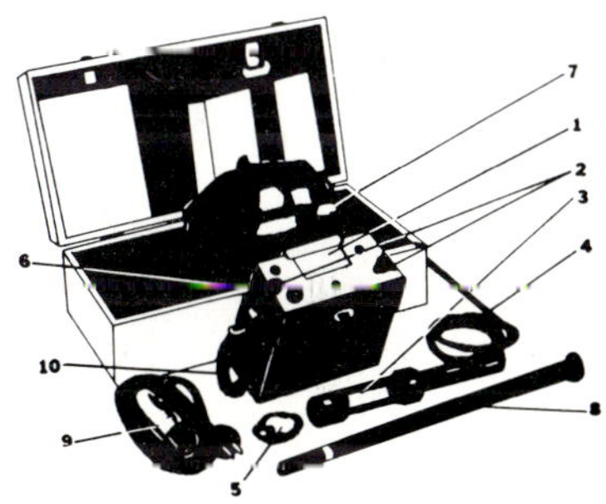

Model DP-66 radiation detection meter **(1)** *measuring scale* **(2)** *power supply* **(3)** *probe* **(4)** *probe cable* **(5)** *miniature headphones* **(6)** *instrument case* **(7)** *extension handle for probe* **(8)** *carrying case* **(9)** *external power supply* **(10)** *carrying strap*

ROMANIA

Radiation Detection Meter Model AD-1

DESCRIPTION
This instrument is entirely Romanian-designed and produced and is used for detecting and measuring gamma radiation. It can also indicate the presence of beta radiation. Using the ion chamber principle it can detect and measure radiation from 0.02 to 600 r/h. The power source is two 1.5-volt cells which, when used at a temperature of about 20°C, provide an operating life of about 100 hours. The instrument weighs 2.35 kg and is 240 mm long, 120 mm wide and 180 mm deep.

STATUS
Probably no longer in production. In service with the Romanian armed forces and possibly with Romanian civil defence organisations.

MANUFACTURER
State factories.

SPAIN

Radiation Detector

DESCRIPTION
This radiation detector is intended for mounting on vehicles, usually under a dashboard with a cable connecting the internal display unit with the external sensor which is contained within a sealed case. Displays on the internal unit show the exposure rate and the accumulated dose and there is an audible alarm combined with a red flashing light that is initiated when the dose rate exceeds 1 r/h.

STATUS
In production.

MANUFACTURER
EISA (Experienias Industriales), Group INISEL, Sor Angela de la Cruz 6-12, Madrid 28020, Spain.

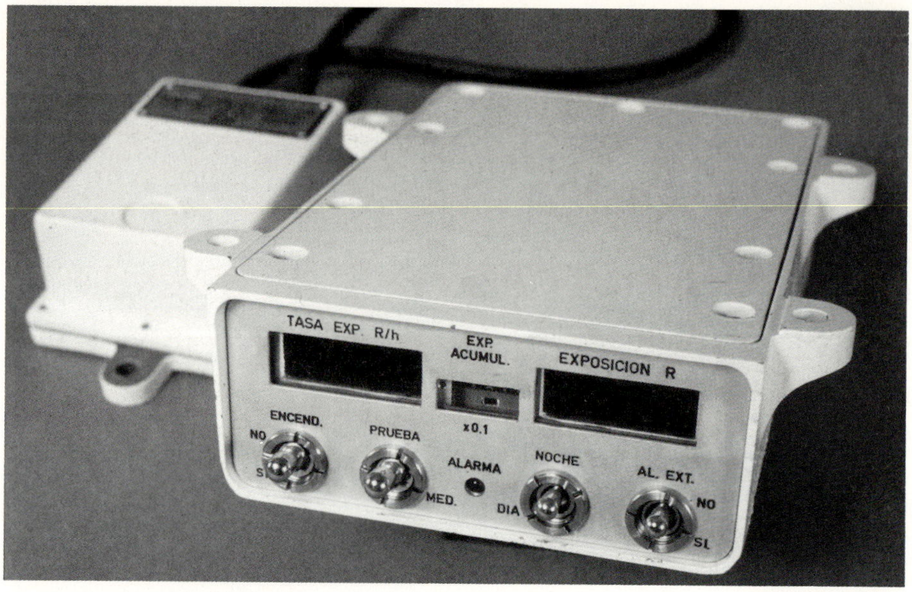

EISA radiation detector for vehicle use

SWEDEN

FFV GAMMA Pod Filtering Device for Air Sampling

DESCRIPTION
The FFV GAMMA Pod filtering device for air sampling in the upper atmosphere was originally developed by the Swedish National Defence Research Institute in co-operation with the Royal Swedish Air Force. The first version was flown during 1957 to collect radioactive debris resulting from tests of nuclear weapons in the atmosphere. Since then the GAMMA Pods have been modified to improve sampling characteristics and an electrically-operated shutter has been introduced to make it possible for the pilot to open and close the units at different altitudes and locations during a sampling flight.

Sweden usually has 18 pods which are flown six at a time by J 32B Lansen fighter aircraft at altitudes up to around 14 000 metres. The pods are loaded with glass fibre filters or, if a more easily dissolvable medium is required, by filters made from an organic material. Depending on altitude and speed up to 10 kg of air per minute is forced through the filter of each open pod. For high sensitivity sampling it is possible to expose all filters simulataneously and, with six pods, collect samples from up to 1000 kg of air. It is also possible to scan a region vertically or horizontally with each individual unit exposed at different altitudes and locations.

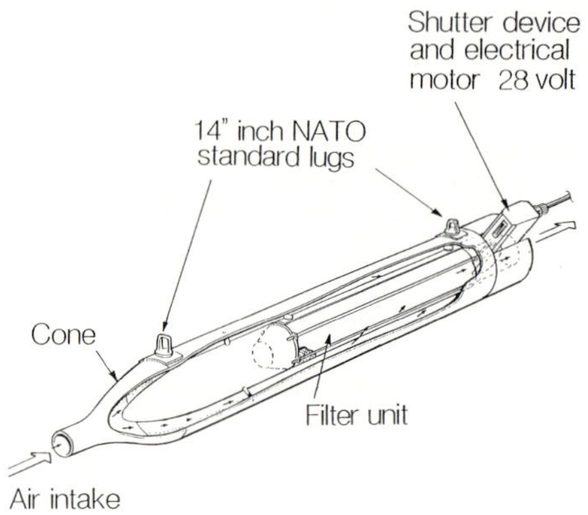

Cross-section drawing of a FFV GAMMA Pod

When a sampling flight has been completed the filters are sent to a central gamma counting laboratory. There the filters are compressed into discs 58 mm across and 10 mm thick ready for placing onto a germanium detector which measures the concentration of all gamma-emitting radionuclides. A day's counting on such a detector means that radionuclides down to less than 0.1 mBg/kg of air can be analysed.

Between 1964 and 1980 the GAMMA Pod system was mainly used to collect samples from radioactive clouds from the then-current series of Chinese atmospheric nuclear tests. The clouds involved usually took between 8 to 10 days to reach Sweden following a test. The system was also used during 1983 when a nuclear-powered radar satellite re-entered the atmosphere and broke up, and again in April 1986 following the Chernobyl reactor incident. Following that incident the system was used during daily flights to scan the south-eastern borders of Sweden at low altitudes. Vertical scans were also made at altitudes up to 12 000 metres.

STATUS
In service with the Royal Swedish Air Force.

MANUFACTURER
FFV Aerotech, S-732 81 Arboga, Sweden.

FFV GAMMA Pods under the wing of a J 32B Lansen

FFV Gamma pod showing filter insert

SWITZERLAND

SIMA-80 Radiation Simulator

DESCRIPTION
The SIMA-80 radiation simulator was developed by the Swiss Autophon concern in conjunction with the Swiss Gruppe für Rüstungsdienste (GRD) to provide a means of training personnel in the use of the RA 73 control and alarm radiation meter, also produced by Autophon under licence from the French firm of SAPHYMO-PHYSIOTECHNIE. The first example was tested during the late 1970s and was made available for service in 1980. Seventy SIMA-80 systems were delivered during 1981 to Swiss civil defence organisations and the Swiss Army ordered 50 for delivery during 1983.

The SIMA-80 radiation simulator uses electromagnetic radiation in place of harmful nuclear radiation sources and is used in conjunction with a special version of the RA 73 meter. The system can be used with any number of the special RA 73 receivers which can be easily identified as they are either yellow or bright blue, not black as is the operational version. The signals from the SIMA-80 transmitter produce simulated readings on the RA 73 receiver scales up to a range of about 2 km using the transmitter's integral aerial; for longer ranges a 6-metre antenna can be raised by one man.

The complete SIMA-80 system consists of a transmitter (the SIMA-80S), a power supply unit (the SIMA-80SG), 25 special RA 73 receiver meters known as the SIMA-80E, a long-range antenna (the SIMA-80FA), a kit of spares, spare programmes and cables. The spare programmes enable training sessions to run for either 1, 2, 4, 8, 12 or 24 hours and ten programmes are supplied with each system. The power supply unit enables the SIMA-80 to be operated on 220-volt ac mains, a 12-volt vehicle battery or internal rechargeable nickel-cadmium cells. If required manual settings may be introduced in place of the preset programmes. The entire system is contained within a sturdy metal casing with the 25 receivers in a similar-sized casing.

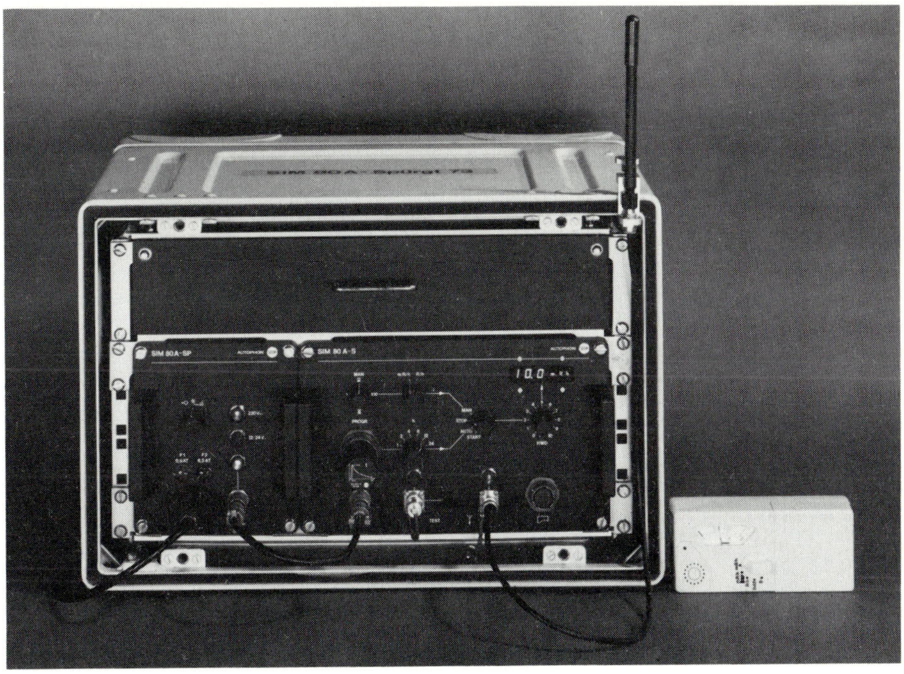

SIMA-80S radiation simulator with SIMA-80E receiver on right

SPECIFICATIONS
SIMA-80S transmitter
Height: 360 mm
Width: 534 mm
Depth: 348 mm
Weight: approx 26 kg
Output: max 1 watt
Range with long-range antenna: 2–5 km
Dose rate adjustable: 1–890 mr/h, 1–890 R/h
SIMA-80E receiver
Dimensions: 165 × 75 × 50 mm
Weight: 0.7 kg
Power supply: 4.5 V cell
Cell life: 24 hours
SIMA-80FA long-range antenna
Height with mast: 6 m
Weight: 8.5 kg

STATUS
Production complete. In service with Swiss civil defence organisations and the Swiss Army (50).

MANUFACTURER
Ascom Radiocom Limited, Military Communications Milcom, Feldstrasse 42, CH – 8036, Zurich 4, Switzerland.

UNION OF SOVIET SOCIALIST REPUBLICS

Radiation Meter, Model DP-5A

DESCRIPTION
The Model DP-5A radiation meter is used to monitor personnel, food and water for low levels of radiation, and has only a very limited application in area monitoring and reconnaissance. The meter is usually worn slung on the chest but the sensing elements are contained in a rod-like probe which can reach into otherwise inaccessible areas. Headphones are supplied for the aural detection of beta radiation. The meter has six separate ranges from 0.05 up to 200 mr/h, and the power is supplied either from three 1.6-volt batteries or any external source providing between 3.5 and 12 volts dc. The meter, which is contained in a fibreglass-reinforced plastic case, weighs approximately 2.1 kg. The whole equipment, including the probe, carrying case, headphones, etc, weighs approximately 7.6 kg.

STATUS
Production complete. In service with the Soviet and Warsaw Pact armed forces.

MANUFACTURER
State factories.

Contamination Survey Meter, Model DP-12

DESCRIPTION
The Model DP-12 is a highly sensitive contamination meter which is primarily intended to monitor the radioactive contamination of personnel, vehicles and equipment, but can also be used in aircraft or helicopters for rapid area surveys. It consists of two main components, the control case and the probe, but a headset for the aural detection of radiation is also provided. The probe unit is connected to the control case by a flexible cable, and consists of a shaft to the end of which is secured a Geiger-Müller tube sensitive to both gamma and beta radiation – the beta radiation can be shielded by a rotating collar over the tube. The shaft itself is telescopic. When in use, the Model DP-12 is carried on the user's chest and the probe unit is held about 25 mm away from the monitored surface, and as the probe is waterproof it may also be used to monitor the contamination of fluids.

STATUS
Production probably complete. In service with the Soviet and Warsaw Pact armed forces.

MANUFACTURER
State factories.

Lightweight Survey Meter, Model DP-62

DESCRIPTION
This lightweight instrument is the smallest and lightest of the Warsaw Pact survey meters and is used mainly for detecting beta radiation in the range from 10 to 500 mr/h. The equipment consists of two components, the meter itself and a hand generator for power. There is no scale as the presence of beta radiation is indicated by a neon bulb flashing behind a lens on top of the instrument. A celluloid window on the bottom of the instrument allows the access of beta radiation and with this window closed by a metal shutter the indication of gamma radiation can be given. The DP-62 lightweight survey meter has been described as 'rugged, lightweight, compact and simple to operate'.

STATUS
Probably still in production. In service with the Soviet and Warsaw Pact armed forces.

MANUFACTURER
State factories.

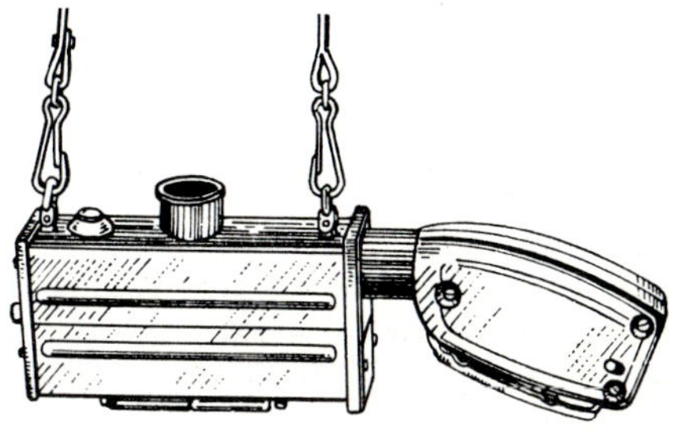

Model DP-62 lightweight survey meter showing hand-operated generator on right

Lightweight Area Survey Meter, Model DP-63A

DESCRIPTION
The Model DP-63A lightweight area survey meter can detect and measure gamma and beta radiation and is widely used by all units of the Warsaw Pact forces. Using the ion chamber principle it is powered by two dry-cell batteries with an operational life of about 50 hours. Normally the instrument is carried in a leather carrying case (weight 0.4 kg) with a shoulder strap and the instrument has two logarithmic ranges selected by push buttons. The bottom scale measures from 0.1 to 1.5 r/h and the other 1.5 to 50 r/h. Beta radiation is measured on the bottom scale only using a beta radiation window. Both scales can be illuminated for night use. The Model DP-63A weighs 0.8 kg and the case measures 165 × 115 × 90 mm.

STATUS
Probably still in production. In service with the Soviet and Warsaw Pact armed forces.

MANUFACTURER
State factories.

Individual Dosimeter Set, Model DP-23

DESCRIPTION
This dosimeter set is contained in a carrying case and consists of a charger-reader unit, 150 Model DS-50 dosimeters packed into three cases each with 50 dosimeters, and 50 Model DKP-50A dosimeters. The case also contains a torch. The Model DP-23 charger-reader unit is used to charge both models of dosimeter, but of the two models, only the Model DKP-50A can be read directly: it is issued to officers and senior NCOs. The Model DS-50 dosimeters can be read only in the charger-reader unit, ie under supervision, and are issued to the rank and file soldiers. Both models read from 0 to 50 r in 2 r steps and both weigh approximately 30 grammes. The Model DS-50 is 120 mm long and 14 mm in diameter and the Model DKP-50A is 130 mm long and 13 mm in diameter. Both can operate in temperatures between −40 and +50°C. The charger-reader unit, which is battery-powered, weighs 3.5 kg and is contained in an aluminium case. The reader function is carried out in increments of 2.5 r on a meter scale.

STATUS
Production probably complete. In service with the Soviet and some other Warsaw Pact armed forces.

MANUFACTURER
State factories.

Individual Dosimeter Sets, Models DP-21A and DP-21B

DESCRIPTION
These individual dosimeter sets consist of a charger-reader unit and a case containing 200 Model DS-50 dosimeters. Each dosimeter can be charged and read only by the charger-reader unit, and each is charged only just before issue. The two models of this set are essentially similar, and are being phased out in favour of the later Model DP-23, but some still remain in service.

STATUS
Production complete. Held in reserve with the Soviet armed forces but may still be in use with some Warsaw Pact forces.

MANUFACTURER
State factories.

Chemical Dosimeter Model DP-70M and Field Colorimeter Model PK-56M

DESCRIPTION
These two equipments are used in conjunction and operate on the colorimeter principle in which any exposure to radiation of a chemical solution will cause it to change colour. This system has the advantage that the dosimeter can be issued without prior charging or any other preparation, with each dosimeter containing a single ampoule of the detecting solution. The field colorimeter Model PK-56M is used to read off the amount of radiation absorbed by the Model DP-70M dosimeter by exposing the ampoule from the dosimeter to a beam of light and comparing it to the colour given by an unexposed ampoule. Colour differences can be compared on a scale and the equipment can be used to determine radiation doses of from 0 to 800 r in 50 r steps. The Model PK-56M is issued with a carrying case and weighs about 1.5 kg.

STATUS
Probably no longer in production. In service with the Soviet and Warsaw Pact armed forces.

MANUFACTURER
State factories.

UNITED KINGDOM

PDRM82 Portable Dose Rate Meters

DESCRIPTION
The PDRM82 range of portable dose rate meters is used to measure gamma radiation and was developed for civil defence and military applications. Over 100 000 instruments have been supplied world-wide and customers included the British Army, Royal Air Force and various other government bodies who use the instrument for post nuclear incident monitoring.

Following the successful introduction of the original PDRM82 a new range of instruments were developed specifically for military applications. The new range includes the portable PDRM82M for mobile survey work, the PDRM82C for use in bunkers or similar environments (this has a remote detector unit connected to the instrument by 11 metres of cable), and the PDRM82D for decontamination operations involving personnel and vehicles and which measures low levels of gamma radiation up to 1 cGy. New instruments are under development.

The PDRM82 is a small rugged instrument with only one control which is large enough to be operated by gloved hands. The simple on/off control ensures that the possibility of operator error during potentially stressful situations is reduced to a minimum. The instrument measures gamma radiation over the range 0.1 to 300 cGy/h. Power is suppled by three standard C batteries which provide a service life of 400 hours. The batteries are contained in a leak proof battery compartment, access to which is gained by removing the control knob. The instrument is normally worn on a neck strap and held in position by a cord around the waist.

Switching the control knob to 'ON' initiates a self-test sequence which covers all aspects of the instrument's operation. Should the self-test sequence reveal a fault the word 'FAIL' will flash on the display and the instrument is discarded.

When the radiation level is rising or falling an arrow is displayed on the liquid crystal display,

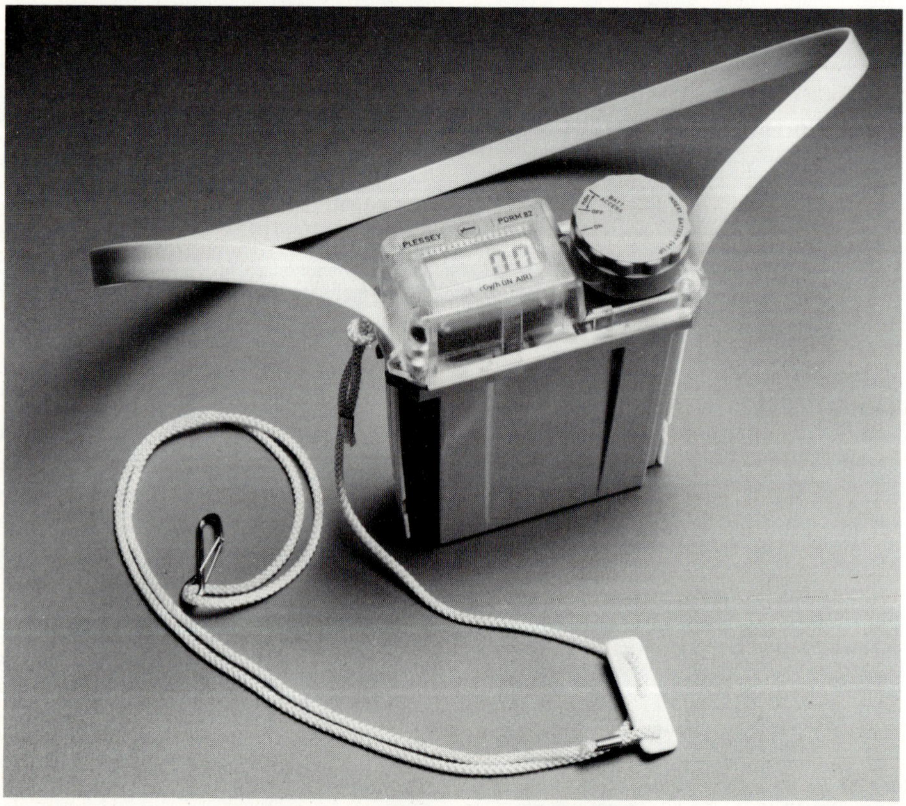

Plessey portable dose rate meter PDRM82

pointing upwards for increasing and down for decreasing. Should the radiation level rise above 300 cGy/h the display will flash '300'. If the battery level is low the word 'BATT' will be displayed. The instrument will then continue to operate for approximately 10 hours.

The PDRM82 requires no maintenance or calibration and has a 20-year storage life. The simple operation procedure is common to all PDRM82 portable or installed instruments, thus keeping training requirements to a minimum.

SPECIFICATIONS
Weight: 560 g
Length: 175 mm
Width: 135 mm
Height: 50 mm
Radiation measurement range: 0.1 to 300 cGy/h
Operating temperature range: –30 to + 45°C

STATUS
In production. In service with the United Kingdom Home Office, British Army, Royal Air Force and a number of overseas government departments.

MANUFACTURER
Plessey Controls Limited, Sopers Lane, Poole, Dorset BH17 7ER, United Kingdom.

Radiac Set MD1 and MD2

DESCRIPTION
The Radiac Sets MD1 and MD2 are portable instruments comprising a control unit, a hand-held probe and a screened interconnection cable. The equipment measures and indicates, on a meter in the probe, gamma radiation over the range 0.1 to 1000 rad/h (MD1) or 0.1 to 1000 mrad/h (MD2). It will also indicate the presence of beta particles provided the level of gamma radiation is not excessive.

The control unit contains self-testing facilities and all power requirements of the equipment are derived from three 1.5-volt primary cells. The equipment is controlled by a four-position switch on the side of the unit; the functions of this switch are:

OFF – batteries disconnected
BATT – battery test
TEST – control unit test
PROBE – set fully operational

A Strontium 90 radioactive test sample is included in the stowage and transit case for testing the complete equipment.

The detector stages comprise two Geiger-Müller tubes, their outputs being paralleled at the input to the head amplifier. When subjected to a radiation field, the tubes generate pulses which are amplified and used to trigger a monostable. The monostable output is shaped and amplitude-stabilised and then fed to a diode pump circuit which, connected to the meter, provides a current proportional to the logarithm of the radiation level. The meter is calibrated in terms of radiation level. When gamma radiation measurements are being made, a sliding screen covers the two windows in the probe face. When the screen is moved to expose the windows, the low energy gamma response is enhanced and the equipment will also detect beta-emitting contaminants.

The probe unit is housed in a sealed casing and the measuring head can swivel through an angle of approximately 270 degrees. A sliding screen is fitted to the probe face. The control unit is housed in a sealed case with a carrying strap at one end.

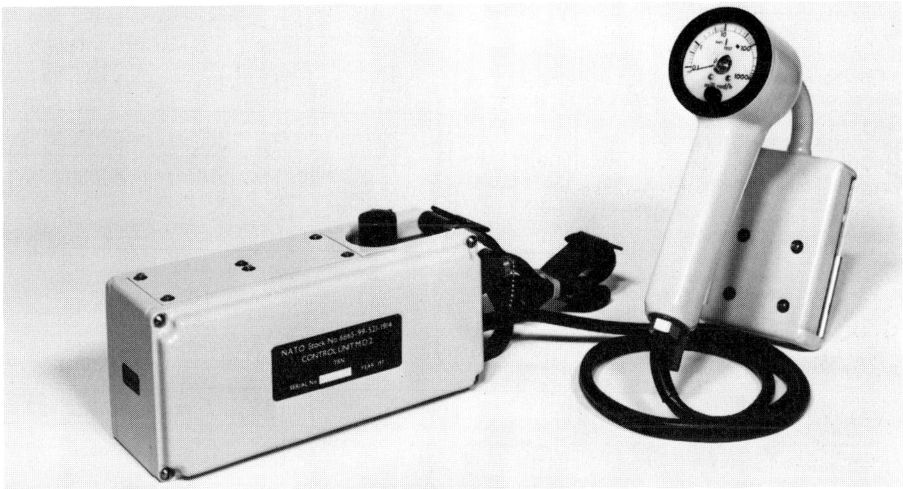
Radiac set MD2

STATUS
Production complete. In service with the Royal Navy.

SPECIFICATIONS
Weight: 2.7 kg
Dimensions:
(control unit) 208 × 89 × 102 mm
(probe) 218 × 77 × 140 mm
(connecting cable length) 920 mm

MANUFACTURER
Jasmin Simtec Limited, Sellers Wood Drive, Bulwell, Nottingham NG6 8UX, United Kingdom.

Radiac Survey Meter No 2

DESCRIPTION
Intended for use by mobile nuclear survey teams, the Radiac Survey Meter is a self-contained device in a grey metal case. It is capable of measuring up to 300 r/h on three scales which can be selected by one of the two operating switches. The lowest scale (white) indicates from 0 to 3, the next (blue) from 0 to 30 and the highest (red) from 0 to 300 r/h. Power for the equipment comes from a single 1.5-volt U2 cell. Normally the equipment is carried in a special pvc-covered cotton haversack designed so that the meter can be read and operated without removing the equipment. For use other than within the haversack, an integral strap handle is provided. Normally the equipment measures only gamma radiation but by removing a panel on the base of the equipment, beta radiation can be detected and measured. Early examples of this meter used thermionic circuitry but most have now been modified to fully-transistorised circuitry. The meter weighs approximately 2.95 kg but complete with the haversack the weight is 4.03 kg. The meter is 232 mm long, 95 mm wide and 149 mm high. The meter interior is sealed and a desiccator is used to keep the contents dry.

STATUS
Production complete. In reserve service with the British Armed Forces and some civilian organisations.

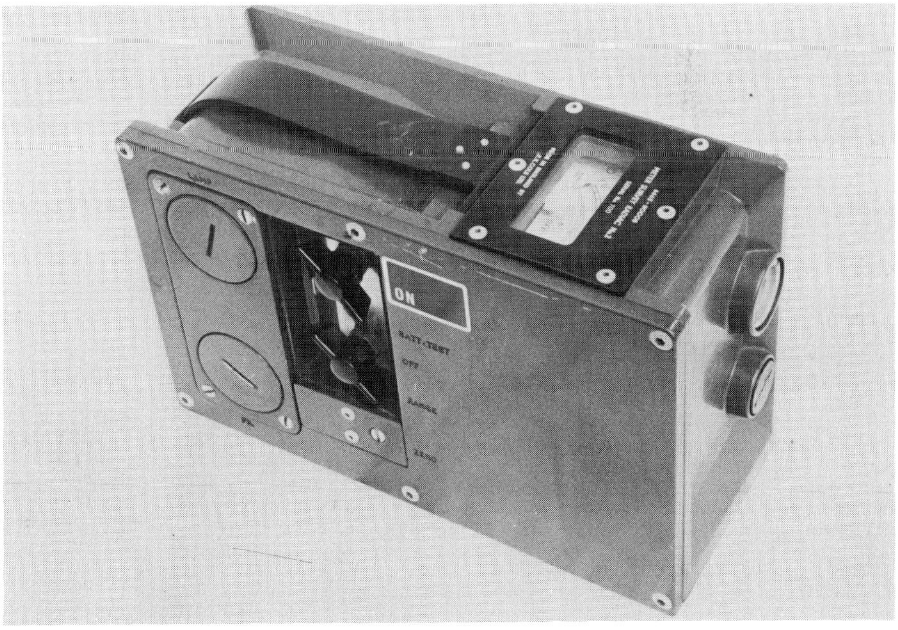

Radiac Survey Meter No 2

Stephen Individual Dosimeter and Charging Unit

DESCRIPTION
Operating on the ionisation chamber/quartz fibre principle, this range of individual dosimeters is manufactured to international standards and can be supplied in a range of types and materials. The standard range, which can be supplied with either brass or aluminium bodies, is 0 to 200 and 0 to 500 mr and 0 to 5, 0 to 50, 0 to 100, 0 to 200 and 0 to 500 r.

Versions outside these ranges can be supplied as needed. All the above models are completely sealed and can be read directly by the user. Operationally they are sometimes carried in small clear plastic bags to prevent nuclear particles alighting on the body and give misleading readings. Normally they are worn like a fountain pen as a pocket clip is fitted to all models. Designed to operate over a wide range of climatic conditions and under all varieties of service use, these dosimeters are stated to have a calibration accuracy of ± 10 per cent, and normal daily leakage is of the order of 0.25 per cent every 24 hours and should never be greater than 2 per cent over the same period. Each dosimeter is 116 mm long and has a diameter of 13.8 mm. Dosimeters with brass cases weigh 40 grammes and those with aluminium weigh 32 grammes.

The above dosimeters can be charged by numerous types of charger unit, but the Stephens charging unit has been ordered by numerous armed forces and other organisations. This unit is powered by a single 1.5-volt U2 cell which provides the charging current and power for the internal light. A spare interior lamp is provided but if this also fails any convenient exterior light can be used by removing the unit base. The charging control device is actuated by a large control dial which can be operated when gloves are worn. In operation, once a dosimeter has been inserted, light pressure can be used for dosimeter reading and a slightly heavier pressure is used for the actual charging operation. The unit measures 120 × 90 × 60 mm and complete with the cell weighs 0.68 kg.

STATUS
In production. In service with the British Armed Forces and many other overseas forces and civil organisations.

MANUFACTURER
R A Stephen and Company Limited, Miles Road, Mitcham, Surrey CR4 3YP, United Kingdom.

Stephen individual dosimeter in use

Plessey Controls Limited Personal Dosimetry System

DESCRIPTION
This personal dosimetry system is designed to record both the instantaneous and residual gamma and neutron radiation dose received by combat personnel. This information allows the significance of any weak link in combat capability to be evaluated, enabling commanders to assess radiation casualty effects, minimise future casualties, predict weapon crew performance, maximise resources and plan correctly.

The system was developed as a result of an international programme between the United Kingdom and the United States, in which Plessey Controls has been the project leader and design authority. The first order for the system was placed in 1982. Since that time over 2800 readers and 300 000 dosimeters have been manufactured for armies around the world.

The system hardware has two main components. One is the detector, the other the reader. The detector is a small dosimeter that resembles a wrist watch and can be worn as one. To detect the dose of gamma radiation the detector depends upon the action of gamma radiation on a silver activated phosphate glass which produces a property known as radiophotoluminescence. As the degree of radiation absorbed by the glass increases so does the degree of photoluminescence and this can be detected under the influence of ultra-violet light. Neutron radiation is detected by a wide base silicon PIN diode inside the detector. Radiation has the effect of creating vacancies in the crystal lattice structure of the diode, increasing the forward voltage of a known current passed through it. This is used to determine the dose received. Thus the detector combines two detection devices in a sealed locket which is worn by the user.

At pre-determined intervals, usually every day, the locket detectors are collected and taken to the reader. Here the lockets are opened individually and placed in a drawer on the front of the reader. This action places the detector in the correct position for reading. In two operations the detector glass is exposed to ultra-violet light and a current is passed through the terminals of the PIN diode. The resultant radiation dose absorbed can then be read once a toggle switch on the reader panel is depressed. Pressing the toggle switch produces the reading on a digital panel meter which stays on for a maximum of 3 seconds. In this way up to 100 detectors an hour can be read. Once the reading is complete the locket is re-sealed and returned to the user.

The reader is powered by a 24-volt vehicle battery or a portable pack. It can be carried in a special field haversack which contains both the reader and a portable battery. The combined dosimeter and reader accuracy at an 80 per cent confidence level is ± 10 cGy ± 10 per cent. The equipment has been designed and tested to meet military and environmental conditions as specified in BDS DEF 133.

SPECIFICATIONS
Reader
Weight: 10.25 kg
Length: 370 mm
Width: 210 mm
Height: 210 mm
Detector
Weight: 75 g
Diameter: 40 mm
Thickness: 12 mm

STATUS
In service with the British Army and US Army and with defence forces in India, Italy and Norway.

MANUFACTURER
Plessey Controls Limited, Environmental Systems, Century Works, Conington Road, Lewisham, London SE13 7LN, United Kingdom.

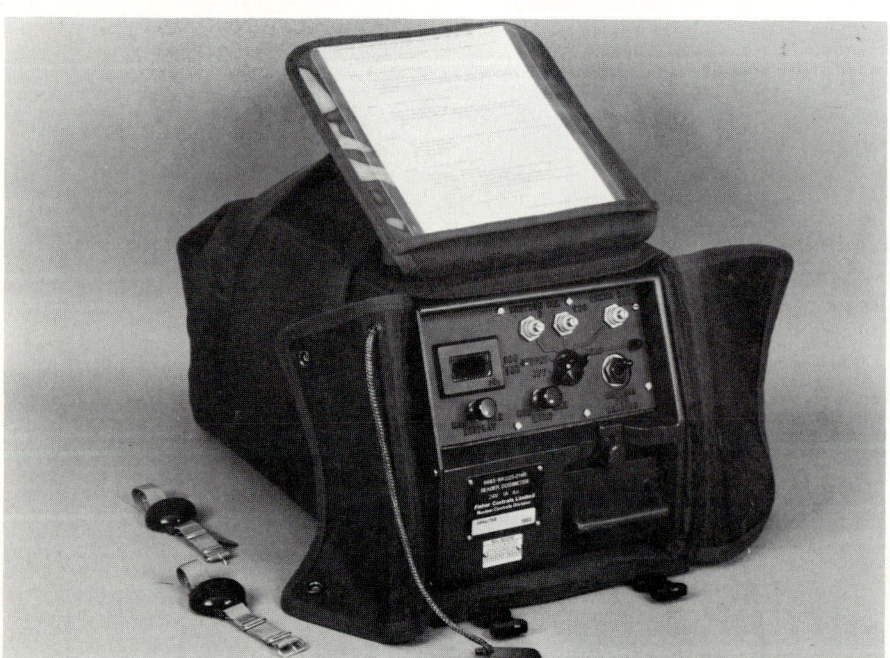

Plessey Controls Limited Personal Dosimetry System with reader in haversack and dosimeters on left

Placing dosimeter into reader

Zengrange ZENBC Hazard Prediction Computer

DESCRIPTION
The ZENBC hazard prediction computer was developed by Zengrange Limited, in consultation with the British Army, to perform as a drill related hazard warning computer. It allows an operator to store the locations and NBC readiness states of more than 20 units within his area of responsibility. Upon receipt of NBC messages (eg, Types 1, 2 or 3) the computer processes the data to determine which, if any, of the units are affected by the NBC hazard. Warnings are updated as necessitated by changes in the weather pattern, new threats or movement of personnel from one location to another. The ZENBC also contains a comprehensive radiation dose-rate and decay calculator.

The ZENBC is housed in the same lightweight, hand-held ruggedised case as the MORZEN and GUNZEN mortar and artillery fire control computers. It has overall dimensions of 200 × 118 × 58 mm and the weight is 0.625 kg.

STATUS
Advanced development.

MANUFACTURER
Zengrange Limited, Greenfield Road, Leeds LS9 8DB, United Kingdom.

The ZENBC hazard prediction computer in use

UNITED STATES OF AMERICA

AN/VDR-2 Radiac Instrument

DESCRIPTION

The AN/VDR-2 radiac instrument is a lightweight autoranging survey meter and dosimeter for radiological surveys and personnel monitoring. It is microprocessor-based and is designed for measuring gamma dose rates from 0.01 mr/h to 10 000 r/h as well as beta dose rates from 0.01 mr/h to 4 r/h.

The instrument functions simultaneously as a dose rate meter and dose meter with independent adjustable alarms that can be set to any level over the entire range of each function. Also provided is measurement of the total integrated dose up to 1000 r. Dose data is independently stored in a non-destructive memory for display on command and may be retained when the unit is switched off. Another feature is a self-test programme which is activated automatically when the unit is switched on. The test sequence includes a calibration check on all ranges, processor and memory test, battery condition test and a detector check.

The AN/VDR-2 comprises a radiacmeter and a probe. The front panel of the radiacmeter, which is contained within a sealed and gasketed aluminium case, contains a liquid crystal display and all operating controls (power on/off; clear/test; rate; dose; light; attenuation; alarm audible/visual). The probe, which is connector coupled to the radiacmeter via a 0.914-metre coil cord, contains two Geiger-Müller tube detectors. A low range tube is housed in the interior of the probe and is active over the gamma range 0.01 millirad/h to 5 rads/h. When the probe window is open the tube mica window is exposed to enable the detection of beta radiation over an equivalent range. The high range Geiger-Müller tube is located in a plug-in housing which rides piggy-back on the probe and covers the range from 5 rads/h to 9999 rads/h. The high range tube does not have a beta detection capability.

When the AN/VDR-2 is used in the area survey mode by dismounted personnel the radiacmeter can be carried in a haversack on a belt or shoulder strap and with pre-set alarms pre-selected for total dose and dose rate levels. The instrument can be used for personnel, food, water and equipment monitoring and was also intended for use as part of the XM87 NBC reconnaissance system, now terminated. Experimental work is being conducted to include the use of the AN/VDR-2 as a communicable monitor used on a remotely-piloted vehicle (RPV). Special carrying racks are available to allow the AN/VDR-2 to be fitted to a number of vehicles including the M577 Carrier, Command Post; M113 series armoured personnel carriers; M151A2 Jeep; M60A1, M60A3 and M1 Abrams tanks; M2 Bradley IFV; and M880 series (4 × 4) trucks.

The AN/VDR-2 is capable of being used as a fixed station monitor with the detection element either remotely located (up to 1524 metres away) or used in the vicinity of the readout. Provision has been made for operation from normal line power and for the mounting of up to 16 equipments to enable widespread coverage of an area or multiple locations. When used in this mode an RS-232 interface unit can be used with the instrument to permit expanded capabilities and operation using computer control, display and recording of radiation information. The RS-232 unit output may be transmitted over long distances using conventional telephone modem units.

AN/VDR-2 radiacmeter and probe unit (foreground) with carrying haversack

Measurement of underwater radiation to depths up to 30.5 metres may be conducted using the AN/VDR-2 and a special underwater probe. The probe is constructed of stainless steel and includes a stainless steel support cable and an easily-decontaminated non-porous signal cable. The AN/VDR-2 may also be used for aerial surveys with ongoing altitude correction data from aircraft altimeters being applied to the AN/VDR-2 and radiation measurements corrected to the 0.914 metre level (ie ground level intensities) will be provided by the instrument read-out.

Other available accessories include a telescopic extender probe up to 3.05 metres long. Probe extensions 30.5 and 61 metres long are available as standard lengths and additional lengths up to 1524 metres are available on request.

SPECIFICATIONS

Weight: less than 1.8 kg
Length:
(instrument) 162 mm
(probe) 159 mm
Width:
(instrument) 102 mm
(probe) 38 mm
Height:
(instrument) 44 mm
(probe) 38 mm
Radiation measured:
(gamma) 0.01 mr/h to 10 000 r/h
(beta) 0.01 mr/h to 4 r/h
Modes of operation:
(ratemeter) 0.01 mr/h to 10 000 r/h
(dosimeter) to 1000 r
Response time:
(above 1 r/h) 2 s
(below 1 r/h) 5 s
Power supply: 3 × 9 V batteries
Battery life: in excess of 100 h
Operating temperature range: −20 to +50°C
Storage temperature range: −40 to +60°C

STATUS

In production. In service with US armed forces.

MARKETING AGENCY

Tradeways Limited, 307-F Maple Avenue West, Vienna, Virginia 22180, USA.

Victoreen Model 450B Survey Meter

DESCRIPTION

The Model 450B survey meter is only one product in a long line of radiation detection and monitoring instruments produced by Victoreen Inc, most of them produced for the medical and nuclear industries. The Model 450B is a small hand-held instrument with the electronics controlled by microprocessor technology and using liquid crystal displays. A liquid crystal analogue/digital display is provided.

The only controls present on the basic instrument are an on/off button and a mode button. The instrument is auto-ranging and auto-zeroing. The top surface of the instrument has an overlay which covers both the display switches to completely seal the surface. Gaskets ensure the instrument is moisture proof and act as shock mounts for the printed circuit boards, liquid crystal displays and the ion chamber assembly. The instrument will remain operational after a drop of nearly one metre onto a concrete floor.

The Model 450B can be supplied calibrated in Röntgens or Sierverts. A FREEZE button is provided to permit the instrument to remember and indicate the highest dose rate from the time the instrument is placed in the freeze mode. A low battery indication is provided when the two 9-volt batteries require changing. Battery life is normally 200 hours of continual use. Warm-up time is less than one minute.

The operating ranges for the Model 450B are: 0 to5 mr/h; 0 to 50 mr/h; 0 to 500 mr/h; 0 to 5 r/h; 0 to 50 r/h; 0 to500 r/h. Accuracy is within 10 per cent of full scale.

A similar instrument, the Model 450P uses a pressurised ion chamber providing an extremely fast response. The Model 450B and 450P can both be fitted with the Model 450-1A communicator that allows the instrument to feed data into a printer or a personal computer.

The Model 450B weighs 0.57 kg, is 200 mm long, 150 mm high and 100 mm wide.

STATUS

In production.

MANUFACTURER

Victoreen Inc., 6000 Cochran Road, Cleveland, Ohio 44139–3395, USA.

Victoreen Model 450B survey meter

AN/ADR-6(V) Aerial Radiac System (ARS)

DESCRIPTION

Developed by the US Army Electronics Research and Development Command (ERADCOM), the AN/ADR-6(V) Aerial Radiac System (ARS) is designed to be carried in any US Army helicopter and allows the rapid survey of large areas of terrain contaminated by radioactive material. The ARS uses sensors external to the helicopter to detect the radiation and then combines the detected radiation signal with an AN/APN-209(V) radar altimeter signal to compute the dose rate on the ground. The result is displayed on the instrument's dose rate meter and recorded on a 'hard' paper copy. The reading may also be transmitted to a ground station or another aircraft. The sensors used are of the compound scintillator type and can be used at altitudes of from 1 to 305 metres. The unit may also be used in fixed-wing aircraft and remotely-piloted vehicles (RPVs). The power required is 28 volts dc with a current requirement of 5 amps at normal operating temperatures and 7 amps below 0°C. The complete installation weighs less than 33 kg. The maximum aircraft dose rate measurement is from 0.04 rads/h up to 400 rads/h. This will indicate a ground dose rate measurement up to 1000 rads/h.

Main components of AN/ADR-6(V) Aerial Radiac System (ARS)

STATUS

All units delivered to US Army.

DEVELOPMENT AGENCY

US Army Electronics Research and Development Command, Combat Surveillance and Target Acquisition Laboratory, Fort Monmouth, New Jersey 07703, USA.

MANUFACTURER

Prime contractor: Autonetics ICBM Systems Division, Autonetics Electronics Systems, Electronics Operations, Rockwell International Corporation, 3370 Miraloma Avenue, Anaheim, California 92803, USA.

Smartbadge Intelligent Personal Radiation Monitor

DESCRIPTION
The Smartbadge intelligent personal radiation monitor has been developed to allow personnel working in a radiation environment to automatically make a number of safety and other calculations relating to radiation exposure and provide a warning when certain pre-determined limits have been exceeded.

Smartbadge is about the size of a hand-held electronic calculator and is carried in a belt pouch. The black plastic case is sealed, shock- and water-proof and is EMP resistant. The keyboard has large keys that incorporate a tactile feed-back for use when wearing NBC gloves. A 32-character liquid crystal display prompts the user to enter basic information on dose restraints. User-friendly software then provides data in response to all likely radiation calculations. For instance, Smartbadge can determine the waiting time necessary before entering a currently too-radioactive area; calculate remaining allowable work time in a radioactive area; predict future exposures, including automatic calculation of radioactive decay; and calculate the residual dose. The instrument continually measures and displays the radiation rate, dose, time remaining until a user-designated dose limit is reached and provides an audible and visual warning of dangerous conditions.

Smartbadge uses a microprocessor-controlled energy-compensated Geiger-Müller tube to measure gamma radiation. The response time is 4 seconds from 0.1 to 1000 r/h and the display updates every second. In an audible 'chirper' mode chirps are produced in proportion to the sensed radiation rate.

Power is provided by a small power pack containing four disposable or rechargeable AA batteries. The display warns when the batteries need changing and there is a built-in self test routine every time the power pack is installed. Dose information, alarm level settings, user identity and other data is stored in a continuous memory.

SPECIFICATIONS
Weight: (with batteries) 0.553 kg
Dimensions: 162 × 102 × 43 mm
Dose rate range: 0.01 to 1000 r/h
Dose range: 1 to 1000 r
Response time: 4 s
Operating temperature range: 0 to +50°C

STATUS
In production.

MANUFACTURER
System Planning Corporation, 1500 Wilson Boulevard, Arlington, Virginia 22209-2425, USA.

Smartbadge intelligent personal radiation monitor

Radiation Alert Monitor 4 Survey Meter

DESCRIPTION
The Radiation Alert Monitor 4 survey meter is a general purpose Geiger-Müller radiation survey meter intended as a handy low cost instrument for use by military and civilian personnel. It can detect alpha, beta, gamma and X-ray radiation.

When using the Radiation Alert Monitor 4 the presence of radiation is displayed by a flashing count light, an audible bleeper that can be switched off for silent operation, and an easy-to-read dual scale analogue meter. The meter has three selectable scales (×1, ×10 and ×100). An internal Geiger-Müller tube eliminates external probes and wires for ease of handling. A single 9-volt alkaline battery can provide up to 2000 hours of operation at normal background radiation levels.

An anti-saturation circuit forces the meter to read full scale in radiation fields as high as 100 times the maximum reading in the highest range. The dual scale meter reads in micro-Sieverts per hour and milli-Röntgens per hour.

The Monitor 4EC is an energy-compensated version of the Monitor 4 that provides a more accurate milliRöntgens per hour reading. The Monitor MC500 detects gamma radiation only and has three selectable ranges up to 5000 μSv/h or 500 mr/h.

An instruction manual is provided with each instrument. Also provided is a padded vinyl carrying case with a wrist strap and belt clip. The instrument measures 145 × 72 × 38 mm and weighs 178 grammes without the battery.

STATUS
In production.

MANUFACTURER
SE International Inc., PO Box 39, 156 Drakes Lane, Summertown, Tennessee 38483, USA.

Radiation Alert Monitor 4 survey meter

Radiation Alert Monitor 5 Survey Meter

DESCRIPTION
The Radiation Alert Monitor 5 is a Geiger-Müller survey meter with a built-in detector capable of detecting alpha, beta, gamma and X-rays. It can be used for the checking of clothing, equipment and working areas and for locating small sources of radiation. It can also be used as a personal monitoring and warning instrument.

The presence of radiation is provided by either a flashing red light, a beeper than can be switched off for silent operation or an analogue meter reading. There are three scales calibrated in counts per minute (cpm) with each range individually calibrated. The three scales are 0 to 500, 0 to 5000 and 0 to 50 000 cpm.

The Monitor 5 is powered by one 9-volt alkaline battery with a life of up to 2000 hours at normal background levels. Each instrument is provided with a padded vinyl bag, a wrist strap and a belt clip. The Monitor 5 measures 145 × 72 × 38 mm and weighs 240.5 grammes with the battery installed.

STATUS
In production.

MANUFACTURER
SE International Inc., PO Box 39, 156 Drakes Lane, Summertown, Tennessee 38483, USA.

Radiation Monitor 5 survey meter

Digilert Nuclear Radiation Monitor

DESCRIPTION
The Digilert nuclear radiation monitor is a personal instrument intended for use by civil and military personnel that can be used to alert the user to high radiation levels, monitor changes in radiation levels and detect radiation leaks. It is based on a Geiger tube radiation detector and can be used to detect and monitor alpha, beta, gamma and X-rays. It has an operating range of from 0 to 19 999 accumulated counts or from 0 to 19 999 counts per minute (cpm).

The Digilert provides a pulsating bleeper alarm once a pre-set level of counts or counts per minute has been reached. The pre-set level is variable in 10-count increments from 10 to 9990. In addition a red LED flashes and the bleeper chirps with each count (the bleeper can be muted). Readings are provided on a clear LCD. The instrument has a dual miniature jack output that can be used to either provide counts to a computer or a data logging device or provide an alert signal to an external warning device.

The Digilert is powered by a 9-volt battery that provides from three to six months of continuous use at normal background levels; it is possible to use an optional ac adapter. The instrument measures 150 × 80 × 30 mm and weighs 270 grammes including the battery.

STATUS
In production.

MANUFACTURER
SE International Inc., PO Box 39, 156 Drakes Lane, Summertown, Tennessee 38483, USA.

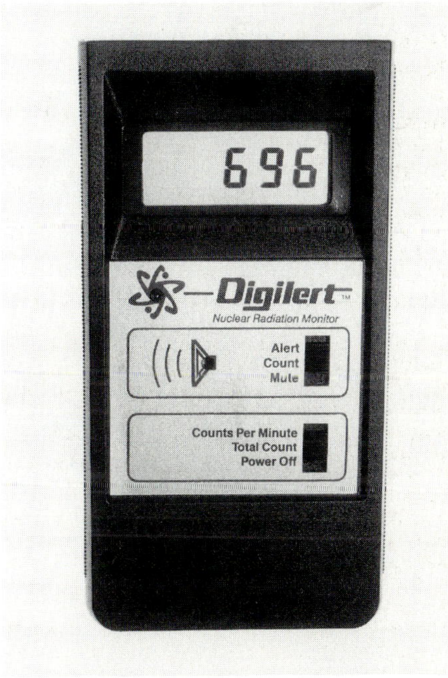

Digilert nuclear radiation monitor

WARSAW PACT

Area Survey Meter, Model DP-1b and Model DP-1c

DESCRIPTION
These two equipments have now largely replaced the earlier Model DP-1a and in their turn are being replaced by later models. Both models can measure gamma radiation and detect beta radiation. The gamma radiation is measured only when the beta radiation window on the bottom of the instruments is closed, and both models utilise the ion chamber principle. The main difference between the two models is that the Model DP-1b has a range of up to 400 r/h and the Model DP-1c up to 500 r/h. Each model has four scales that can be selected by a switch. The East German version of the Model DP-1b is known as the 'Röntgenmeter Type B'. The Model DP-1c can be converted for training by replacing the left-hand compartmented unit by a special unit which multiplies the sensitivity of the bottom two ranges by 3000 and the third by 1500. Range 4 is not used.

STATUS
Production complete. Still in limited service with the Soviet and Warsaw Pact armed forces.

MANUFACTURER
State factories.

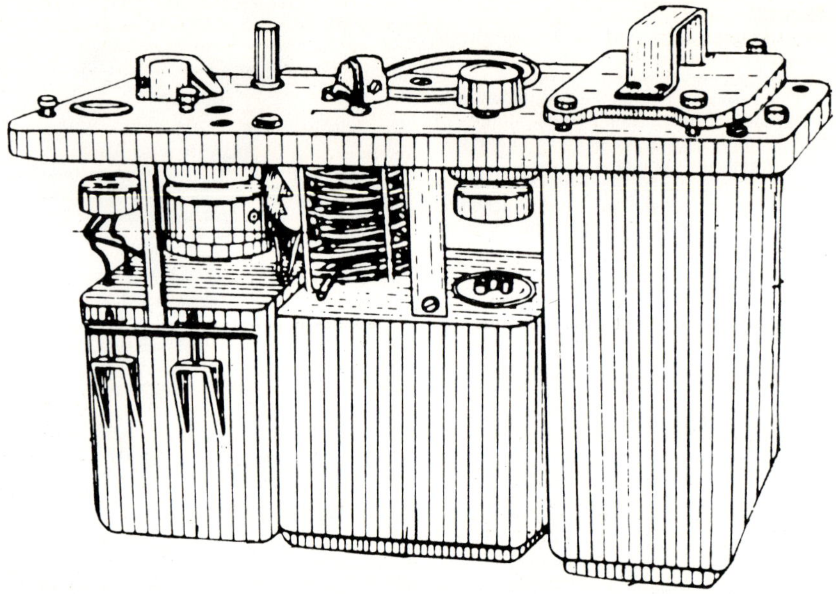

Model DP-1b survey meter without metal casing

Area Survey Meter, Model DP-2

DESCRIPTION
The Model DP-2 area survey meter is used by ground survey teams for detecting gamma radiation and employs the ion chamber principle. It is normally carried in a case slung around the user's neck, and it can measure radiation levels of up to 200 r/h on three scales that are calibrated logarithmically. The scales can be illuminated by an interior lamp if required. Batteries supply the operating power for the instrument and the hermetically-sealed aluminium case measures 240 × 170 × 130 mm. Weight is approximately 3.5 kg.

STATUS
No longer in production. Still in limited service with the Soviet and Warsaw Pact armed forces and civilian defence organisations.

MANUFACTURER
State factories in USSR and East Germany.

Radiation Survey Meter, Model

DP-3
DESCRIPTION
This instrument is standard equipment for employment with most Warsaw Pact radiological and chemical reconnaissance vehicles but can also be fitted to other motor vehicles, ships and aircraft. It is normally bolted to a bulkhead or wall and power usually comes from a vehicle battery or a similar source. The Model DP-3 can measure gamma radiation from 0.1 up to 500 r/h on four scales selected by a single switch. Weight of the Model DP-3 is approximately 4.5 kg.

STATUS
Production probably complete. In service with the Soviet and Warsaw Pact armed forces.

MANUFACTURER
State factories.

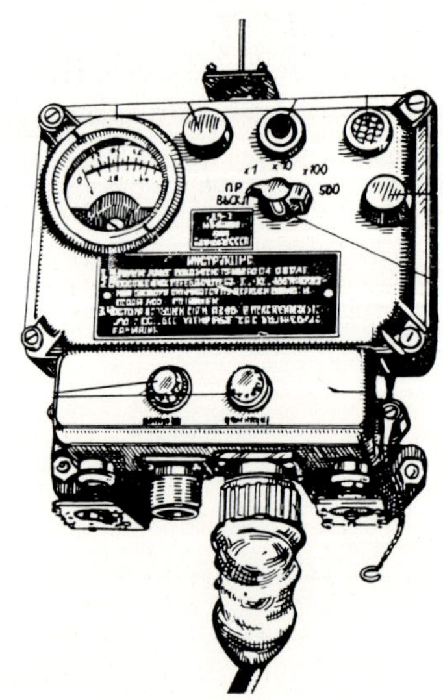

Model DP-3 radiation survey meter

YUGOSLAVIA

UPNE Nuclear Blast Detection and Analysing Equipment

DESCRIPTION
The UPNE nuclear blast detection and analysing equipment is intended to provide a definite confirmation that a nuclear detonation has taken place and provide as much relevant data as possible relating to that blast for subsequent planning purposes.

The equipment consists of two main components plus interconnecting cables. The UPNE sensor consists of a 360-degree azimuth sensor (elevation detection limits are − 15 degrees and + 30 degrees) under a hardened transparent dome supported on a metal tripod. The related computer is based upon a microprocessor and uses data from the sensor unit to display to the operator the real time and magnitude of a nuclear burst, its azimuth and elevation from the sensor unit and also the distance of the centre of the fireball from the sensor. Nuclear blasts with powers from 0.01 kT up to 1 MT can be analysed with a detection range of over 20 000 metres for a 0.1 kT burst. The computer, contained in a hardened case, has an automatic data store for up to ten nuclear bursts and the resolution between successive bursts is less than 10 seconds for bursts of less than 100 kT.

The UPNE takes two men less than 20 minutes to install and may be powered by a 220-volt, 50-Hertz supply or by two 12-volt batteries. The integral real time clock is powered by an independent lithium battery with an operational life of one year. The UPNE has a built-in automatic test feature for all functions.

The UPNE sensor unit weighs approximately 80 kg together with its transport packaging and the computer weighs approximately 25 kg. The operating temperature range is − 25 to + 60°C.

STATUS
In production. In service with the Yugoslav armed forces.

MANUFACTURER
Federal Directorate of Supply and Procurement (SDPR), 9 Nemanjina Street, 11105 Belgrade, Yugoslavia.

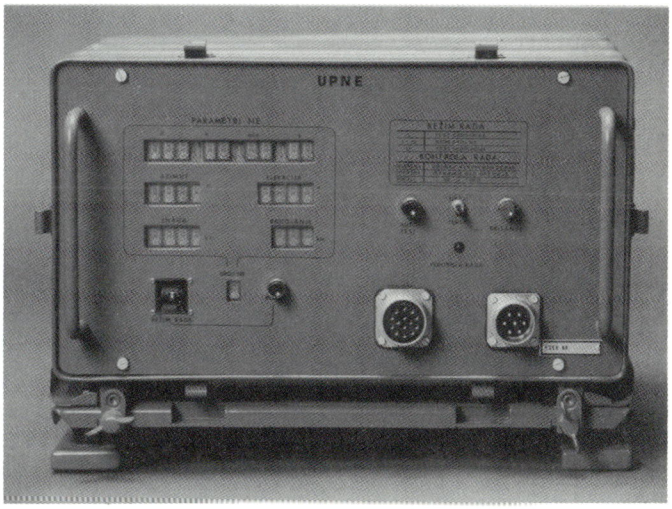

UPNE nuclear burst detection and analysing equipment computer

UPNE nuclear burst detection and analysing equipment sensor unit

MZ-10 Gamma Radiation Alarm Monitor

DESCRIPTION
The MZ-10 gamma radiation alarm monitor is a portable instrument designed to provide visual and audible warnings when the ambient level of gamma radiation rises above a preset level. It is intended for use in a variety of environments from nuclear laboratories and installations to nuclear shelters and may also be used in vehicles.

The MZ-10 weighs 2.5 kg and is contained in a waterproof case. It has its own integral sensor but may be provided with an optional detector probe that can be used up to 100 metres from the instrument. The MZ-10 may be preset within a measurement range of 10 to 9999 pC/kg/s which corresponds to 14 to 140 mr/h – other ranges can be produced on request. The preset alarm level may be set at selected fixed intervals within this measurement range and when it is exceeded the four-digit LED display will start to flash. A small loudspeaker on the instrument will also begin to emit sound pulses, as will the device on the optional probe, if used.

The MZ-10 can be powered by a normal 220-volt, 50-Hertz mains supply or by a 12-volt battery. There is an automatic overload protection device for use when the MZ-10 enters a high level radiation field.

STATUS
In production. In service with Yugoslav armed forces and some civilian authorities.

MANUFACTURER
Federal Directorate of Supply and Procurement (SDPR), 9 Nemanjina Street, 11105 Belgrade, Yugoslavia.

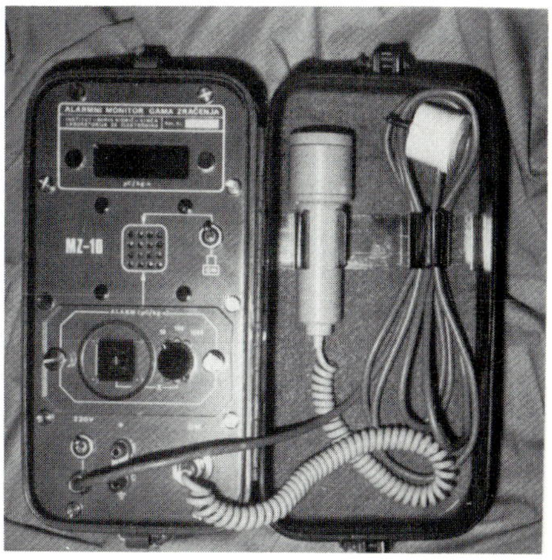

MZ-10 gamma radiation alarm monitor

Radiation Detector and Alarm M81

DESCRIPTION
The Radiation Detector and Alarm M81 is a fully automatic system that can monitor radiation and provide an alarm when either the radiation dose or absorbed dose rate reaches a pre-set level. It is intended for installation at military and other special locations and provides real time detection and monitoring of hazards from gamma radiation. It can also be used to automatically switch on filter and other ventilation systems when necessary.

The system consists of a measuring unit, up to three Type MS2 sensor probes, a connection box Type SK2, inter-connecting cables and maintenance accessories. The three Type MS2 sensor probes may be located up to 200 metres away from the central measuring unit.

The system can measure a radiation dose from 0.1 to 999.9 cGy and an absorbed dose rate of from 0.01 to 990 cGy/h. The alarm threshold level can be pre-set within the absorbed dose rate range so that when a sensor probe detects contamination at that pre-set level, visual and aural alarms are activated and filter and ventilation systems may be switched on. LED displays are used to provide information on radiation levels at each sensor probe.

STATUS
In production. In service with the Yugoslav armed forces.

MANUFACTURER
Federal Directorate of Supply and Procurement (SDPR), 9 Nemanjina Street, 11105 Belgrade, Yugoslavia.

Radiation detector and alarm M81 ready for installation

DRBIV M81 Radiation Detector for Combat Vehicles

DESCRIPTION
The DRBIV M81 radiation detector is intended for use in combat and reconnaissance vehicles. It can be used to measure the gamma radiation dose and dose rate and can transmit data through the vehicle's radio network.

The DRBIV M81 consists of an external sensor unit and an internal display unit together with the associated cables. The sealed sensor unit is mounted in an appropriate position on the exterior of the user vehicle. The wall-mounted internal display unit uses two LED arrays to display the dose and dose rate. The dose range is from 0.1 to 999.9 cGy and the dose rate from 0.01 to 990 cGy/h. A socket on the display unit may be used to connect the unit to the vehicle radio to transmit data to a central monitoring point.

The DRBIV M81 operates off the vehicle dc supply which may be between 10.8 to 30 volts dc. The complete equipment weighs less than 3 kg with the sensor unit measuring 58 × 200 mm; the display unit measures 199 × 194 × 51 mm. The operational temperature range is from −25 to +55°C.

DRBIV M81 radiation detector for combat vehicles

STATUS
In production. In service with the Yugoslav armed forces.

MANUFACTURER
Federal Directorate of Supply and Procurement (SDPR), 9 Nemanjina Street, 11105 Belgrade, Yugoslavia.

Personal Dosimeter DL-M3 Dosimeter Reader M-3 (CDL-M3)

DESCRIPTION
The Personal Dosimeter DL-M3 works on the colorimeter principle, resembles a conventional fountain pen and is worn as such. It consists of a body, a pocket clip and a cap with a washer. Inside the body there is a glass ampoule filled with a chemical sensitive to radiation. Unirradiated ampoules emit a yellow to yellow-orange colour. When exposed to radiation the colour changes to orange, orange-red or red and red-violet depending on the radiation dose absorbed. These colours are compared with a circle of 33 ampoules contained in the Dosimeter Reader M-3 (also known as the CDL-M3).

The reader consists of a body containing a barrel with the ring of 33 ampoules – the visible part of the barrel is the dose indicator. There are telescopic shields for the eyepiece and telescopic sliding shields to protect the illuminating aperture. The dosimeter is inserted into a plug at the far end of the reader body from the eyepiece. In use the exposed dosimeter has its cap removed and it is plugged into the dosimeter reader which is held or placed in a horizontal position. Prior to this insertion the reader is removed from its carrying case and the telescopic shield is pulled out to its maximum extremity. A light filter is then placed so that the maximum amount of light is obtained. The light source can vary from candlelight to full sunlight although full sunlight can be shielded by the insertion of an extra white or yellow sliding filter.

When reading the colour of the dosimeter ampoule is matched with one of the ring of 33 ampoules in the reader. The ring of ampoules in the reader is rotated until a close colour match is found and a value is read off from the knob at the front of the reader used to rotate the ampoule barrel. By the use of repeated readings and comparisons an overall dose rate reading can be obtained. The dosimeter can be used over a range of from 0 to 1500 rads while the reader can be used over a range of from 0 to 1200 rads. Maximum accuracy is obtained within the range from 200 to 1000 rads.

Training versions of these equipments are available. The training dosimeter is known as the DL-M4 and the training reader is the M-4 (or CDL-M4). The training dosimeter is used over a range of from 0 to 15 rads and the training reader over the range of 0

Reader M3 in use with Personal Dosimeter DL-M3 inserted at end of body

to 12.2 rads. Both are otherwise identical to the operational versions.

Dosimeters are delivered in batches of 300 packed in cardboard boxes. Readers are delivered packed individually, also in cardboard boxes. The reader is normally carried in a polythene bag inside a strong plastic case complete with cover and a shoulder strap. Some spare parts, colour comparison charts and five spare dosimeter caps are also carried in the case.

SPECIFICATIONS

Item	**Dosimeter**	**Reader**
Designation	DL-M3	M-3 (CDL-M3)
Range	0–1500 rads	0–1220 rads
Weight	20 g	1.4 kg with case
Dimensions	13 × 135 mm	80 × 90 × 160 mm
Shelf life	10 years	10 years

STATUS

In production. In service with the Yugoslav armed forces.

MANUFACTURER

Federal Directorate of Supply and Procurement (SDPR), 9 Nemanjina Street, 11105 Belgrade, Yugoslavia.

NBC RECONNAISSANCE VEHICLES

FRANCE

Renault VAB Reco NBC Reconnaissance Vehicle

DESCRIPTION
The prototype of the Renault VAB Reco NBC reconnaissance vehicle was first shown during January 1988 and is a 4 × 4 variant of the Renault VAB armoured personnel carrier (for details of the basic VAB refer to *Jane's Armour and Artillery 1988-89*, pages 279 to 282).

The VAB Reco was developed for the French Army by the French Ministry of Defence with the conversion work being carried out at Roanne Arsenal. The vehicle has a crew of two and is equipped with nuclear fallout sensors and a series of devices for the detection, collection and analysis of chemical agent and fallout samples. The crew will also transmit warnings of NBC contamination and will probably carry marker devices to indicate contaminated areas. The vehicle is equipped with a comprehensive suite of land navigation systems.

Two prototypes of the VAB Reco were evaluated by the French Army during 1988. Production is expected to commence during 1992.

STATUS
Prototypes.

MANUFACTURER
(Vehicle) Renault Véhicules Industriels and Creusot-Loire Industrie under the leadership of Renault VI. Marketing by Société des Matérials Spéciaux Renault VI-Creusot-Loire Industrie, 316 Bureaux de la Colline, 92213 Saint-Cloud, Cedex, France.

GERMANY, WEST

Transportpanzer 1 'Fuchs' NBC Reconnaissance Vehicle

DESCRIPTION
This wheeled NBC reconnaissance vehicle is built upon the components and hull of the Transportpanzer 1 armoured personnel carrier and for details of the basic 6 × 6 vehicle refer to *Jane's Armour and Artillery 1988-89*, pages 296 to 299.

The NBC reconnaissance version of this vehicle is known to the West German Army as the ABC-Abwehrtruppe (ABC-AbwTrp) and 140 have been ordered; the first NBC variants entered service in February 1988.

Following the US Army's withdrawal of funding from the XM87 NBC reconnaissance system in early 1988 it was announced on March 16 1988 that the US Army was to adopt the Transportpanzer 1 'Fuchs' NBC reconnaissance vehicle as the sole NBC reconnaissance vehicle for the Army. Initially the US Army acquired and tested five vehicles, renamed the Fox, using Congressionally-appropriated NATO comparative test funds. The US Army originally had plans to acquire 280 vehicles at a cost per vehicle of $1.17 million, with total programme costs estimated at $790 million by increasing the quantity to 576 vehicles.

The original plans were that first vehicles to be fielded, beginning in December 1990, will be interim examples using the NBC reconnaissance suite fitted to West German Army vehicles. A product improvement programme was planned to modify all US Army Fox vehicles to a standard that will incorporate stand-off chemical agent detection, automatic sampling, automatic data processing and jam-proof data transmission systems.

In July 1988 the US Congress directed the US Army to cancel its planned acquisition of the Fox and initiate a competition with not less than two different prototypes. Thyssen Henschel have teamed with General Dynamics Land Division for this competition – see entry under *United States* for further details.

For nuclear radiation detection the Transportpanzer 1 Fuchs carries a set of ASG 1 nuclear tracing equipment with a print-out facility that can

The double wheel sampling unit at the rear of the Transportpanzer 1 NBC reconnaissance vehicle with one wheel raised to the MM-1 mass spectrometer external probe

Transportpanzer 1 Fuchs NBC reconnaissance vehicle

Transportpanzer 1 Fuchs NBC reconnaissance vehicle

provide a hard copy of detection location co-ordinates when coupled to the vehicle's vehicle orientation system. For the analysis of soil and other samples the Fuchs carries a MM-1 mobile mass spectrometer (see entry in *Germany, West* section under *Chemical and biological detection and identification equipment*[1]). This is associated with a double wheel sampling unit at the rear of the vehicle. Two sampling wheels with silicone tyres can be run along the ground surface and lifted up to an external probe connected to the MM-1 mobile mass spectrometer for analysis. This probe is extendable for use as a ground probe. Also at the rear of the vehicle is a work hatch through which a sampling device can be manually operated to take soil and other samples for later analysis.

Other NBC equipment carried includes a chemical agent detection and monitoring unit based on the American M43A1, and NBC marking equipment to mark contaminated areas. The markers are ejected through a work hatch at the rear of the vehicle. The vehicle has an NBC air filtering unit capable of providing up to 180 cubic metres an hour.

SPECIFICATIONS
Configuration: 6 × 6
Weight:
(empty) 15 300 kg
(loaded) 17 000 kg
Max load: 1700 kg
Length: 7.3 m
Width: 2.98 m
Height: 2.43 m
Ground clearance: 0.406 m
Track:
(front) 2.54 m
(rear) 2.56 m
Wheelbase: 1.75 m + 2.05 m
Max speed: 105 km/h
Range: approx 800 km
Fuel capacity: 390 litres
Max gradient: 60%
Vertical obstacle: 0.6 m
Trench: 1.08 m
Engine: Mercedes-Benz Model OM 402A V-8 liquid-cooled diesel developing 320 hp at 2500 rpm
Transmission: ZF model 6 HP 500 6-speed automatic with torque converter
Steering: hydraulic, recirculating ball
Turning radius: 8.5 m
Electrical system: 24 V
Batteries: 4 × 12V, 125 Ah

STATUS
140 NBC vehicles on order for the West German Army. The US Army has five NBC vehicles for evaluation. The West German Army has more than 1000 examples of the basic Fuchs vehicle and orders have been received from Venezuela and the Netherlands.

MANUFACTURER (vehicle)
Thyssen Industrie AG Henschel, Postfach 10 29 69, D-3500 Kassel, Federal Republic of Germany.

TM 170 NBC Reconnaissance Vehicle

TM 170 NBC reconnaissance vehicle

DESCRIPTION
The TM 170 NBC reconnaissance vehicle was produced as a private venture by Thyssen Maschinenbau GmbH in co-operation with DECO-DELTA GmbH of Stuttgart. It is based on the TM 170 armoured personnel carrier which was developed from the chassis of a Daimler-Benz Unimog U 1700 L (4 × 4) cross country vehicle. For full details and specifications of the TM 170 range of vehicles refer to *Jane's Armour and Artillery 1988-89* pages 300 and 301.

The TM 170 NBC reconnaissance vehicle is designed to operate in an NBC-contaminated environment and measure, monitor and mark affected areas. All operations, including gathering soil samples, can be performed from within the vehicle.

The armoured superstructure is all-welded and provides protection against armour-piercing 7.62 mm ammunition and shell splinters. The same level of protection is provided by the armoured windows. The ventilation system is equipped to filter out chemical and bacteriological agents and special non-absorbent paint is used to facilitate decontamination. Other safety measures include the provision of fire warning and extinguishing systems, additional protection for the fuel supply systems and an emergency air supply to the engine. The fuel tank is an integral part of the structure. For ease of maintenance the superstructure can be detached from the chassis within minutes by removing five connecting elements. Run-flat tyres are fitted as standard.

An FNA 5 navigation system is fitted. This indicates the vehicle position, both in analogue and digitial terms, in UTM co-ordinates and shows the direction of the vehicle's longitudinal axis. The system can be used with map scales of 1:20,000 and 1:250,000.

The vehicle crew consists of the commander, seated on the right with the driver on the left. One further crew member, a tracer, is accommodated in the main compartment together with all the specialist NBC equipment. If required the vehicle can be modified to operate with a crew of four. In action the crew would wear NBC protective clothing and would be able to operate outside the vehicle if necessary.

Specialist NBC equipment carried includes an ASG 1 automatic nuclear tracing device measuring gamma radiation in the 20 mrad/h to 1000 rad/h range and photon energy from 83 keV to 2 MeV. For biolgical warfare agent detection a sampling device is fitted at the rear of the vehicle to take ground samples; these are despatched to a laboratory for analysis. A MM-1 mobile mass spectrometer (for details refer to entry under *Germany, Federal Republic* in *Chemical and biological detection and identification equipment* section) is carried for identifying organic substances in the atmosphere or found in samples collected by a double-wheel sampling unit under the rear of the vehicle. Also provided is a M43A1 chemical nerve agent detector which can be removed from the vehicle for dismounted use and to enable internal and external readings to be taken and compared as a check on the protective ventilation system.

The vehicle is also fitted with various meteorological sensors for temperature, relative humidity, wind speed and direction and barometric pressure. Data provided by the sensors can be recorded on a printer.

Other equipment carried includes contaminated area markers and various instruments and personal radio equipment for crew members operating outside the vehicle.

The TM 170 NBC reconnaissance vehicle has a combat weight of 11,650 kg and a power-to-weight ratio of 14.4 hp/tonne.

STATUS
Ready for production.

MANUFACTURER (vehicle)
Thyssen Maschinenbau GmbH, Stockumerstrasse 28, D-5810 Witten-Annen, Federal Republic of Germany.

TM 170 NBC reconnaissance vehicle with meteorology mast raised

UNITED STATES OF AMERICA

Nuclear, Biological and Chemical Reconnaissance System

DEVELOPMENT/DESCRIPTION

The US Army has a requirement for an NBC reconnaissance system (NBCRS) which will locate, mark and report in a timely manner any contaminated or uncontaminated terrain to facilitate troop, supply and equipment movements. The XM87 NBC reconnaissance system was under development to meet this requirement with initial development involving the M113A2 armoured personnel carrier. Future development would have involved a vehicle from the M998 HMMWV series and the M2/M3 Bradley IFV. Six prototypes of the M113A2 version were scheduled to be delivered by July 1989 with system development and operating trials commencing during the third quarter of 1988. It was expected that the first XM87 tracked vehicles would have been fielded in Europe in June 1992. The TRW Defense Systems Group of Redondo Beach, California, were the prime contractor.

In 1988 the US Army proposed dropping the development of the XM87 programme in favour of purchasing the off-the-shelf Transportpanzer 1 Fuchs NBC reconnaissance vehicle from the West German company Thyssen Henschel (see entry in this section for details). The US Army acquired and tested five vehicles, re-named the Fox, using Congressionally-appropriated NATO comparative test funds. The US Army originally had plans to acquire 280 vehicles at a cost per vehicle of $1.17 million, with total programme costs estimated at $790 million by increasing the quantity to 576 vehicles.

The original plans were that first vehicles to be fielded, beginning in December 1990, would be interim examples using the NBC reconnaissance suite fitted to West German Army vehicles. A product improvement programme was planned to modify all US Army Fox vehicles to a standard that will incorporate stand-off chemical agent detection, automatic sampling, automatic data processing and jam-proof data transmission systems.

However in July 1988 the House of Representatives directed that a competition be held between US and international companies for 'off-the-shelf' NBCRS vehicles; not less than two different types were to be involved. A request for proposals was released during late 1988 and bids were received from the TRW Defense Systems Group teamed with General Motors of Canada Limited, and from General Dynamics teamed with the West German Thyssen Henschel concern.

The vehicle involved in the TRW/General Motors of Canada bid is a modified version of the LAV-L (8 × 8) 2470 kg armoured logistic carrier (for full details of this vehicle refer to *Jane's Military Logistics 1989* page 580). The General Dynamics/Thyssen Henschel entry is the Fox vehicle originally requested by the US Army.

A 'shoot-off' competition between the two was expected to be conducted throughout the Summer of 1989; TRW/General Motors of Canada had delivered two vehicles by March 1989. The original schedule was that a 'winner' was to be announced by 29 September 1989 and a contract was to have been issued to the winning competitor for the delivery of 48 interim vehicles plus a research and development programme to product-improve the system to meet US Army requirements. The contract will also provide for the serial production of 279 units.

WARSAW PACT

Chemical and Radiological Reconnaissance Vehicles

DESCRIPTION

All the Warsaw Pact armed forces have vehicles devoted primarily to chemical and radiological reconnaissance. The specific vehicle in service with any particular army usually depends on the types of vehicle already in service but few of these special vehicles have any drastic modifications for their role other than the addition of lane-marking equipment to the rear. However, they nearly all carry a standard equipment kit consisting of one Model GSP-1 or GSP-1M nerve agent and radiation detector-alarm or one Model GSP-11 nerve agent detector-alarm and a separate radiation alarm, one Model VPKhR chemical agent detection and identification kit or the equivalent, one Model PPKhR automatic chemical agent detection and identification kit, one DP-3 area survey meter, one DP-5A radiation meter, one Model KPO-1 biological warfare sampling kit, SKhT signal cartridges, RDG smoke grenades and lane flag markers and emplacers.

The main vehicles involved are the UAZ-69-rkh, the UAZ-469-rkh, the BRDM-1 RKh, the BRDM-2 RKhb and RKha, and the Hungarian FUG (OT-65). The last three vehicles all have two 20-flag marker emplacers carried at the rear, but the UAZ-69-rkh may have either one 10- or 12-flag emplacer, or a flag rack from which the flags have to be emplaced manually.

The BDRM-2 RKhb carries the GSA-12 automatic chemical indicator and has a special device for firing warning signs into the ground using explosive charges, even when the vehicle is moving at speeds of up to 20 km per hour. This device has 19 barrels mounted on a plate, rather like a Gatling gun, and can be swung through an axis of 270 degrees. The vehicle is also equipped with the TNA-3 inertial navigation device for accurate course and position finding. The combat weight of the BRDM-2 RKhb is 7090 kg and it is armed with two 7.62 mm machine guns. An essentially similar vehicle, the BRDM-2 RKha, is armed with a single turret-mounted 14.5 mm KPVT machine gun and lacks some of the sensors carried by the BRDM-2 RKhb.

A more recent addition to the Soviet range of chemical reconnaissance vehicles is the RKhM, based on the chassis of the MT-LB multi-purpose armoured vehicle. Modifications to produce the RKhM include repositioning the engine and the provision of a box-like structure on which a machine gun cupola and the commander's cupola are installed. The upper part of the track is covered by a three- or four-part skirting. At the rear is a lane flag dispenser.

Hungary employs a chemical and radiological

UAZ-469-RKh chemical and radiological reconnaissance vehicle dispensing lane flags

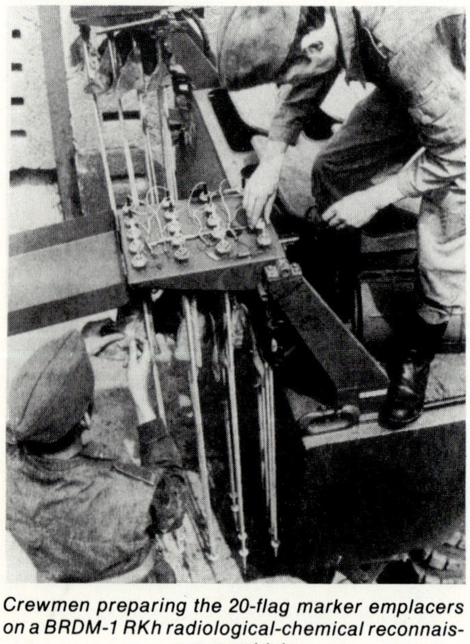

Crewmen preparing the 20-flag marker emplacers on a BRDM-1 RKh radiological-chemical reconnaissance vehicle

RKhM NBC reconnaissance vehicle based on chassis of MT-LB multi-purpose armoured vehicle

Hungary employs a chemical and radiological reconnaissance variant of the PSZH-IV armoured personnel carrier known as the FUG.

STATUS
All the above are in service with the Warsaw Pact and Soviet armed forces.

BRDM-2 RKha NBC reconnaissance vehicle in travelling configuration with 14.5 mm machine gun turret and warning marker dispensers covered

BRDM-1 RKh team wearing combined and protective suits and Model ShM masks, with soldier in foreground using VPKhr chemical agent detection and identification kit and soldier behind using DP-5A radiac survey meter

COLLECTIVE PROTECTION SYSTEMS AND ASSOCIATED COMPONENTS

CHINA, PEOPLE'S REPUBLIC

Prefilter, Particulate, M-75/2000

DESCRIPTION
The M-75/2000 particulate prefilter has an air flow capacity of 2000 cubic metres an hour and an air flow resistance at full capacity of up to 290 Pa. Using artificial dust for testing at full capacity the filtration efficiency is up to 95 per cent.

The M-75/2000 weighs 35 kg and measures 650 × 650 × 570 mm.

STATUS
In service with the Chinese armed forces.

DEVELOPMENT AGENCY
Research Institute for Chemical Defence, PO Box 1044, Beijing, China.

MANUFACTURER
Xinhua Chemical Factory, Taiyuan, Shanxi, China.

Prefilter, particulate, M-75/2000

Filter, Gas-particulate, M-77/500

DESCRIPTION
This filter has an air flow capacity of 500 cubic metres an hour and an air flow resistance at full capacity of less than 686 Pa. The filtration efficiency is stated to be 99.9999 per cent against aerosols when operating at full air flow capacity. Life when operating at full capacity against 2 milligrammes per litre droplets of cyanogen chloride is over 2 hours.

The M77/500 weighs 160 kg and measures 920 × 604 × 604 mm.

STATUS
In service with the Chinese armed forces.

DEVELOPMENT AGENCY
Research Institute for Chemical Defence, PO Box 1044, Beijing, China.

MANUFACTURER
Xinhua Chemical Factory, Taiyuan, Shanxi, China.

Filter, gas-particulate, M-77/500

Filter, Gas, M-73/1000

DESCRIPTION
The M-73/1000 gas filter has an air flow capacity of 1000 cubic metres per hour and an air flow resistance of less than 343 Pa at full capacity. Life when operating at full capacity against 2 milligrammes per litre of cyanogen chloride is over 2 hours at 20°C.

Weight of the M-73/1000 is 240 kg. Dimensions are 1000 × 700 × 670 mm.

STATUS
In service with the Chinese armed forces.

DEVELOPMENT AGENCY
Research Institute for Chemical Defence, PO Box 1044, Beijing, China.

MANUFACTURER
Xinhua Chemical Factory, Taiyuan, Shanxi, China.

Filter, gas, M-73/1000

FRANCE

AMF 80 Modular NBC Shelters

Exterior view of emplaced 100-man AMF 80 series semi-hardened modular shelter

DESCRIPTION

The AMF 80 modular NBC shelter is produced by a combination of two companies, Société des Tuyaux Bonna and Sofiltra-Poelman. The AMF 80 as furnished to the French armed forces shelters up to 60 occupants. It consists of standard prefabricated 2 and 2.5-metre diameter modules. The comfort and protective equipment is factory assembled. All modules are connected by elastomer (neoprene) joints. The shelter is usually constructed in a trench using conventional equipment and covered with a layer of earth. A basic AMF 80 can be constructed in about two days by a small construction team. The reinforced structure contains sanitary and filtering equipment providing for a lengthy stay. Airlocks at each entrance/exit are equipped with a steel 1.2-metre diameter armoured door and a cast aluminium air tight door of the same diameter.

The ventilation system consists of air inlet blast dampers, filter housings equipped with approved standard filters, fans driven by crank-handle or electric motors with a combined output of 480 cubic metres. Overpressure is maintained in the shelter and monitoring equipment measures the overpressure in the air locks and shelter toxic free area.

As a consequence of its modular design, the shelter can be adapted to accommodate any number of persons or any final use by changing the number and size of the elements.

The basic AMF 80 does not permit ingress/egress during NBC attack. An AMF 80 derivative with upscaled 3.2-metre diameter contains a contamination control area (CCA) where personnel can safely doff and don NBC clothing and equipment in a positive air flow environment. The CCA is capable of multiple line processing at the rate of six people every five minutes assuming that the decontamination procedures can support this rate. This shelter configuration provides protection for 100 persons and is under production for the Canadian Forces Europe as the CF CPS 100 modular semi-hardened shelter. This shelter is provided with emergency exits for the work and rest areas. In the CF CPS 100 configuration it is possible to have a personnel ingress/egress rate of about 40 persons per hour. The shelter is self-supporting for seven days.

The protective capability of the basic AMF 80 against a conventional warfare attack meets the criteria of 6 metres effective miss distance of a 1000 lb general purpose bomb from the side wall of the shelter (US Air Force trial).

The shelter sustains a static overpressure of 4.2 bars (blast effect of a simulated nuclear explosion, Direct Course Test, October 1983, the shelter being located at 138 metres from point zero).

The shelter provides protection against the following agents in vapour, liquid and aerosol form at threat concentrations: GB, GD, TGD, VX, HD, THD, HL, CX, AC, biological agents and toxins.

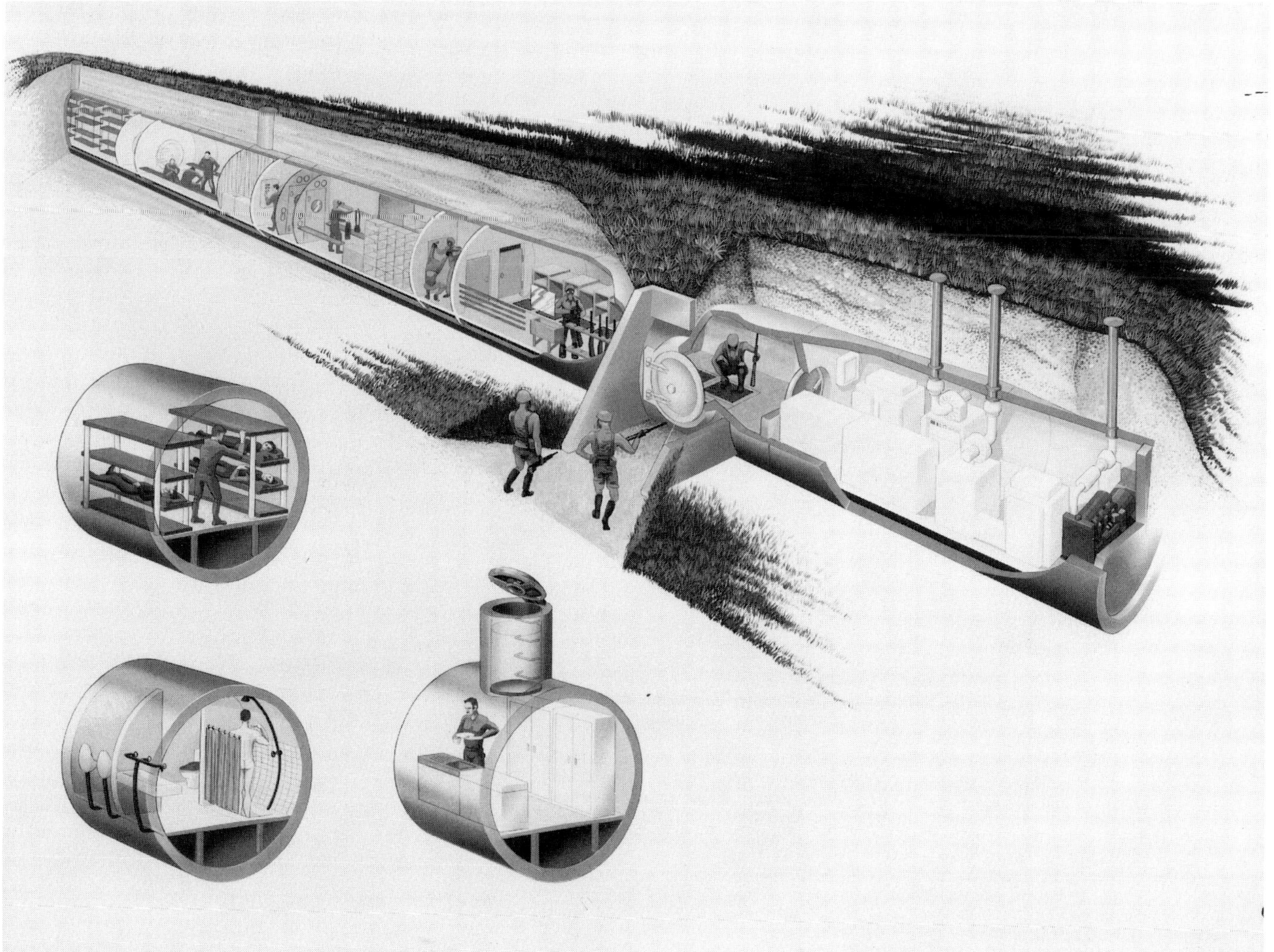

Artists impression of a AMF 80 modular shelter for 100 persons

SPECIFICATIONS

Type	**Model 1**	**Model 2**	**Model 3**
Nominal diameter	2 m	2.5 m	3.2 m
Internal diameter	2 m	2.5 m	3.2 m
External diameter	2.36 m	2.95 m	9.78 m
Wall thickness	0.18 m	0.225 m	0.29 m
Useful length	3.3 m	2.35 m	2.31 m
Weight per metre	3220 kg	4930 kg	7150 kg

STATUS
In production. In service with the French armed forces. Both models ordered for US Air Force trials in Europe and USA. Ordered by the Canadian Forces Europe.

MANUFACTURERS
Société des Tuyaux Bonna, 91 rue du Faubourg St Honoré, 75008 Paris, France.
Sofiltra-Poelman, 71 boulevard National, 92250 La Garenne-Colombes, France.

Decontamination area of CF CPS 100 modular semi-hardened shelter

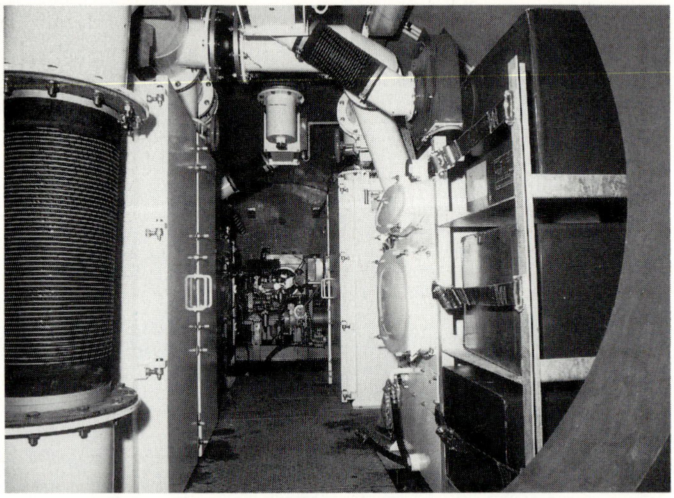

Technical services area of CF CPS 100 modular semi-hardened shelter

AP 60 Modular Semi-hardened Shelter

DESCRIPTION
The AP 60 modular semi-hardened shelter follows the same general construction lines as the AMF 80 shelter (see previous entry) and is constructed from cylindrical reinforced concrete elements. It has a main agent-free area closed at each end by a light alloy airtight door with locks on the inside and outside. Each of these doors provides access to an L-shaped air lock which gives access to the outside.

The main agent-free area can accommodate up to 60 persons, 50 sitting and 10 laying down. The area is equipped with a lighting system, a ventilation system, benches, stretchers, shelves for stores, and a toilet unit. The NBC filtration and ventilation unit consists of two 240 cubic metres per hour units. The filtration system and a power generator are located in an air lock with exhaust gases expelled from the shelter via a flexible hose.

The AP 60 can be self-supporting for approximately 48 hours, depending on the stores contained inside the shelter.

STATUS
In production for the French Air Force.

MANUFACTURERS
Société des Tuyaux Bonna, 91 rue du Faubourg St Honoré, 75008 Paris, France.
Sofiltra-Poelman, 71 boulevard National, 92250 La Garenne-Colombes, France.

Emplacing an AP 60 modular semi-hardened shelter

Main area of an AP 60 modular semi-hardened shelter

Bachmann-TMB NBC Shelters

DESCRIPTION
Bachmann-TMB produce a range of NBC shelters for field use. A typical structure is a field repair shelter capable of accommodating vehicles up to the size of a main battle tank. The main workshop area consists of an enclosed space formed by placing a layer of impervious fabric over a structure formed from arc-shaped bows. Vehicles to be repaired enter the structure by raising the bows at one end. Personnel enter through a shelter built onto one side of the main area. The shelter acts as an air lock and also has space for the main air filtration and conditioning plant.

A multi-purpose NBC tent structure can be formed by erecting a series of inflatable modules formed from coated compound fabrics that are impervious to chemical agents. The tents can have either inflatable frames or the modules can be erected within rigid metal frames.

Bachmann-TMB also produce a NBC protection system for aircraft shelters. The shelter involved is protected by an external covering over the entire structure roof that is impervious to chemical agents. The cover may also be camouflaged. A system of impervious flexible doors is arranged under the entrance arch for aircraft exits and entrances, and an impervious flap is provided for the aircraft engine exhaust. There is an internal air lock for personnel and further protected personnel entrances and exits are provided within the structure. The protective structure is pressurised by a combined air ventilation and filtration system. This protected shelter can be supplemented by impervious housings for aircraft parked outside the shelter and by easily

Bachmann-TMB inflatable NBC tent erected inside a rigid frame and with a 900 cubic metres per hour air filtration unit

decontaminated munitions and other trailers associated with aircraft servicing.

STATUS
In production.

MANUFACTURERS
Bachmann SA, 69 avenue Danielle-Casanova BP15, 94200 Ivry-sur-Seine, France.
Techniques Michel Brochier SA, ZI – Route de Chartinières, Dagneux 01120 Montluel, France.

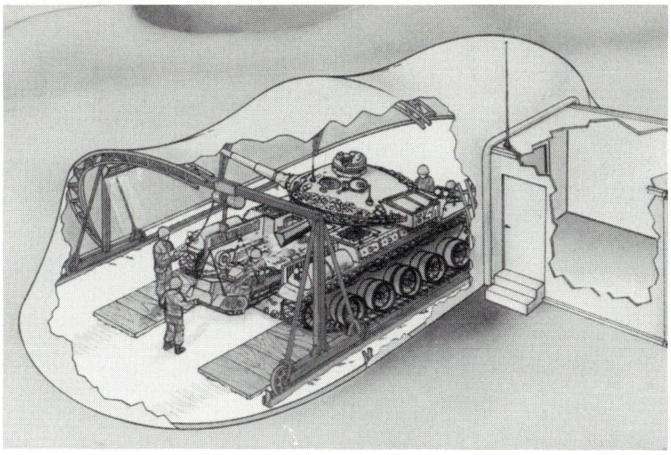

Artists' impression of a typical Bachmann-TMB NBC field repair workshop

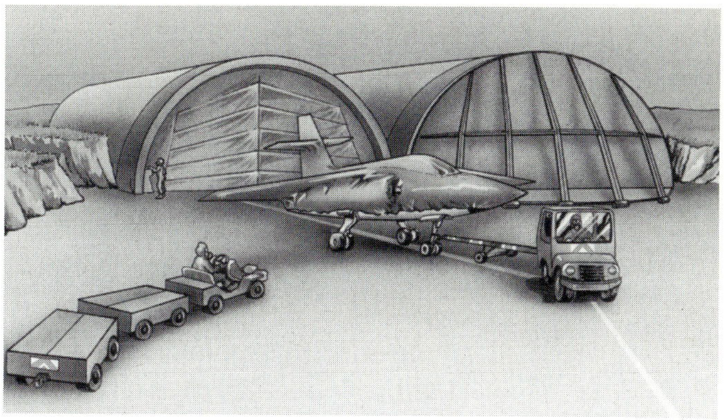

Artists' impression of a typical Bachmann-TMB NBC-protected aircraft shelter

FMGC High Capacity Composite Filter

DESCRIPTION
The Sofiltra-Poelman FMGC (Filtre Mixte de Grande Capacité) high capacity composite filter has a capacity of 1000 cubic metres per hour and is designed to prevent the ingress of aerosols or vapours into the air systems of collective shelters and bunkers, etc.

For use the unpacked filter is placed in its storage state at the location where it is to be used with the filter caps hermetically closed. The FMGC is emplaced on four floor bolts and is not connected to the air system until required. The caps are removed before use and the filter is connected to the air supply.

The filter has two cylindrical rings containing the filtering media and is kept within the filter body. The composite filter is composed of a circular particulate filter consisting of a microglass fibre paper media, pleated and spaced with glass threads. The paper media is sealed and pleated using a material formed from a vinyl plastic or polyurethane. The vapour filter has two perforated cylinders covered by a polyester cloth. The cylinders are supported at the base by a viscoelastic sealant and are compressed between two plates by eight threaded tension rods. The charcoal filter mass (25 litres in volume and in a layer 140 mm thick) is loaded under continuous vibration and is located under the upper plate.

Overall dimensions of the filter are 1.64 × 0.78 × 0.78 metres and the approximate weight without packing is 390 kg. The filter has a storage life of 15 years and can withstand a typical attack after effective ventilation for a one-year period without chemical attack or two typical attacks after a one-year period of non-ventilated storage followed by a one-month period of effective ventilation in the absence of chemical attack.

FMGC high capacity composite filter

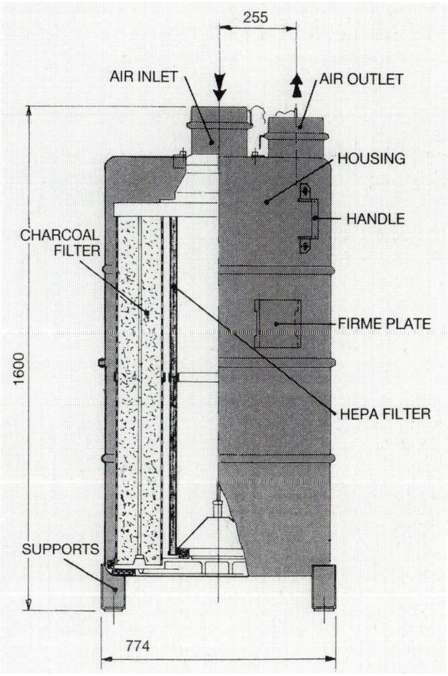

Cross section diagram of FMGC high capacity composite filter

STATUS
In production for the French Army.

MANUFACTURER
Sofiltra-Poelman, 71 boulevard National, 92250 La Garenne-Colombes, France.

Sofiltra-Poelman NBC Filters

DESCRIPTION
Solifiltra-Poelman produce three types of NBC filter. One type is a composite filter capable of providing protection against aerosols and vapours. The other two types are proof either against aerosols or against vapours.

Of the composite filters all are rectangular in cross section apart from the FMPC 30 which is circular. All the composite filters have painted steel frames apart from the FMMC 225 which uses an aluminium alloy and stainless steel frame. The efficiency of the anti-aerosol stage against a 0.12 micron monodispersed sodium fluorescein aerosol is stated to be up to 99.995 per cent.

The anti-aerosol and anti-vapour filters are all rectangular in cross section.

STATUS
Composite filters are at prototype or pre-production stage, depending on model. Anti-aerosol and anti-vapour filters in production for the French Army.

MANUFACTURER
Sofiltra-Poelman, 71 boulevard National, 92250 La Garenne-Colombes, France.

SPECIFICATIONS
COMPOSITE FILTERS

Model	Flow rate	Dimensions L × W × H	Filtering area	Charcoal volume	Initial pressure drop	Weight
FMPC 30	30 m^3/h	255 diam × 155.5 mm	0.75 m^2	4.4 dm^3	700 Pa	5.5 kg
FMPC 90	90 m^3/h	566 × 254 × 126 mm	1.5 m^2	8.1 dm^3	1600 Pa	22 kg
FMPC 180	180 m^3/h	605 × 384 × 156 mm	3.6 m^2	18 dm^3	1600 Pa	30 kg
FMMC 225	225 m^3/h	461 × 461 × 328 mm	2.9 m^2	900 dm^3	900 Pa	42 kg

ANTI-AEROSOL FILTERS

Model	Flow rate	Dimensions L × W × H	Filtering area	Weight
FCAA 130 MF1	130 m^3/h	600 × 230 × 55 mm	2.2 m^2	4 kg
FCAA 250 MF1	250 m^3/h	600 × 130 × 202 mm	4.4 m^2	6.5 kg
FCAA 250 MF2	250 m^3/h	600 × 130 × 202 mm	5.7 m^2	6.5 kg
FCAA 900 MF1	900 m^3/h	610 × 305 × 292 mm	14 m^2	16 kg
FCAA 1800 MFI	1800 m^3/h	610 × 610 × 292 mm	30 m^2	27 kg
FCAA 1800 MF1M	1800 m^3/h	610 × 610 × 292 mm	30 m^2	27 kg

ANTI-VAPOUR FILTERS

Model	Flow rate Dimensions L × W × H	Charcoal volume	Filtering area	Weight
FCAV 90 MF1	90 m^3/h	600 × 230 × 105 mm	9.8 dm^3	15.5kg
FCAV 150 MF1	150 m^3/h	461 × 461 × 328 mm	48 dm^3	51 kg
FCAV 150 MF1M	150 m^3/h	461 × 461 × 328 mm	48 dm^3	52 kg
FCAV 150 MF2	150 m^3/h	461 × 461 × 328 mm	27 dm^3	37 kg

NBC Collective Protection Units for Surface Vessels

DESCRIPTION
Sofiltra-Poelman produce three models of NBC collective protection units for naval surface vessels. All three units have been derived from a 900 cubic metres per hour unit and have capacities of 1000, 1500 and 3000 cubic metres per hour respectively.

All three units are NBC pressurisation and filtration units for the protection or personnel and material inside a ship's pressurised area or citadel. Each unit is housed inside a stainless steel housing containing one or two high efficiency filters, FMMC 255 composite filters (4, 6 or 12 according to capacity), a washable pre-filter, a washing device for the pre-filter, up to 3 anti-blast valves, an electric heater and a high pressure motor fan.

The 1000 cubic metres per hour unit is intended for use with small or medium tonnage naval vessels. The 1500 cubic metre per hour unit is intended for medium tonnage naval vessels while the 3000 cubic metres per hour unit is intended for use with large tonnage naval vessels. The latter unit was designed for use with the nuclear aircraft carrier *Charles de Gaulle*.

Weight of the 1000 cubic metre per hour unit with unpacked filters is 1432 kg. Figures for the two larger units are 1730 kg and 2576 kg respectively.

STATUS
Pre-production.

MANUFACTURER
Sofiltra-Poelman, 71 boulevard National, 92250 La Garenne-Colombes, France.

Typical Sofiltra-Poelman NBC collective protection unit for naval vessels

ITALY

Moldip Igloo NBC Tent

DESCRIPTION
The Igloo NBC tent was developed by Moldip SpA and Sekur SpA working in close collaboration and is intended to be a special module that can be fitted onto the standard Igloo pneumatic tent manufactured by Moldip SpA.

The standard Igloo tent is an inflatable tent made of an impermeable rubber-coated fabric. Each tent measures 1.4 × 0.65 × 0.35 metres when packed and weighs 138 kg. When erected the tent is of semi-cylindrical shape with a maximum height of 2.3 metres and a base of 4.6 × 4.6 metres. All joints are air-tight. The tent is supported by inflatable tubular members divided by internal membranes to form separate inflation circuits. The outer skin is made of olive green rubber-coated fabric with a weight of 800 grammes per square metre while an inner skin uses close-weave nylon. The outer skin is temperature resistant and has an infra-red reflectivity range close to that of the natural environment. The material is resistant to chemical agents and can be easily decontaminated.

The standard Igloo tent has sealable apertures at both ends through which ventilation tubes can be inserted. Extra tents can be assembled in a modular fashion to produce larger structures. There are fittings for the metal frame for extra support and carrying handles are provided to move the tent when inflated.

For the NBC role a lobby module can be added to the Igloo tent. Accessories that can be used inside the module include an air lock entrance with a double flap, a NBC filtration unit and associated air ducting units to produce an internal overpressure, and a shower assembly for the decontamination of personnel entering the tent.

The Igloo NBC tent has two modules, one measuring 4 × 4 metres and the other 6 × 5 metres. An indefinite number of modular elements can be added to provide shelters of varying dimensions and various associated filter unit capacities can be employed.

STATUS
In production.

MANUFACTURERS
Moldip SpA, Via Milano 8, 20038 Seregno (MI), Italy.
Sekur SpA, Via di Torrespaccata 140, 00169 Rome, Italy.

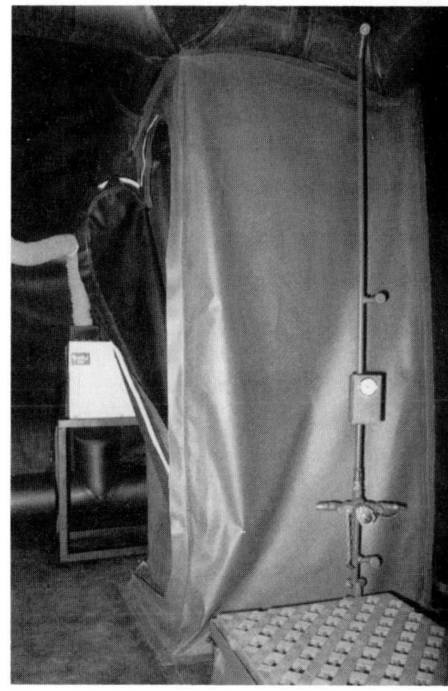

Decontamination unit inside Igloo NBC tent lobby module

Sekur NBC Filter Units

DESCRIPTION
Sekur SpA produce a range of NBC filter units to suit a variety of purposes ranging from the protection of building interiors to mobile installations on vehicles and warships.

All the Sekur NBC filter units follow the same basic operating pattern. There is a blast-proof valve that may also be used as an outlet valve. The first filter stage is a pre-filter to remove dust and other coarse particles. Pre-filtering can be obtained by the use of dynamic filtration (cyclone) or by using a static unit using washable synthetic foam and up to 95 per cent effective. For naval use a special pre-filter to retain sodium chloride particles is available. Then follows an aerosol filter unit that retains particulates larger than 0.3 micron. The efficiency of this stage is not less than 99.995 per cent. The pleated filters used at this stage consist of a special fibre-glass paper with resistance against heat and chemical agent corrosion. The next stage is an activated charcoal layer to remove remaining chemical agents and radioactive particles. All these stages are contained in stainless steel or galvanised steel casings.

Accessories available for use with the Sekur systems are air intake and delivery pipelines, intercept valves, air flow dampers and electric fans.

Models available are as follows:
SP-35. A compact filter system for use on mobile shelters or containers and small vehicles. The unit has a capacity of approximately 35 cubic metres per hour. Included in the one-piece stainless steel unit (dimensions 780 × 380 × 350 mm) are an electric

Sekur SP-180 NBC filter unit as fitted to Ariete main battle tank

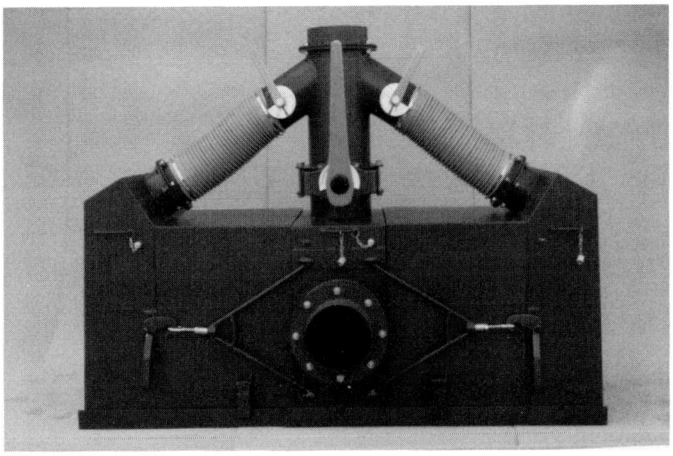

Sekur SP-1200 NBC filter unit for large structures

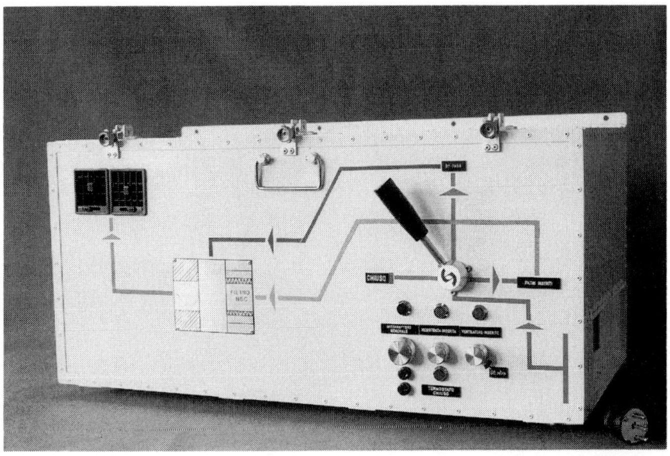

Sekur SP-35 NBC unit for use in mobile shelters or small armoured vehicles

fan, an air flow regulator, a silencer, a rapid-operating bypass valve and an air diffuser. The activated charcoal element is in a stainless steel container.
SP-85. This unit is intended for use in underground shelters and bunkers. It has a nominal capacity of 85 cubic metres an hour and is contained in a cylinder 400 mm in diameter and 700 mm high. Weight is approximately 40 kg.
SP-120 and SP-180. These units are intended for use on armoured vehicles and main battle tanks. The SP-180 is used on the Ariete main battle tank.
SP-1200. Intended for use in large underground and other structures the SP-1200 has a capacity of 1200 cubic metres per hour. Construction uses section and sheet steel and the activated charcoal elements are held in stainless steel containers. Weight is approximately 300 kg and dimensions are 1.7 × 0.75 × 1.25 metres.

In static installations, for air flow capacities of between 85 and 1200 cubic metres an hour, or higher than 1200 cubic metres per hour, SP-85 and SP-1200 filter units can be assembled and connected in parallel.

Emergency units for single users, such as the SP-7 filter unit, are also available.

STATUS
In production.

MANUFACTURER
Sekur SpA, Via di Torrespaccata 140, 00169 Rome, Italy.

NETHERLANDS

Bronswerk NBC Collective Protection Filter Unit

DESCRIPTION
This filter unit was developed for collective protection in mobile shelters to be used in conjunction with air conditioning and heating equipment also produced by Bronswerk BV. It can also be used in other mobile systems, vehicles, naval vessels or permanent installations such as bunkers.

The unit uses a cyclone pre-filter and a changeable composite filter formed from pleated aerosol media and activated carbon. The filter complies with NATO AQAP-1 requirements. In the filtration mode and using a two-speed fan the unit has a capacity of 90 cubic metres an hour.

The unit is powered by a 24 volt dc power supply. Total weight is approximately 70 kg and overall dimensions are 652 × 441 × 620 mm.

In association with AAF International BV, Bronswerk BV also produce a compact, vibration resistant NBC filter cassette suitable for use in transport systems. No pre-filter is used and the absorption stage uses impregnated activated carbon bonded in panel form to provide a 100 mm deep carbon bed. The filter has a rated capacity of 90 cubic metres an hour and has a filter efficiency of 99.997 per cent. Dimensions are 600 × 300 × 170 mm.

STATUS
In production.

MANUFACTURER
Bronswerk BV, Brabantsestraat 10, PO Box 28, 3800 HC Amersfoort, Netherlands.

Bronswerk NBC collective protection filter unit

SWEDEN

JP SHELTEC Shelter Equipment

DESCRIPTION
The first Swedish air raid shelter was built over 40 years ago and shelter construction still continues to the stage where the number of shelter places available in Sweden has reached 5.5 million. This allows over 70 per cent of Sweden's population access to a shelter place at home and work. Every year between 150,000 to 200,000 new shelter places are produced.

The shelters constructed fall into four main categories. These include multi-storey shelters for use as civil defence control centres. Typical of these is the two storey Type D control centre which houses a staff of from 60 to 90. Public shelters are often large constructions built into rock or using reinforced concrete structures. Some of these shelters can hold up to 15,000 people and may have peace-time uses varying from public gymnasiums to workshops. There are a number of smaller public shelters, typical of which is the TB 78 holding a maximum of 180 persons. The TB 78 can be built into public buildings such as schools and hospitals and various numbers of these shelters can be constructed together in groups to provide extra accommodation. A typical TB 78 shelter could be a ground floor school room in normal use. In an emergency windows could be blocked off with concrete elements and extra ventilation could be installed. The installation time of such items may not exceed 48 hours and many components such as shelter doors remain in place permanently.

Also provided are shelters for important community functions (such as telephone and computer centres) and commodities such as oil storage.

Shelter installation policy is determined at government level and shelter construction is mainly carried out by commercial concerns. Research and development on civil defence systems is continuously carried out by the Swedish Civil Defence Administration in close collaboration with the Swedish National Defence Research Institute, the Swedish Fortifications Administration, the Royal Institute of Technology, the National Board of Health and welfare and various industrial and consulting companies.

JP SHELTEC AB produce a range of shelter components. They can also produce special designs and will supervise installation when required. Included in their product range are NBC-protective doors for

Air purification plant for a large shelter

JP SHELTEC NBC protective door installed in an underground shelter

personnel and equipment. These doors can provide protection against blast, splinters, electro-magnetic pulses (EMP) and NBC agents.

Also produced are protective valves which are available in varying protection classes and for various air flow requirements. The let-through impulses of these valves are extremely small, even at loads over 100 bars. The valves can be equipped with EMP wave traps, prefilters, etc, depending on installation requirements. Valves may be provided in stainless steel for use in corrosive conditions or where diesel exhaust fumes may be present.

SHELTEC air purification units are based on modular concepts. Installations involved include sealed gas and aerosol filters, filter stands, fans, pre-heaters and gas-tight dampers, all designed to use a minimum of space. The units are protected against ground shocks and can meet the highest military standards. No special tools or other equipment are required for maintenance and the systems are designed for ease of handling. Each installation includes all the auxiliary items required.

JP SHELTEC also produce shock absorbers for shelters to provide them with protection against ground explosions. The shock absorbers involved will restrict acceleration to 2 g vertically and 0.2 g horizontally, thereby making it possible to use non-hardened equipment in shelters so protected. The SHELTEC range includes shock absorbers designed for static loads of from 2 tons up to hundreds of tons. No servicing or maintenance is required by the shock absorbers throughout their life times.

STATUS
All the above components are in production and are in service within Sweden.

MANUFACTURER
JP SHELTEC AB, Box 1163, S-14124 Huddinge, Sweden.

SWITZERLAND

BERICO Shelter Components

DESCRIPTION
All Swiss civil defence constructions have to utilise components that have been authorised by the Bundesamtes für Zivilschutz (BZS), the Swiss civil defence authority. BERICO AG of Niederglatt hold such an authority and produce a range of armoured doors, gates, covers and sliding walls for shelters and compression doors for use inside structures. They also produce ventilation plant for shelters of varying sizes.

The armoured doors and similar components are 200 mm thick to withstand one atmosphere excess pressure. Stronger components to withstand up to three atmospheres excess pressure are 250 mm or 350 mm thick. All components have standard height and width dimensions, eg a typical armoured door (Model PT1) is 1.85 metres high and 0.8 metre wide. Also available are sealable armoured windows.

Three basic models of NBC ventilation equipment are produced. They are the VA 40, the VA 75 and the VA 150, with the numbers referring to the filter air flow capacity in metres per hour. The VA 40 is intended for shelters holding between 8 and 13 people, the VA 75 for shelters holding 14 to 25 people and the VA 150 for shelters holding between 26 and 50 people. All three models can be either hand operated or powered by a three-phase electrical supply.

STATUS
In production.

MANUFACTURER
BERICO AG Niederglatt, 8172 Niederglatt, Switzerland.

BERICO AG filtered ventilation system for NBC shelter

UNITED KINGDOM

Barton APD Shelters

DESCRIPTION
Intended to be used buried underground the Barton APD Shelter is of modular construction designed to withstand heavy blast effects and may provide protection from NBC agents. The shelter can be supplied as a single unit or alternatively the modular design enables the shelter to be built up into underground complexes of any number of configurations. Once installed the internal construction enables the shelters to be used for a variety of purposes ranging from simple shelters to large command, control and communication complexes.

The main shelter module consists of nine octagonal main frames manufactured, as are most components, from galvanised steel. The frames are joined together by tubular distance pieces to create a main structure 3.6 metres long. On each frame a set of end covers are attached, one consisting of two outer plates and a centre plate. The other cover consists of two outer plates and a bottom plate providing an opening 1.8 metres high and 0.76 metre wide to which a conning tower is connected. Holes are provided in each end cover to take the air inlet and exhaust pipe flanges. These are normally set at opposite ends. At the air inlet end a bilge pump bracket is located together with its outlet. The main shelter is covered with pre-stressed galvanised sheets which are lap jointed and bolted to the outside of the main structure. Where joints occur in the covers, including the end covers, a rubber sealing strip is placed between mating surfaces. Fibre washers are placed under all bolt heads. Vent and exhaust pipes are connected to the end covers by a flange which is supplied complete with a sealing gasket.

Internally a bracket is supplied to house a wash basin. Hinged brackets support four bunks which are supplied with foam mattresses; each bunk is 1.8 metres long. The softwood timber floor is supported on 100 x 50 mm joints at 450 mm centres. Other items supplied for the standard interior include a hand-operated air pump, a chemical toilet and a hand-operated bilge pump.

The standard shelter has a conning tower leading to the surface bolted onto the main structure at one end. The tower is 0.76 metre wide and 3.2 metres high with an inward-opening hatch at the top. The hatch has a diameter of 450 mm and has an 8 mm armoured lid. When a number of shelter modules are combined the conning tower may be replaced by an entrance module which can serve up to four shelters in a cruciform arrangement. Bolted construction is used for the entrance module which is manufactured from 4 mm plate flanged and reinforced with 4 mm flats. It uses two ladders.

The Barton APD shelter has been tested by the British Army.

Barton APD shelter showing entry conning tower

SPECIFICATIONS
Basic unit length: 3.6 m
Cross-section: 2.4 m across octagonal points
Conning tower entrance: 0.76 m^2
Total weight: (including fittings) 2130 kg
Entrance module
Total weight: 2300 kg
Height: 2.75 m
Plan size: 2.75 m^2

STATUS
In production.

MANUFACTURER
BC Barton and Son Limited, 1 Hainge Road, Tividale, Warley, West Midlands B69 2NJ, United Kingdom.

Barton APD shelter showing cross-section shape

Heywood Williams Field Shelter MEXE Mk III

DESCRIPTION
The Field Shelter MEXE Mk II (or Mexeshelter) was developed by the Military Engineering Experimental Establishment at Christchurch (now the Royal Armament and Research Establishment Christchurch), with production and marketing being undertaken by Heywood Military Engineering, a division of Heywood Williams Limited. The shelter provides protection against conventional attack, near miss nuclear attack and protection against chemical and bacteriological attack. Protection against the first of these two categories is given by the standard shelter with protection against chemical and bacteriological attack being provided by the kit described below.

The Mk III was originally conceived as an underground shelter for Command Posts (CPs), Observation Posts (OPs) and Regimental Aid Posts (RAPs), for use mainly by the Army in forward and rear Observation Posts (OPs) and Regimental Aid Posts (RAPs), for use mainly by the Army in forward and rear area tactical situations, but it is now receiving consideration for the underground protection of personnel on air stations and in areas not directly related with the tactical battlefield itself. In its present configuration, it is being used on a much wider basis which includes a casualty clearing and first aid station in conjunction with triage; storage for medical supplies, including plasma, drugs, dressings; fuel, food and other combat support items, and as temporary magazines for ordnance items. In the Command Post usage it has moved into battalion and divisional use.

The Mk III consists of four components plus accessories – three metal structural members, pickets, spacers, arches, plus flexible revetting material to clothe the metal structural members and which is held in place by earth backfill. All components are packaged in man-portable kits and can be installed by unskilled personnel with no special tools; all work can be accomplished with a pick and shovel. With the use of mechanical equipment such as a bulldozer, scraper or shovel, approximately two hours are required to fully install the Mk III. In addition to the mechanical equipment, approximately 15 man-hours of troop labour are required.

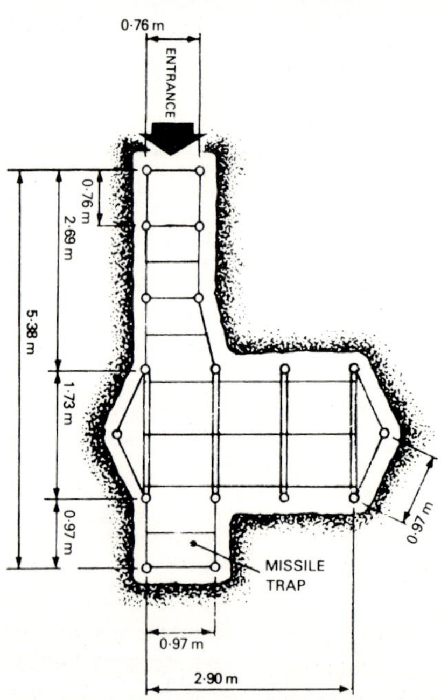

Command post MEXE Mk III modular shelter built with 18 pickets, 4 arches, 29 spacers and 4 rolls of revetting material

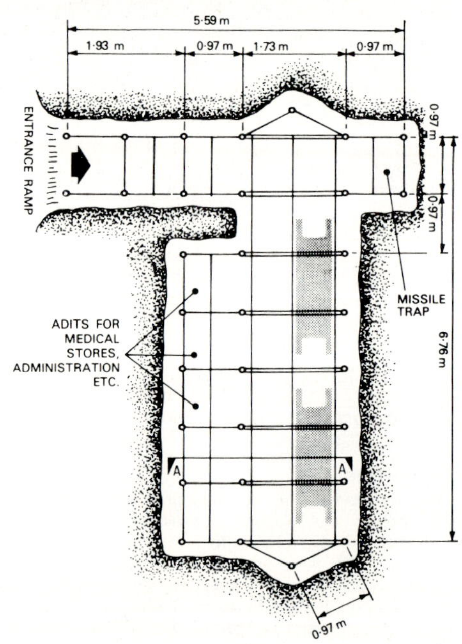

Regimental first-aid post built from 32 pickets, 8 arches, 55 spacers and 7 rolls of revetting material

The flexible revetting material used to line the walls and roof of the shelter is a pvc-coated jute fabric reinforced with wire. The pvc layer, being impermeable, provides an obstacle to chemical agent penetration, and, when this is combined with the additional barrier of 450 to 600 mm of soil over the shelter roof, it is likely that there will be no hazard in the shelter even with heavy chemical agent contamination on the ground outside. Modifications necessary to give chemical protection without the use of respirators and NBC clothing are the installation of an NBC filtration unit and liner as outlined below.

Normalair-Garrett Limited of Yeovil, Somerset, has produced an NBC air filtration package for use with the MEXE shelter (qv).

In service with the British Army and the Royal Air Force is the L1A1 air filtration unit (qv). The associated liner, including the airlock, is constructed of translucent reinforced plastic and folds down in similar fashion to a tent. It is self-supporting, even when there is no inflation from the AFU, and takes approximately six minutes to inflate. The airlock has easy access 'fold away' doors and adjustable ports to vary the liner/airlock pressure. The liner was developed by Driclad Limited of Sittingbourne, Kent and liners for other configurations of shelter can be made available.

Various tests have been run to evaluate the protection provided by the Mk III in both a projected nuclear and a conventional weapon environment. In the case of a nuclear blast, the results record the protection provided against heat (thermal radiation), nuclear radiation (immediate) and nuclear fall-out. The shelter components have been designed to withstand, to a given level, the effects of a 'near miss' weapons blast. The tests were undertaken by the American, West German and British armed forces.

STATUS
In production. In service with British, Canadian, Kuwait and Saudi Arabian armed forces. Under active evaluation by Belgium and the USA (40 sets purchased).

MANUFACTURER
Heywood Military Engineering, Birds Royd Works, Brighouse, West Yorkshire HD6 1NG, United Kingdom.

MEXE shelter covering prior to earth backfill

Assembling MEXE shelter and showing general construction method

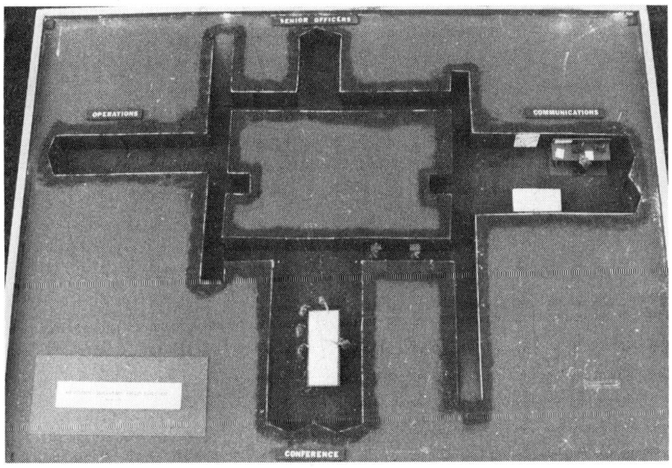

Cutaway model showing layout of MEXE shelter in use as an underground command post

Interior bay of MEXE shelter showing impermeable pvc layer for protection against chemical agents

Beaufort NBC Field Shelter

DESCRIPTION
The Beaufort NBC shelter is based on the Mark III Mexe frame and has been updated by using high tensile chemical agent resistant materials. It can be used as an underground command post, observation post or regimental aid post, or for a range of other purposes.

The Beaufort NBC shelter can be installed by unskilled personnel with no special tools being required. Using mechanical equipment an excavation for a shelter can be produced within two hours with an additional 15 man hours required to complete the installation. The shelter uses three metal com-

Interior of Beaufort NBC shelter

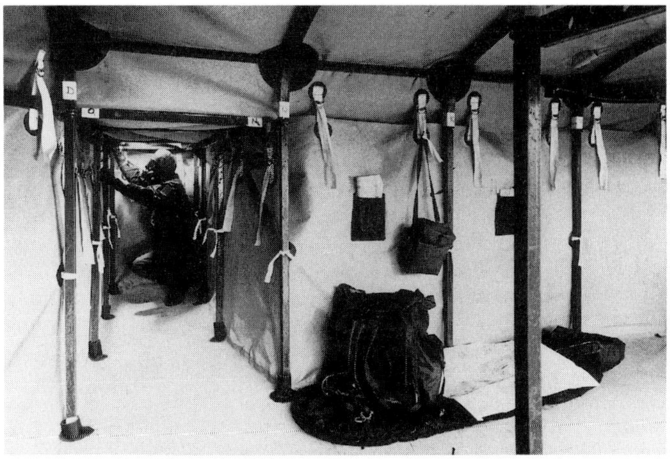

Interior of Beaufort NBC shelter

ponents, pickets, spacers and archers which assembled form the skeleton of the shelter. There are three fabric components; roof section, bottom section and floor panel/template. The latter serves as a template for the excavation and is afterwards placed on the excavation floor for protection against sharp rocks or protrusions. The two-part fabric cover is held in place by stainless steel 'D' rings with an overlapping top cover giving the side walls double layer protection. The final component is the glass reinforced plastic hatch which has watertight seals.

Ancillary equipment includes webbing straps for cross lashing to prevent bowing of the fabric after backfilling. All components are packed in man portable units. For full blast protection earth backfill should be a minimum of 450 to 500 mm at the apex of the structure.

STATUS

In production. In service with the British Armed Forces.

MANUFACTURER

Beaufort Air-Sea Equipment Limited, Beaufort Road, Birkenhead, Merseyside L41 1HQ, United Kingdom.

M.D.H. NBC Collective Protection Equipment for Armoured Vehicles and Ships

DESCRIPTION

M.D.H. Limited has been designing, developing and producing their Microflow range of air filtration products for British and overseas defence ministries for over 20 years. The range of products now covers vehicle collective protection equipment involving two stage filtration systems, three stage NBC filtration systems and three/four stage by-pass NBC/ventilation filtration systems. The latest group of filters and systems designed for vehicles such as the Warrior APC, Challenger and M1A1 Abrams MBTs, Challenger ARRV and the Stormer/Starstreak HVM carrier are much reduced in volume compared to earlier systems.

M.D.H. Limited (at one time Microflow Limited) has designed hybrid collective protection equipment (HCPE) for the United States Department of Defense. The systems involved include units with 130, 255 and 510 cubic metres an hour.

The two stage filtration system was designed for use in riot control vehicles and gives protection to the vehicle crew from high dust concentrations, smoke, and some riot control gases. It incorporates a prefilter to retain coarse particles and prolong the life of the second stage. At the second stage a high efficiency filter removes fine particles. This type of unit has a nominal flow capacity of up to 180 cubic metres an hour, a power consumption of 25 amps and a weight of 50 kg. A typical application for this type of system is that used on the Timoney/BDX (Vickers Valkyr) riot control vehicle.

A typical four stage NBC filtration system (the No 7 Mark 1) is that fitted to the British CVR(T) range of vehicles (Scorpion, Samaritan, Spartan, Sultan, Scimitar and Samson). The unit features multiple air connections and facilities for a remotely located fan and control box. There are four stages of filtration, three for particle removal and a final stage for the removal of toxic chemical vapours. The nominal flow capacity is 170 cubic metres per hour, power consumption is 35 amps and weight is 52 kg.

The Microflow four stage by-pass NBC/ventilation

No 6 Mark 2 NBC pack fitted to Chieftain MBTs

A 680 cubic metres per hour NBC unit produced by M.D.H. Limited for shipborne use

Challenger MBT with a No 15 Mark 1 NBC pack housed in the turret bustle; the pack was designed and developed by M.D.H. Limited

Spartan APC with No 7 Mark 1 NBC pack visible inside hull wall

filtration system was designed at the request of the British Ministry of Defence for installation in the Chieftain MBT. The pack involved (the No 6 Mark 2) utilises four stages of filtration, three for particle removal and a fourth stage for the removal of toxic chemical vapours. However the unit has an additional facility which allows the third and fourth stages to be by-passed, thus preserving these filters until subjected to an NBC contaminated environment. The first and second stages would be used for extracting coarse particulate material from the ventilation air. Facilities to monitor particulate filter life are embodied and vehicle over-pressure can be checked and adjusted. This unit is available as a retrofit to the Centurion MBT and as an option on the Vickers Main Battle Tank. The nominal flow capacity of this unit is 340 cubic metres an hour, power consumption is 65 amps and weight is 152 kg.

A more compact Microflow by-pass unit has a nominal capacity of 187 cubic metres an hour. A generally similar unit (the No 15 Mark 1), designed and developed by M.D.H. Limited for the Challenger MBT, has a nominal flowrate of 340 cubic metres an hour, a power consumption of 55 amps and a weight of 135 kg.

The latest family of Microflow NBC filtration systems are considerably reduced in volume. This has been achieved by improving the efficiency of the first (inertial separator) stage, allowing the deletion of the second (particulate prefilter) stage and by providing the third (particulate high efficiency) and fourth (anti-vapour) stages with a combined housing. The redesigned system attains the former high degree of filtration yet can be serviced, in the case of a 170 cubic metre per hour unit, by the replacement of a single composite filter canister. This type of system can be provided with a by-pass facility.

M.D.H. Limited also produce associated equipment such as NBC filtration/ventilation control boxes, NBC control panels, respirator hoses for connections to a main NBC system, eyeball outlets, hose connections, cooling units, pressure relief valves, duct discharge points, bleed valves and hoses. A full range of support equipment for testing and training is available, including environmental chambers for testing complete systems.

STATUS
In production. In service world-wide.

MANUFACTURER
M.D.H. Limited, Walworth Road, Andover, Hampshire SP10 5AA, United Kingdom.

On the FV 435 WAVELL tracked vehicle the NBC pack is carried behind the armoured door in the side of the hull; an M.D.H. environmental control unit, a generator and other components are carried inside the housing over the hull

Typical Microflow composite NBC filter module; the module shown has a capacity of 170 cubic metres an hour

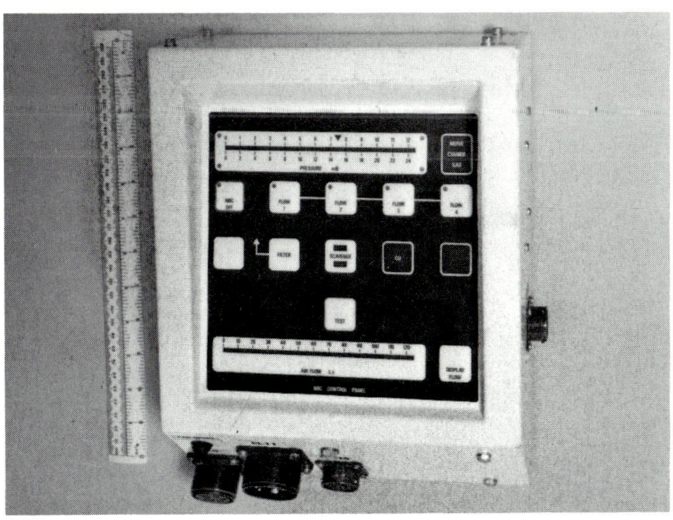

Electronic NBC control box produced by M.D.H. Limited with a membrane keyboard for ease of decontamination

No 15 Mark 1 NBC pack fitted to the Challenger MBT and being retrofitted to Chieftain MBTs to replace the No 6 Mark 2; both packs were designed and developed by M.D.H. Limited

M.D.H. Survivaire Shelter Filtration Unit

DESCRIPTION
The Survivaire shelter filtration unit was developed specifically for smaller fallout shelters for both civil and military applications. A three-stage filtration system is employed and is available in various capacities. The system can be supplied in manual and manual/electrical versions. A fully electrical version is also available.

The manual version has been designed to require a minimum of physical effort to the extent that the system can be employed by children. The manual/electrical version can be powered by an electrical supply taken from a 12-volt vehicle battery. The latter version involves a circuit breaker and a two-stage fan. When using manual operation the number of turns of the operating handle is about 19 rotations per minute.

The Survivaire is supplied complete with a blast valve and a pressure relief valve which allows stale air to be released. When not in use in the NBC mode the second filtration stage may be by-passed and the system may then be used for general ventilation. An airflow indicator is provided on the unit for the operator to see that an adequate airflow is being achieved for the size of shelter and the number of occupants.

STATUS
In production.

MANUFACTURER
M.D.H. Limited, Walworth Road, Andover, Hampshire SP10 5AA, United Kingdom.

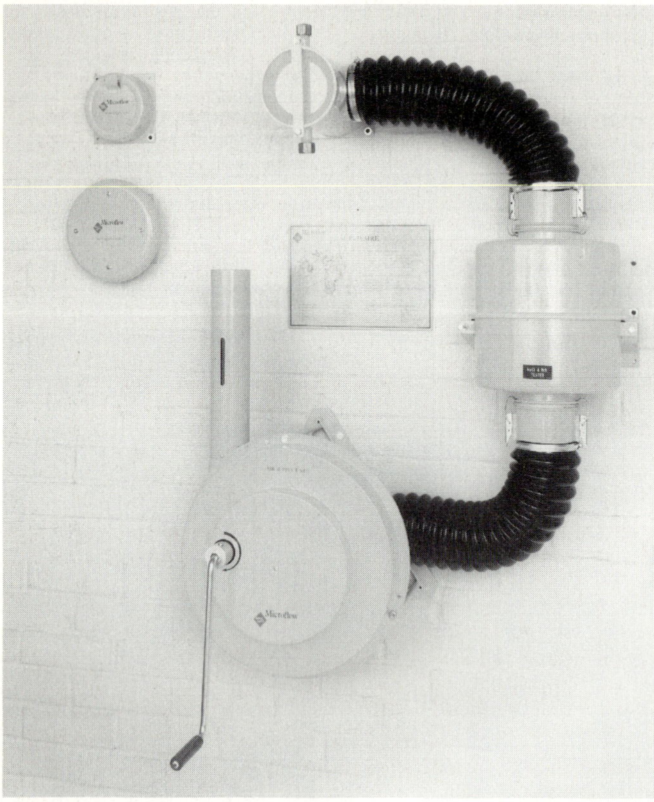

Manual version of a Survivaire shelter filtration unit with a capacity of 42 cubic metres an hour

Electrical version of a Survivaire shelter filtration unit with the filter unit by-passed for general ventilation purposes

NGL Filtration Division NBC Air Filter Unit

DESCRIPTION
NGL (Normalair-Garrett Limited) Filtration Division produce a NBC air filter package intended to provide NBC protection for 25 to 50 persons inside structures such as underground troop shelters, field hospitals and suitable surface structures. The package is man-portable (by two men) and consists of an air filtration unit and a liner. For larger shelters additional numbers of units and liners can be employed.

The air filter unit is intended to be installed at the furthest point from the shelter airlock. The air filtration unit is a compact quiet running filter/blower unit that provides an uncontaminated airflow of 170 cubic metres an hour; a 370 cubic metres an hour unit is also available. It also provides positive pressure to inflate the associated liner. The unit is electrically powered and is suitable for either 24 to 28 volts dc or 110 or 240-volt ac voltages derived from mains supplies or generators.

The air filtration unit contains a pre-filter, a particulate filter and a vapour filter, all designed to conform to the NATO filtration efficiency requirements for service respirators. The unit body is constructed of high strength glass fibre/resin material and can be installed in any suitable space either above or below ground. Filtered air is passed to the liner through a 101.6 mm (4-inch) flexible hose.

The liner, which folds down similar to a tent, is manufactured from translucent flexible reinforced plastic material and has an easily detachable airlock. It is self-supporting even when no inflation is supplied from the air filtration unit. The liner was designed in a modular form such that it can be extended to provide protection in multiple shelter installations. Liners are available for other standard shelters and can be produced to suit any reasonable building shape. Modules can be connected together using Drilock closures. The liner inlet port iuced to suit any reasonable building shape. Modules can be connected together using Drilock closures. The liner inlet port is fitted with flow indicato tapes and a grid. Adjustable vents to ensure the flow from the liner to the airlock to ambient are fitted to the liner and airlock walls adjacent to the access doors. Airflow is controlled by means of adjustable velcro flaps fitted across the vents. Tests have demonstrated that the liner can be installed and inflated complete with an airlock in a Mexe-type shelter in approximately 6 minutes.

The air filtration unit is 954 mm long and has a diameter of 508 mm. Weight is 56.8 kg.

NGL Filtration NBC air filtration unit

STATUS
In production.

MANUFACTURER
Normalair-Garrett Limited, Hy-Mech Product Group – Filtration, 11 Garrett Road, Lynx Trading Estate, Yeovil, Somerset BA20 2TH, United Kingdom.

L1A1 Portable NBC Filtration Unit

DESCRIPTION
The L1A1 portable NBC air filtration unit is used to deliver up to 47.2 litres per second of filtered air to an inflatable chemical agent-proof liner erected under a conventional canvas tent, a Mexe Mark 111 shelter (qv), or within a suitable building. The air flow inflates the liner at a pressure slightly above atmosphere. Air flow within the liner is controlled by vents in the liner.

The air filter unit is electrically powered using a 24-volt dc supply or a mains ac supply. Filtration involves two pre-filters, one for coarse particulates and one for fine, an ultra-high efficiency particulate filter and two activated charcoal anti-vapour filters. The filter unit is suspended by anti-vibration mounts within a carrying frame. The unit can be carried by two men and weighs 55 kg. Dimensions are 975 × 590 × 500 mm.

Supplied with the unit are two hose reels. One carries 100 metres of three-core cable on a portable hand reel for use with ac mains supplies. The other reel has 10 metres of two-core cable for dc supplies; two terminal clips are provided for connections to batteries. Also supplied are two hose assemblies. A single hose is used to connect the air filter unit to the liner. The hose is 4 metres long and has a chemical resistant terylene and butyl rubber cover. A spiral spring steel reinforcement is provided to allow the hose to regain its shape if run over by a wheeled vehicle. Each hose weighs 6 kg.

STATUS
In production. In service with the British Army and Royal Air Force.

L1A1 NBC air filtration unit in use

MANUFACTURER
Howden Aircontrol Limited, Hawthorn Road, Staines, Middlesex TW1 3AY, United Kingdom

Howden Aircontrol Limited NBC Filtration for Container Bodies

DESCRIPTION
Howden Aircontrol Limited produces a range of NBC filtration kits for use on military container bodies and shelters and soft-skinned vehicles. The externally-mounted four stage filtration pack will supply 47 litres per second of clean air into a container where a pre-set pressure relief valve will maintain an overpressure to overcome leaks in the container. Control and monitoring of the system is carried out from within the container using a pressure indicating unit and an electrical control box.

The NBC filtration pack contains a two speed fan and four stages of filtration. The first filtration stage is a self-cleaning inertial separator for particulates. The second stage uses a washable particulate pre-filter while the third stage uses a disposable ultra-high efficiency particulate filter. The fourth stage uses 2 disposable activated charcoal anti-vapour filters. Air flow on the normal setting is 32 litres a second; on high it is 47 litres a second. Pack dimensions are 858 × 490 × 470 mm and the weight is 64 kg.

A heater/attenuator unit has a thermostaticaly-controlled air heater which is switched on automatically when air entering the container from the filtration pack falls to 0°C. The heater also acts as a supply air sound attenuator. When a heater is not required a sound attentuator of identical dimensions is fitted. The unit weighs 2.65 kg.

The system is completed by the electrical control box (weight 1.9 kg), a pressure indicating unit (weight 2.35 kg) and a pressure relief valve (weight 0.25 kg).

STATUS
In production. In service with the British Armed Forces.

Complete NBC filtration system for container bodies

MANUFACTURER
Howden Aircontrol Limited, Hawthorn Road, Staines, Middlesex TW18 3AY, United Kingdom.

Howden Aircontrol Limited Collective NBC Protection for Armoured Vehicles

DESCRIPTION

Howden Aircontrol Limited designed and manufacture environmental control systems for armoured vehicles and have been chosen to participate as specialists in the Warrior infantry combat vehicle programme.

For Warrior they produce an environmental control system which provides NBC collective protection plus crew heating and forced draught cooling. In both the ventilation and NBC modes the air flow is 165 litres a second, providing an internal pressure of 7.5 mbar. The heating capacity is 3.5 kW. System controls include a four-speed fan control, a direct reading internal pressure gauge (with a facility for checking the particulate NBC filter condition), an interlock and instant NBC operation control, and a pressure relief valve.

Power consumption in the NBC mode is 47.5 A at 28.5 volts dc. Weight complete is 170 kg and the dimensions are 1400 × 660 × 520 mm.

STATUS

Warrior systems in production for the British Army.

MANUFACTURER

Howden Aircontrol Limited, Hawthorn Road, Staines, Middlesex TW18 3AY, United Kingdom.

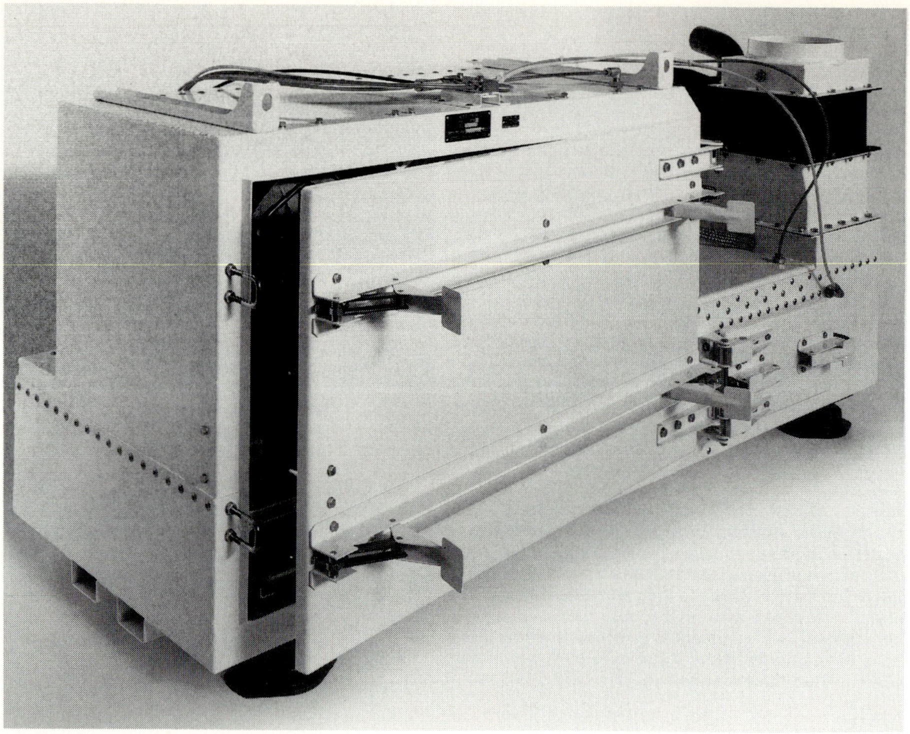

Typical Howden Aircontrol Limited NBC collective protection unit for armoured vehicles

UNITED STATES OF AMERICA

Survivable Collective Protection System (SCPS)

DESCRIPTION

The Survivable Collective Protection System (SCPS) is a modular, semi-hardened structure intended to protect military personnel against conventional, biological and chemical weapons. The SCPS has a fully integrated life support system to allow command and control functions to proceed in the event of an enemy attack. A single SCPS will provide 168 personnel (84 per 12-hour shift) with four days of continuous life support in a chemical or biological warfare environment, or 336 survivable man days in its present configuration.

To enter the SCPS personnel pass through T-sections and process through steel blast doors which aid in protecting shelter occupants and equipment from the effects of conventional munition blast. A processing area, a Toxic Free Area (TFA) and an equipment room make up the major segments of the shelter. Personnel process through air locks en route to the TFA. A filtration sub-system provides a sufficient and continuous flow of toxic-free air within the air locks. Electrical power is provide by an engine/generator unit which is an integral part of the shelter. The shelter also incorporates sanitary facilities, food, fuel, water storage and sleeping accommodation.

A complete SCPS incorporates the following components:

- A 120/208-volt three phase engine/generator
- A 300-gallon (1364-litre) fuel tank
- Three 600 cfm filter blower units
- A 5000 cfm exhaust fan
- Air locks
- Changing booths
- Blast doors and valves
- Blast T-sections
- 60-watt lighting with dimmers
- Automatic emergency lighting
- Sanitary facilities
- A central electrical control
- Food and water (420 gallons/1910 litres) storage
- A heat and smoke detection system
- A transfer switch for an external power capability
- A two-way inter-communication system with master and sub-stations
- Sleeping accommodation with litters for 84 occupants
- Retaining walls
- Drainage provision

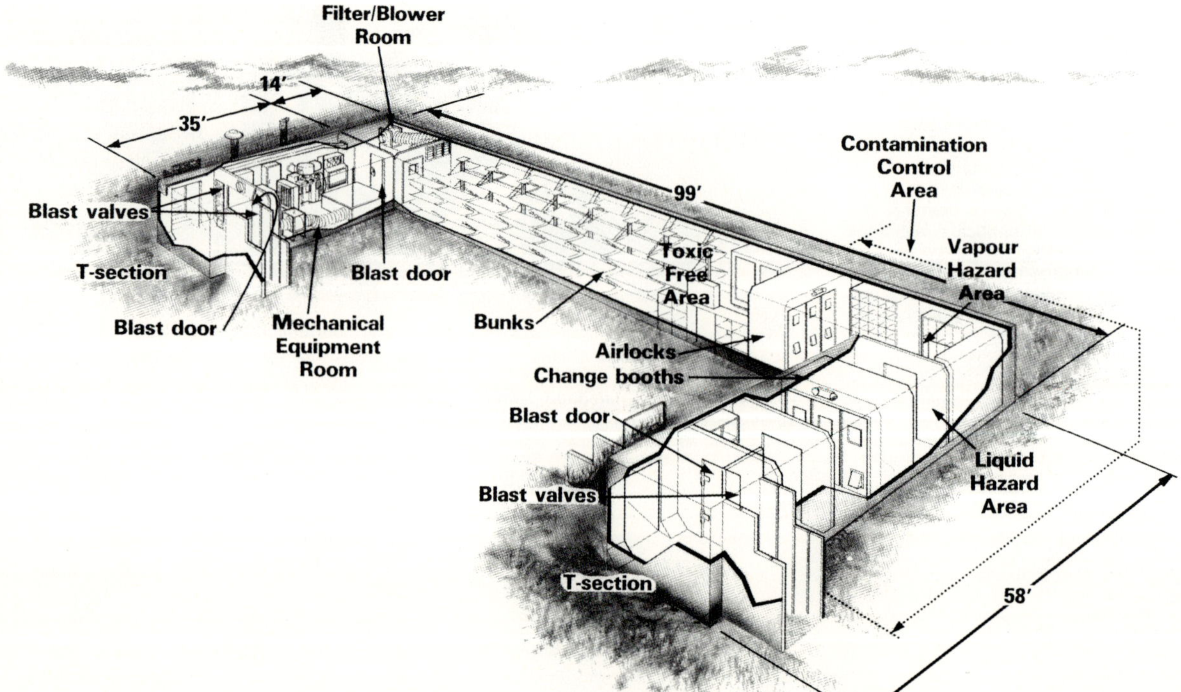

Layout of a typical Survivable Collective Protective System (SCPS)

Storage cabinets
Fullers earth troughs
A shelter manager's desk
Chemically resistant mirrors
Fire extinguishers
Optional equipment includes an air conditioning system, a 120 000 BTU heating system, a pile and cap foundation, and an exterior communication facility.

STATUS
In production for the US Air Force since 1985.

MANUFACTURER
BMY Defense Systems Operations (A Division of Harsco Corporation), PO Box 1512, York, Pennsylvania 17405-1512, USA.

AirResearch Environmental Control/Nuclear, Biological and Chemical (EC/NBC) Protection System for M1A1

Allied-Signal Aerospace Company's AirResearch Los Angeles Division based at Torrance, California, is producing the Environmental Control/Nuclear, Biological and Chemical (EC/NBC) protection system for the US Army's M1A1 Abrams main battle tank. AirResearch is producing the system under contract to General Dynamics Land Systems Division, the prime contractor for the M1A1.

The M1A1 Abrams is the first of any of the US Army's armoured vehicles to be equipped with an EC/NBC system. In 1977 the US Congress passed Public Law 95-79 requiring the US Army to begin equipping its main battle tanks, mechanised infantry vehicles, armoured personnel carriers and similar combat vehicles with such a system to protect personnel assigned to them.

AirResearch began full-scale development engineering of the EC/NBC system in 1982 and successfully completed field testing in mid-1984. As a result of these tests, which were performed from Fort Greeley, Alaska, to the Panama Canal Zone, AirResearch was awarded an initial production contract and began production deliveries in early 1985. Current deliveries are being made under subsequent follow-on contracts.

The AirResearch EC/NBC system provides the M1A1's four-man crew with conditioned air for breathing as well as personal heating and cooling as required, while they are wearing their protective suits and masks. Additionally it provides positive air pressure within the tank hull to prevent the entry of any NBC contaminants. The system uses an aviation-type air cycle refrigeration unit in which bleed air from the tank's gas turbine engine provides the energy required for powering the air cycle machine. The system has five major components: a precooler, a high pressure water extractor, an air-bearing cooling turbine, a primary heat exchanger, and an NBC filter assembly plus valves and controls. The bulk of the system is packed into a 152 mm high sponson tool box below the turret. The filter canisters, which remove contaminants from the air by physical and chemical means, are furnished by the US Government.

STATUS
In production. In service with the US Army.

MANUFACTURER
Allied-Signal Aerospace Company, AirResearch Los Angeles Division, 2525 West 190th Street, PO Box 2960, Torrance, California 90509-9500, USA.

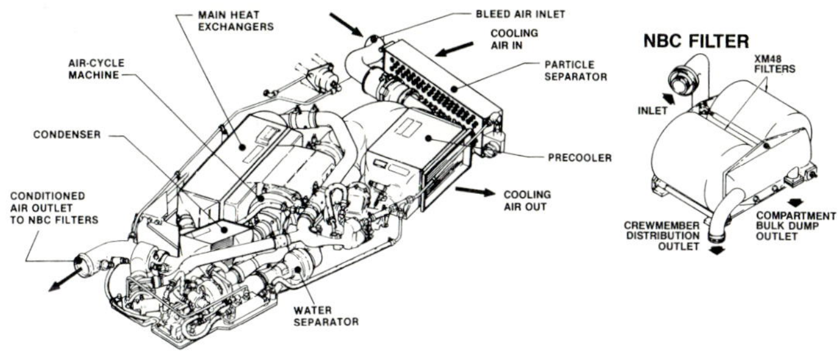

AirResearch Environmental Control/Nuclear Biological and Chemical (EC/NBC) protection system components for M1A1 Abrams MBT

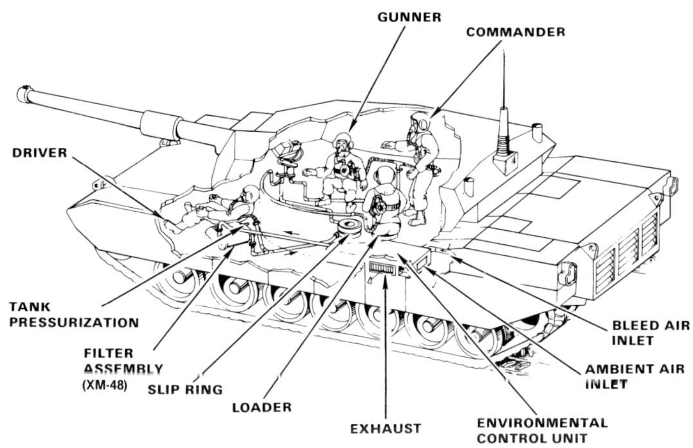

Schematic layout of AirResearch EC/NBC protection system for M1A1 Abrams MBT

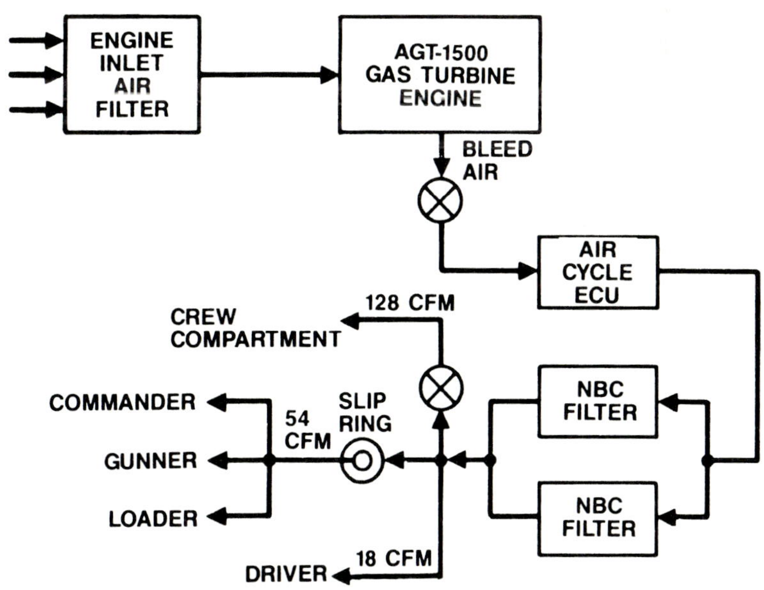

Layout of AirResearch EC/NBC protection system in M1A1 Abrams MBT

MEDICAL AND LIFE SUPPORT EQUIPMENT

AUSTRIA

SAB-JAB 87 Ventilation Unit for NBC Protective Clothing

DESCRIPTION
The Blaschke SAB-JAB 87 ventilation unit for NBC protective clothing is intended to provide cooling air to enable NBC suit users to remain operational for longer periods and prevent much of the debilitating stress caused by body heat being unable to escape through impermeable fabrics. The unit has a small electrically-driven impeller fan that blows air into an NBC suit through a short length of corrugated hose. Air is drawn into the unit through two filter canisters and the unit is carried slung from a strap. The unit causes a slight over-pressure within the protective suit and thus prevents the ingress of harmful agents due to suit damage.

There are two models of ventilation unit. One is the SAB-JAB 87/1L which is powered by four 1.5-volt LR20 batteries; the weight is 2.05 kg. The other unit is the SAB-JAB 87/2L powered by three 1.5-volt LR20 batteries; the weight of this version is 1.94 kg. Both units are 280 mm long, 230 mm wide and 108 mm deep. On both models the batteries provide power for up to 40 hours of uninterrupted use.

STATUS
In production.

MARKETING AGENCY
BMS Group of Companies, Braung. 44, A-1170 Vienna, Austria.

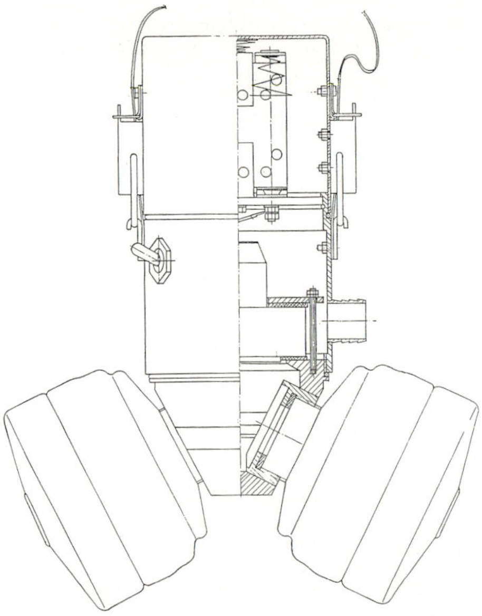

Cut-away drawing of SAB-JAB 87 ventilation unit for NBC protective clothing

CANADA

NBC Casualty Bag

DESCRIPTION
Manufactured from the same material as that used in the Canadian Protective Coverall, the NBC casualty bag is a container with a zip-fastener along one side and half-way across each end. Laminated plastic windows are provided at each end and the bag incorporates a metal frame to hold the bag away from the patient's face. The bag completely encloses the body and is olive green in colour. Carrying handles made from webbing material are incorporated.

The casualty bag weighs approximately 2 kg, is 2.51 metres long and 0.91 metre wide.

STATUS
In production. In service with the Canadian Armed Forces.

MANUFACTURER
Anachemia Canada Inc, Case postale 147, Lachine (Québec), Canada H8S 4A7.

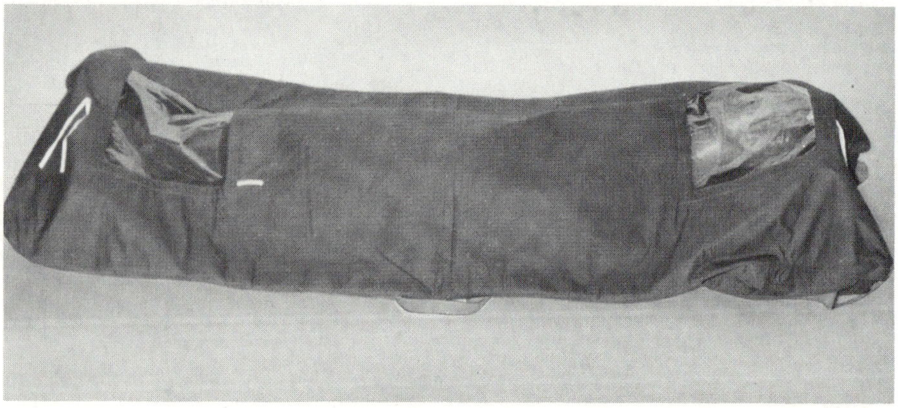

Canadian protective casualty bag

FRANCE

NBC Casualty Bag

DESCRIPTION
This NBC casualty bag for recumbent wounded is provided with an air filter generator using a bank of standard filtration canisters, the same cansiters used with the Protective Mask, Model ANP 51 M 53 (see entry in *Protective masks and clothing* section). The canisters provide the casualty with protection for up to 10 hours. The casualty bag is provided with a zip fastener for access to the casualty and an interior pocket is provided to contain medical care instructions.

STATUS
In production. In service with the French armed forces.

MANUFACTURER
Techniques Michel Brochier SA, ZI – Routes des Chartinières, Dagneux 01120 Montluel, France.

SA TMB NBC casualty bag

Polyinject Syringe

DESCRIPTION
Poutrait-Morin produced their first autoject syringe for the French armed forces in 1967. The latest version is the Polyinject, produced in association with the Institut Mérieux, which can contain up to three different nerve agent or other chemical agent antidotes for self-administration in one operation.

The Polyinject is a multi-chambered, pressure activated self-injector syringe with the antidote solutions contained within the main body of the syringe. The chambers are separated by a system of rubber valves with the volume and number of compartments variable according to the antidote solutions and their volume. The maximum volume for two solutions is 5.5 millilitres. Each addition compartment or chamber decreases the total capacity by 1 millilitre. A double bevelled needle is fitted with an external diameter of 0.9 mm and an internal diameter of 0.6 mm. The exposed length of the needle after injection is 20 mm and it is capable of penetrating several layers of protective or other clothing.

Each Polyinject is issued sealed in an individual rigid plastic tube with each syringe bearing a simple illustrated instruction label. After unwrapping, a safety cap is removed and the Polyinject is placed against the thigh muscles. A button on the end of the syringe is then pressed and the automatic injection of the various antidote solutions is made without the solutions mixing in the syringe. The entire operation takes between 4 to 6 seconds.

The Polyinject weighs 48 grammes and is 152 mm long; the diameter is 21 mm.

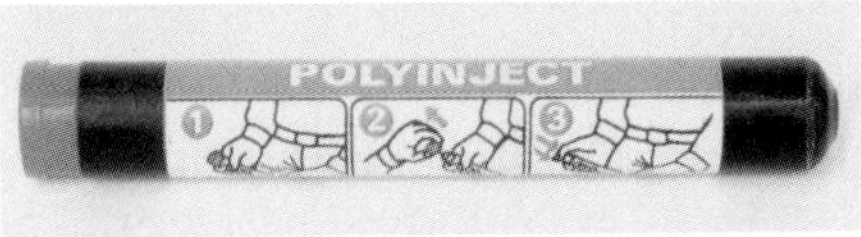

Poutrait-Morin Polyinject syringe

STATUS
Under test in various countries.

MANUFACTURER
Poutrait-Morin, 25 rue de la Courneuve, 93300 Aubervilliers, France.

GERMANY, EAST

Protective Kit, Medical, Model MSP-18

DESCRIPTION
The Model MSP-18 kit is issued to every East German soldier for self-treatment of the effects of chemical and biological warfare. It is issued in a small case measuring 100 × 90 × 30 mm and contains several different types of medical drug administrants and an instruction sheet. One half of the case contains six squeeze-type hypodermic syringes (syrettes) for treating nerve agents, some cotton pads, and a package of six tablets for use against lung irritants. The other half of the case holds a red plastic eye dropper for flushing nerve agents from the eyes, a blue plastic syrette containing an antidote for hydrogen cyanide poisoning, a phial containing ampoules for inhaling after hydrogen cyanide (HCN) poisoning, and a colourless syrette containing a morphine-based pain killer. The kit is said to be effective against the agents for which it is issued.

STATUS
In service with East German armed forces.

MANUFACTURER
State factories and laboratories.

GERMANY, WEST

Kärcher NBC Ventilation Kit for NBC Protective Suits

DESCRIPTION
One disadvantage of NBC protective suits providing up to 100 per cent protection is that temperatures inside the suit increase until heat stresses prevent further activities by the wearer. Even at temperatures around 20°C the working time for a fully-protected wearer is limited to a maximum of one hour. This problem has been addressed by the Kärcher NBC Ventilation Kit for NBC protective suits.

The Kärcher Ventilation Kit is a lightweight fan which draws air through normal standard NBC filter canisters and blows it through a hose connection to inside the protective suit. A second hose can be connected to the protective mask. The filtered air effects the temperatures inside the suit and helps to extend its period of use by a factor of approximately three during normal temperatures.

The ventilation kit is driven by rechargeable batteries with an operating time per battery of approximately 4 to 6 hours. The kit is fixed to a belt worn over the protection suit. Special valves incorporated in the protection suit allow internal ventilation and build up an interior overpressure that provides extra protection in the event of slight damage to the suit exterior.

SPECIFICATIONS
Weight: (including battery) approx 2.1 kg
Length: 220 mm
Width: 180 mm
Height: 170 mm
Air flow rate:
(with 3 filters) 250 l/min
(with 2 filters) 160 l/min
Battery pack: 9 V, 3.6 Ah

STATUS
In production. Under test by several countries.

MANUFACTURER
Alfred Kärcher GmbH & Co, Alfred-Kärcher-Strasse 28-40, D-7057 Winnenden, Federal Republic of Germany.

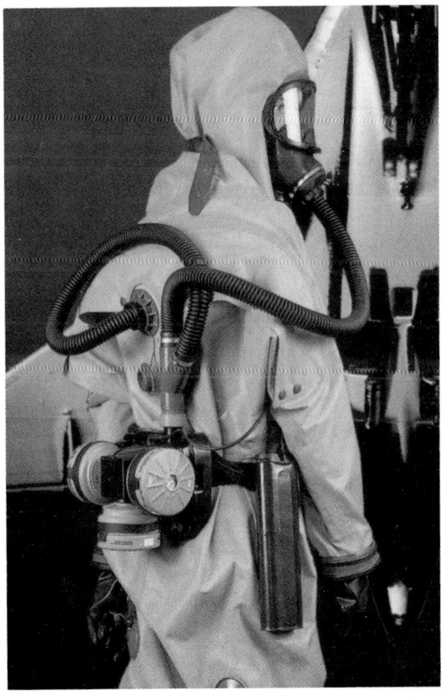

Kärcher NBC ventilation kit for NBC protective suits adapted for use with the Kärcher NBC protective suit

HUNGARY

ReaJector Autoinjector

DESCRIPTION
The ReaJector autoinjector was designed for single-hand personal use by personnel in stress situations. It delivers a dose of 2 millilitres of a nerve agent antidote or some other medication through protective or other clothing.

The ReaJector has a cylindrical plastic body containing the medical solution, a needle and a spring. Pressing a button on one end of the device releases the spring to drive the needle into the body, usually the thigh, and automatically inject the solution. The ReaJector is then discarded.

STATUS
In production.

MANUFACTURER
TECHNIKA Foreign Trade Company, Budapest-Hungary X, Salgótarjáni.20, 1475 Budapest, POB 125, Hungary.

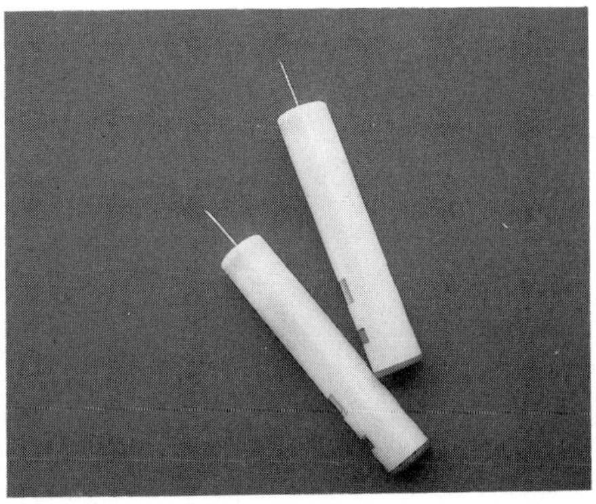

ReaJector autoinjectors

NETHERLANDS

AtroPen, ComboPen and MultiPen Nerve Agent Antidotes

DESCRIPTION

The terms AtroPen, ComboPen and MultiPen are registered trade names for auto-injectors. They are hypodermic devices meant for self-administration and are usually issued to personnel likely to encounter nerve agents.

The AtroPen contains atropine only and counters the symptoms of nerve agent poisoning. The device is 90 mm long and 12 mm in diameter.

The ComboPen contains a solution of water soluble medicaments (atropine and oxime). The device is 145 mm long and 20 mm in diameter. (The ComboPen is in service with the British Armed Forces as the Autoject (ComboPen) Nerve Agent Antidote L2A1. The L3A1 is a training device.)

The MultiPen may contain up to three medicaments stored separately and, for example, makes it possible to combine the water-insoluble diazepam with other nerve agent antidotes. The device is 170 mm long and 22 mm in diameter.

All three devices are used in the same manner. The safety cap is removed from the end and the injector is pressed against the outer thigh muscles. Further pressure releases a needle which can penetrate combat clothing and the body skin to introduce the atropine-based nerve agent antidote into the body. The needle is held in place for five seconds and the device can then be discarded. If nerve agent symptoms persist the self-treatment is repeated after 15 minutes. Instruction for use are printed on the body of the device.

If required the device can be fitted with a tablet holder to contain an anti-convulsant tablet such as 5 mg diazepam (Valium).

For training, an inert version with no needle or medical content is available. In place of the needle a blunt plastic button is projected and the device can be re-cocked with a special cap attachment.

Various types of filling other than nerve agent antidote can be used, and if required the ComboPen can be reloaded using a cartridge that contains the nerve agent antidote and a needle.

In June 1987 Duphar were awarded a contract worth $4 508 905 to supply 1 610 323 atropine injectors to the American Defense Logistics Agency by July 1988.

STATUS

AtroPen and ComboPen: In production and in service with several NATO nations including the United Kingdom and the United States.
MultiPen: In production.

MANUFACTURER

Duphar BV, Medical Devices Division, Drentestraat 11, 1083 HK Amsterdam, Netherlands.

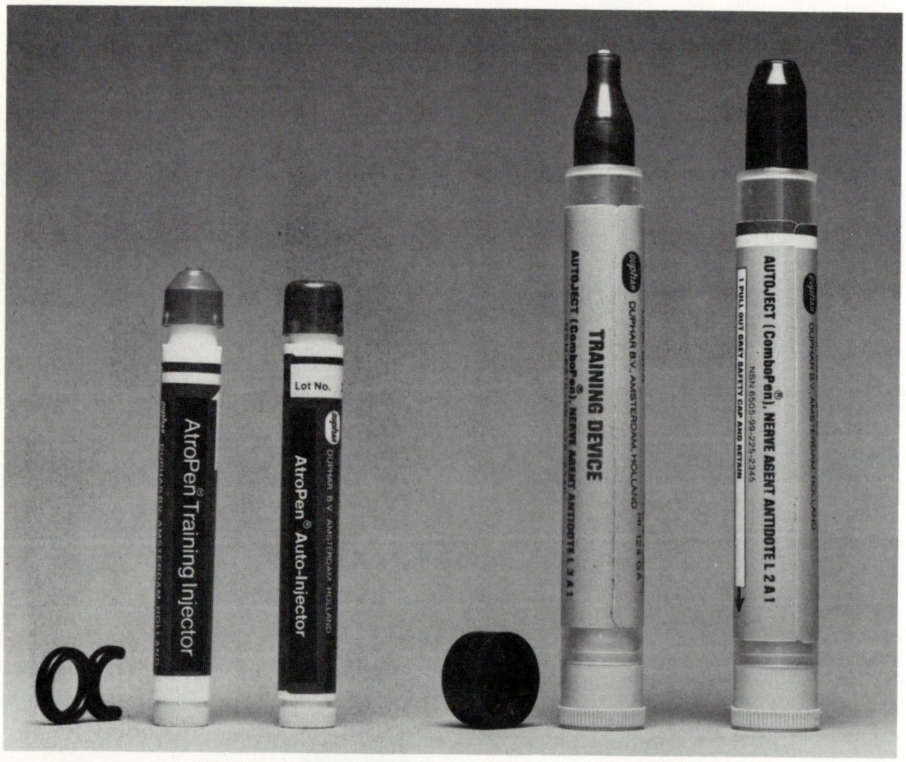

AtroPen (left) and ComboPen (right) together with their training devices

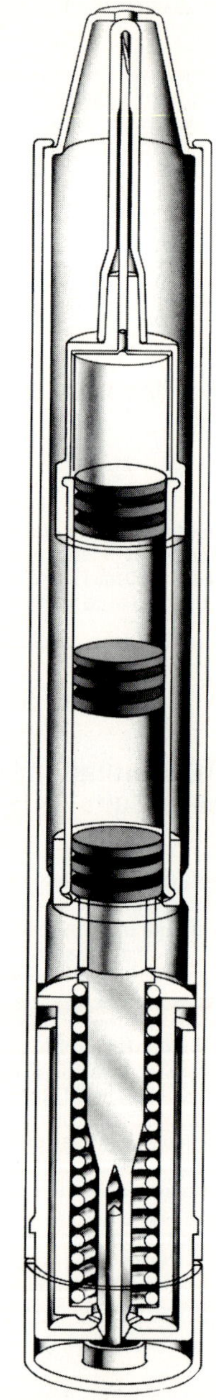

Cutaway drawing of MultiPen showing separate antidote agent compartments

Duphar Nerve Agent Poisoning Prophylaxis

DESCRIPTION

The most advanced therapy for nerve agent poisoning is the immediate (self) injection of atropine, oxime and diazepam. This therapy is efficient against most nerve agents but not against Soman (GD). The only efficient and available antidote against Soman is pyridostigmine given prophylactically and followed by the usual therapy. This drug is available as a 30 mg tablet that should be taken orally three times a day. The standard packing is a blisterpack containing 21 tablets but other packagings are available on request.

STATUS

In service with several NATO nations.

MANUFACTURER

Duphar BV, Medical Devices Division, Drentestraat 11, 1083 HK Amsterdam, Netherlands.

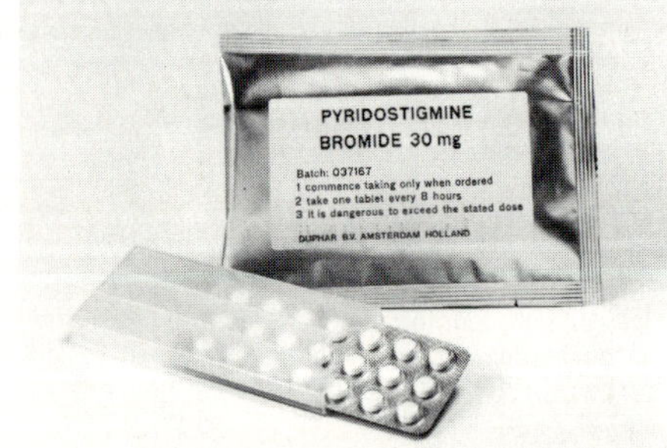

Blisterpack of pyridostigmine tablets

UNION OF SOVIET SOCIALIST REPUBLICS

Atropine Injector

DESCRIPTION
This injector is issued to all Soviet troops and is used as a self-administered treatment for nerve agent poisoning. The injector is a squeeze-type hypodermic syringe which contains and dispenses a 1 cc dose of 0.1 per cent atropine sulphate solution from a flexible plastic capsule. The injector needle cover is removed and a piercing wire already in the needle is pushed into a membrane that seals the needle from the capsule. The wire is discarded, the needle pushed into an outer thigh muscle and the capsule squeezed to inject the solution. This injector is simple and relatively cheap and can be used to administer the atropine solution in about 20 to 25 seconds, but it is now likely to have been replaced by an automatic self-inject item.

STATUS
Production status uncertain. In service with the Soviet and Warsaw Pact armed forces.

MANUFACTURER
State factories and laboratories.

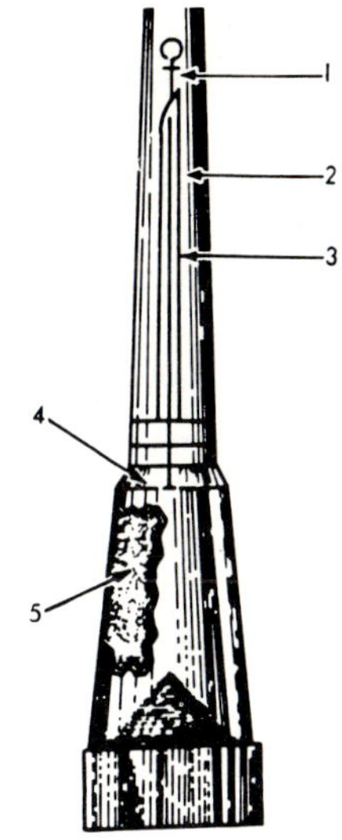

Atropine injector **(1)** *piercing wire* **(2)** *cap* **(3)** *injector needle* **(4)** *internal seal* **(5)** *liquid antidote*

Head-wound Protective Mask, Model ShR

DESCRIPTION
There are two known versions of this mask which differ only in the positioning of the extra outlet valves. One is virtually identical to that used on the Model ShM (see entry under *Protective masks and clothing*) and the other has an outlet valve on the right cheek, but both have the inlet valve at the level of the wearer's nose rather than at the normal mouth level which helps to prevent clogging of the air system by body liquids. The nose-level system also permits the easy supply of oxygen when necessary, but the usual anti-dim deflection of air across the eye-pieces as used in the Model ShM is not incorporated. The same hose and filter canister as on the Model ShM are used but the hose can be fitted with a T-piece to allow two filter canisters to be fitted to ease breathing. The Model ShR covers the whole head and it is secured in place by rubber straps.

STATUS
In production. In service with the Soviet and Warsaw Pact forces.

MANUFACTURER
State factories.

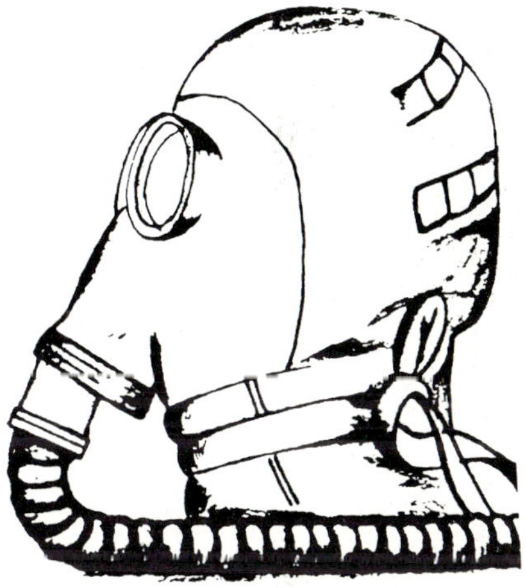

Model ShR head-wound protective mask

UNITED KINGDOM

NBC Casualty Bags

DESCRIPTION
There are two main types of British NBC casualty bag. The first is full length for use with stretchers and is intended to cover the stretcher fully with provision being made for stretcher handles. The second is for walking wounded and consists of a bag covering the top half of the body down to the knees and secured at the waist. There are sleeves and a transparent film panel in front of the face. Velcro fasteners are fitted to the access panels of both bags.

Both types of casualty bag are made using an outer layer of flame resistant cloth twill modocrylic and nylon yarn with silicone showerproofing. The inner material is flame retardent with a charcoal coating. The underside of the full length casualty bag uses a groundsheet formed from a 0.3 mm thick polyethelene laminate weighing 200 grammes per square metre and consisting of two outer layers of 0.05 mm low density polyethelene film bonded to central plain weave high density polyethelene.

The full length casualty bag weighs 1.58 kg and when packed measures 330 × 270 × 62 mm. When

NBC casualty bag for use with stretchers

NBC casualty bag for walking wounded

packed the half-length bag measures 240 × 120 × 62 mm and weighs 0.65 kg.

Under development at the Chemical Defence Establishment (CDE) at Porton Down in Wiltshire, is a nursing casualty bag that will provide a degree of access to a protected casualty by medical personnel.

STATUS
In production. In service with the British Armed Forces.

MANUFACTURERS
Compton Webb Group, Herbert Road, Newport, Gwent NP9 7XN, United Kingdom.
J. & S. Franklin Limited, Franklin House, 151 Strand, London WC2R 1HL, United Kingdom.
Remploy Limited, 415 Edgware Road, Cricklewood, London NW2 6LR, United Kingdom.
James North and Sons Limited, PO Box 3, Market Street, Hyde, Cheshire SK14 1RL, United Kingdom.

pneuPAC Ventilator/Resuscitator Equipment

DESCRIPTION
The pneuPAC ventilator/resuscitator equipment was produced orginally for medical and industrial use in situations where breathing is impaired or stopped after an accident, or after the sudden collapse of a casualty. The equipment is also of great assistance to chemical warfare casualties where breathing may become difficult or erratic following a chemical warfare attack and especially following exposure to, or inhalation of, nerve agents when muscular control of the breathing process may become uncontrollable or non-existent.

The basic process used with the pneuPAC involves placing a mask over the casualty's nose and mouth and supplying either oxygen, an oxygen and air mixture or pure air to the casualty at a predetermined and variable rate and up to a limited pressure, all controlled automatically by a unit known as the pneuPAC control module.

The pneuPAC control module is powered by gas and requires no battery or other electrical supply. It operates immediately when connected to a source of gas under pressure which may be a cylinder of oxygen or air, piped gases or a compressor. The gas drives the pneuPAC control module which generates the required breathing pattern using pneumatic logic components within the module. The control module weighs 1.3 kg and measures 90 × 60 × 180 mm. The pneumatic logic components are contained inside a foam plastic case that is strong enough to withstand being driven over by a vehicle. It can be safely dropped from a height of 1 metre and can be immersed in water. The control module can operate continually for many hours, requires no routine maintenance and normally requires a factory service check after every three years. It has a storage life of 5 years.

The gas from the pneuPAC control module passes through a flexible, reinforced, non-kink hose to the casualty valve. The valve is manufactured from reinforced plastics. Located on the valve is a transparent mask covering the casualty's nose and mouth. In the inspiration phase ('gas on') gas is delivered to the casualty and in the expiration phase ('gas off') the valve allows the casualty to exhale to the atmosphere. The valve, with a switching force of 0.5 kg, is not easily jammed, is simple to clean and cannot be misassembled. An audible relief valve is fitted to prevent excess pressure in the casualty's airway and warn of obstruction.

The pneuPAC control modules have a single protected control knob which can provide a continuous variation of the ventilation parameters (tidal volume and frequency of respiration) to suit the casualty. Four models of control module are available, as follows:

Model 1. Adjustable; for infants and children up to 5 years old

Model 2R. Adjustable; for adults down to 5 year old children with a click stop setting for adult cardiopulmonary resuscitation (CPR)

Model 4R. Pre-set; for adults only at CPR setting

Model 4R(R). Similar to Model 4R but adapted for adult rescue operations so that accidental separation of the mask from the casualty is prevented.

The models most applicable to military applications are Models 4R and 4R(R).

The casualty valve on the mask can be supplied with the option of air entrainment, ie mixing and augmenting the supplied oxygen with air. Most resuscitation operations, especially following exposure to nerve agents, commence with the supply of 100 per cent oxygen. After satisfactory breathing control is established the entrainment valve can be switched to draw in atmospheric air so that the casualty can receive the more suitable level of 45 per cent oxygen for long term ventilation. This facility can extend the life of a supply cylinder by a factor of three. When the entrainment valve is used in an atmosphere contaminated by chemical warfare agents the entrainment valve intake port can be covered by an activated charcoal filter.

pneuPAC control module with air entrainment valve on left

Air entrainment valve in position on casualty valve on nose and mouth mask

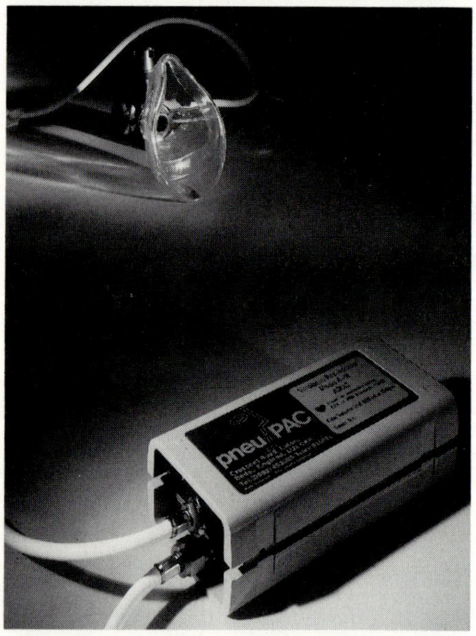

pneuPAC control module (foreground) with casualty valve and mask

The basic units for the system are the pneuPAC control module and the casualty valve set. All four modules of control module and casualty valve can be supplied in portable self-contained resuscitation sets with their own cylinder. Two typical sets are the Compact Instant Action Set which includes as standard a pneuPAC aluminium cylinder which gives, on CPR setting, up to 17 minutes resuscitation or 50 minutes with air entrainment. The larger Comprehensive Instant Action Set is designed for use with a pneuPAC aluminium 'D' cylinder which gives, on CPR setting, up to 25 minutes resuscitation, or 75 minutes with air entrainment.

The control modules may be incorporated into control consoles mounted in ambulances or static installations. For instance, the British Army uses two pneuPAC control modules mounted behind a single control panel mounted inside the rear com-

SPECIFICATIONS
Performance specifications at different settings, Model 2R, 4R and 4R(R)

Factor	Minimum	Maximum	CPR
Tidal volume	340 ml (+10%/−20%)	1450 ml (±)10%)	900 ml (±11%)
Frequency (breaths/min)	21 (+20%/−10%)	11 (±10%)	13 (±10%)
Inspiratory time	0.5 s	2.2 s	1.35 s
Expiratory time	2.35 s	3.3 s	3.2 s
Inspiratory/Expiratory ratio	1/5	1/1.5	1/2.4
Minute volume	7.1 litres	16 litres	11.7 litres

partment of a 1-tonne Land Rover ambulance, each module capable of resuscitating a casualty.

A wide range of accessories is available. A test set is available to check the pneuPAC control module calibration, tidal volume, frequency and relief valve settings. Other accessories include oxygen therapy sets, aspirator systems, patients' airway pressure monitoring systems, different relief and other valves, carrying pouches, mounting brackets and knapsacks. Portable and other compressors are available and a portable high pressure oxygen compressor capable of producing oxygen from atmospheric air has been introduced.

STATUS
In production. In widespread use with many NATO armed forces, including the United Kingdom (Army and Royal Air Force) and US Air Force, and in countries in North Africa, the Middle and Far East.

MANUFACTURER
pneuPAC Limited, Crescent Road, Luton, Bedfordshire LU2 0AH, United Kingdom.

pneuPAC Automatic Lung Ventilators

DESCRIPTION
These automatic lung ventilators are intended for the automatic resuscitation/ventilation of casualties following NBC situations. The ventilators are of modular construction and can provide a range of options to meet user requirements and operation logistics, from field environments to hospitals.

On all models there is an option of a facility which monitors automatically the level of breathing of the patient and changes from automatic controlled ventilation to demand as required by the patient. There exists also the option of an adjustable relief valve with an audible alarm, a pressure gauge to indicate airway pressure, and an air entrainment system which reduces driving gas consumption by two thirds. Special arrangements are required to remove toxic elements in the ambient entrained air.

The basic units, which differ mainly in their ventilatory controls, are as follows:
rescuePAC. This is the simplest model having a single ventilatory parameter control knob operating in a manner similar to the pneuPAC ventilator-resuscitator (see previous entry). A single control knob controls the frequency of breathing and tidal volume.
paraPAC. This model has two ventilatory parameter control knobs which vary the tidal volume and frequency.
transPAC. On this model the two ventilatory parameter control knobs vary the frequency and minute volume.
ventiPAC. This model has three ventilatory control knobs to control the inspiratory time, the expiratory time, and flow.

The internal control elements are housed in vibration isolator mountings within a shock resistant rigid casing capable of withstanding abusive use. The average weight of the units is 2 kg and the dimensions are 92 × 220 × 162 mm.

All units can be operated using pneuPAC airPAC compressors or the oxyPAC range of high pressure oxygen concentrators (see next entry).

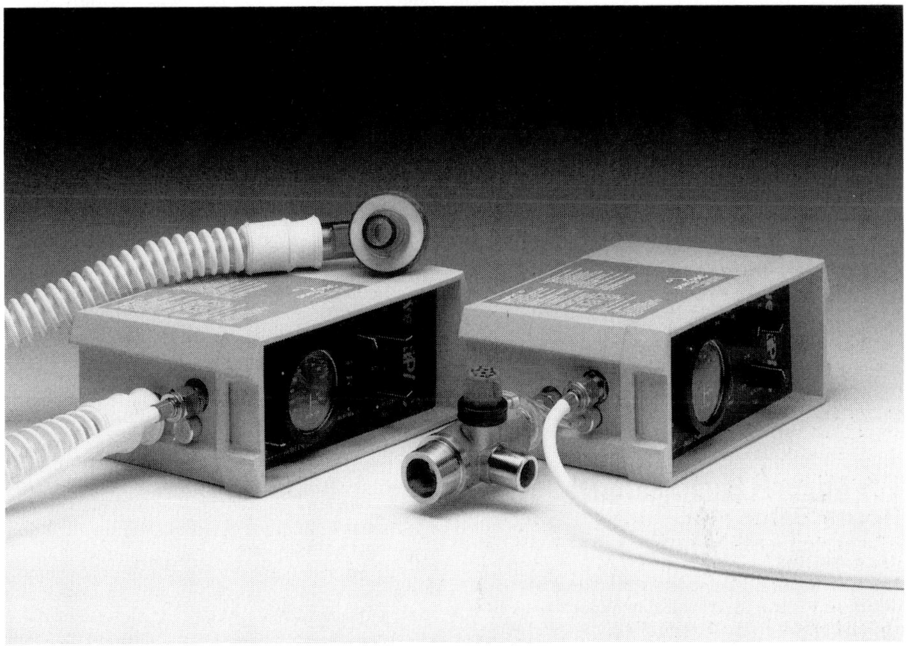

pneuPAC ventiPAC automatic lung ventilator

STATUS
In production.

MANUFACTURER
pneuPAC Limited, Crescent Road, Luton, Bedfordshire LU2 0AH, United Kingdom.

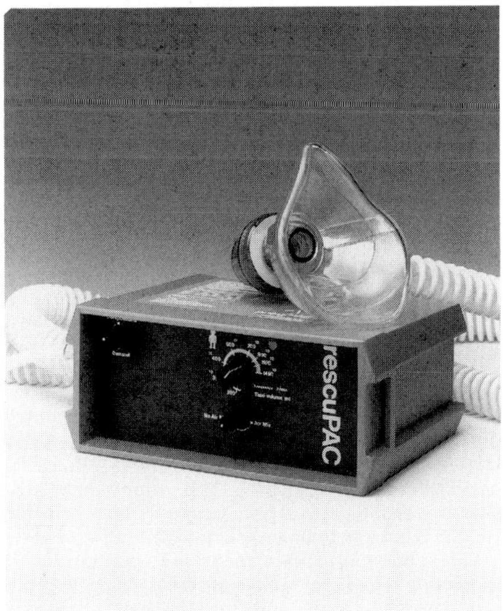

pneuPAC rescuePAC automatic lung ventilator

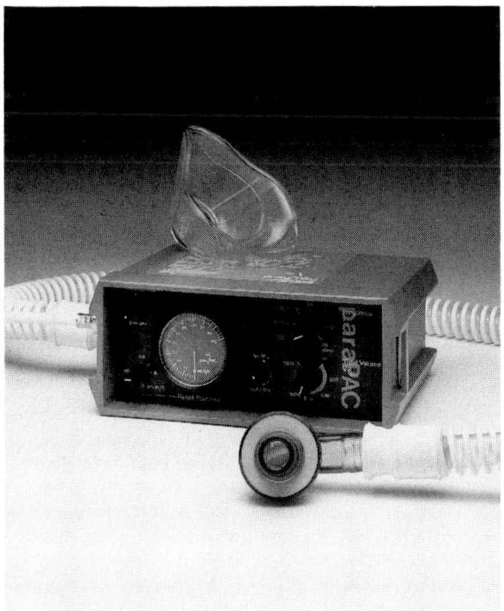

pneuPAC paraPAC automatic lung ventilator

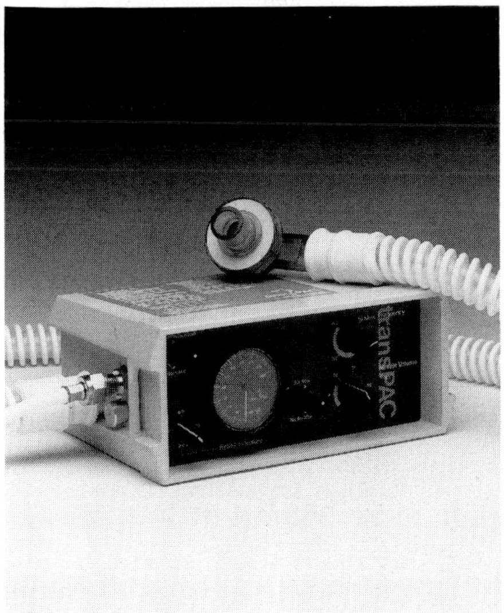

pneuPAC transPAC automatic lung ventilator

pneuPAC Oxygen and Air Supplies for Lung Ventilation

DESCRIPTION

To provide the necessary gas drive for lung ventilators pneuPAC Limited produce the airPAC range of air compressors. They also produce the oxyPAC range of oxygen concentrators for resuscitation in NBC situations.

The airPAC compressors take in ambient air which is filtered to remove harmful substances. The air is then compressed to provide the required drive gas for ventilators. The compressors can be driven from dc and/or ac power supplies. A commercial unit incorporates a non-moving part cooler working on the Peltier effect principle which cools the compressed air and removes moisture. This ensures that there is no bacterial growth and colonisation within the lung ventilator unit.

The oxyPAC range of high pressure oxygen concentrators take in ambient air which is filtered and harmful substances are removed using a pressure swing adsorbtion system. The oxygen is then separated and supplied to the patient. Oxygen can be supplied at pressures above 3.5 bar together with filtered air to drive ventilators and provide oxygen supplementation at levels above 30 per cent. The system is of a modular construction and can be built into vehicles in simple elements. In vehicles the air supply can be taken in from a vehicle compressor or from a separate compressor incorporated into the unit.

STATUS

In production.

MANUFACTURER

pneuPAC Limited, Crescent Road, Luton, Bedfordshire LU2 0AH, United Kingdom.

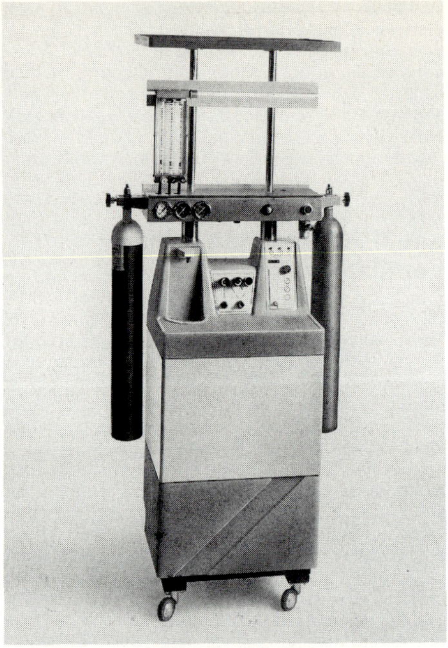

pneuPAC oxyPAC 5 oxygen concentrator

Life Support Engineering IA Resuscitator

DESCRIPTION

The Life Support Engineering IA resuscitator was designed for use in irrespirable atmospheres (IA). The equipment is manual and all equipment involved is housed in a strong steel carrying case.

The set comprises a cylinder filled with oxygen or air and the cylinder is fitted with a regulator. A medium pressure hose leads from the regulator to a sensitive demand valve fitted to a resuscitation bag. Two resuscitation masks, three guedel airways and head harnesses complete the unit. The carrying case measures 501 × 342 × 131 mm.

Operation is by means of a simple on/off valve. When the resuscitation bag is squeezed and then released, it opens the demand valve and draws oxygen or air direct from the cylinder into the resuscitation bag. As the bag is squeezed it pushes uncontaminated air or oxygen directly into the patient. The 2-litre steel cylinder contains approximately 400 litres of oxygen or air to provide a duration of 20 to 30 minutes.

If the casualty begins to breathe normally during resuscitation, his inspiration will automatically operate the demand valve.

STATUS

In production.

MANUFACTURER

Life Support Engineering Limited, Robell Way, Storrington, West Sussex RH20 3DN, United Kingdom.

MARKETING AGENCY

Denley Instruments Limited, Health & Life Support Division, Natts Lane, Billingshurst, West Sussex RH14 9EY, United Kingdom.

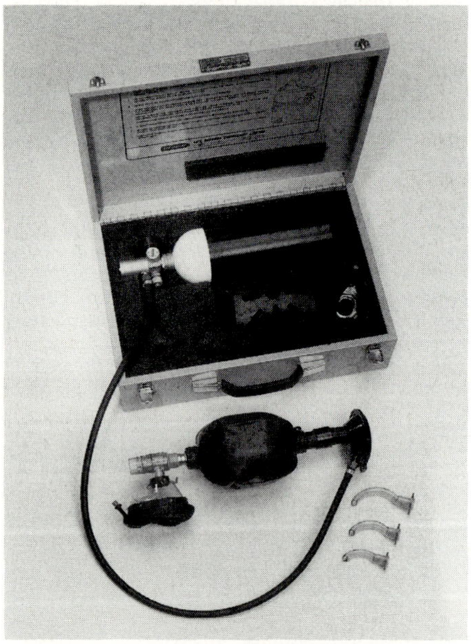

Life Support Engineering IA resuscitator set

Resuscitator NBC Portable Mark 2

DESCRIPTION

The Resuscitator NBC Portable Mark 2 is intended for the first aid treatment of respiratory failure in a contaminated environment. It is manually-operated, simple, robust and designed in such a way that it can be used with limited training.

The resuscitator consists of four main components; a rubber facepiece, an exhalation valve unit, an air supply unit and a filtration canister. All metal parts are anodised aluminium and all rubber parts are high quality with a life of at least ten years. The components are joined together by simple threaded joints. The whole apparatus can be sterilised by autoclaving or by boiling.

The rubber facepiece consists of an orinasal mask designed to make a pressure fit on as wide a range of faces as possible. The facial fit is achieved by a large anatomically shaped pneumatic cushion which is formed integrally with the wall of the mask. The air cushion is maintained at atmospheric pressure by means of a special pressure release plug situated on the left cheek of the mask. Attached to the front of the mask is a female screw adaptor which facilitates the inclusion of the filtration canister and is so positioned as to ensure that the apparatus is at the correct angle to thrust the casualty's jaw upwards to ensure a clear airway.

The exhalation valve unit is a slightly concave moulded rubber diaphragm with a central circular orifice which is closed by a hinged rubber flap. In its resting position the diaphragm is lightly pressed against a circular metal seating on the bellows side of the unit. When the bellows are compressed the diaphragm is raised off its original seating by the air pressure and seals against a larger circular seating on the facepiece side of the unit, and air passes through the central orifice into the facepiece, via the canister. At the end of the compression stroke the valve returns to its resting position and the casualty exhales freely to the atmosphere through a gap around the periphery of the unit. The hinged flap over the central orifice in the diaphragm ensures that the air will not leak back past the diaphragm, and thus that no exhaled air can be drawn into the bellows. The flap is reinforced by a larger rubber disc to prevent it from becoming irreversibly inverted. The diaphragm valve is contained in a flat circular metal housing which has a male screw adaptor at one side and a female screw adaptor at the other. This prevents the unit from being assembled incorrectly.

The air supply unit consists of circular rubber bellows with a male screw adaptor at one end for connection to the exhalation valve unit. At the outer end of the bellows, mounted on a metal plate, is the inlet valve. Also mounted on the metal plate is a spring loaded safety valve. The safety valve prevents over-inflation of the lungs. When used for resuscitation the bellows will deliver approximately 2.5 litres of air per stroke. Incorporated in the safety valve is a whistle to give audible warning that the casualty is being ventilated too vigorously or that air is not reaching the lungs.

In order to ensure a supply of uncontaminated air a filtration canister is placed between the exhalation unit and the facepiece. This is a modified version of the standard respirator filter canister.

The entire apparatus weighs approximately 0.9 kg and is contained in a rigid box measuring 254 × 152 × 152 mm. The box is fitted with a carrying handle.

STATUS

Production complete. In service with the British Armed Forces.

Interspiro Spiromatic High Performance Breathing Apparatus

DESCRIPTION

The Interspiro Spiromatic high performance breathing apparatus is designed for use by personnel operating in highly toxic environments. When used in conjunction with normal NBC protective clothing and a one-piece protective NBC overall the apparatus provides the user with a nominal protection factor in excess of 1,000,000.

The Spiromatic 324 is a typical model being a 300 bar twin 4-litre cylinder model that can provide air for up to 60 minutes with a facility to extend this to 4 hours or more for most situations. The equipment is comfortable and easy to use and offers no resistance to inhalation, thereby reducing wearer fatigue. The equipment is easy to service in the field using simple test equipment. Air cylinders are coated with zinc to protect against agressive chemical corrosion.

The Spiromatic face mask is designed to provide maximum safety in all environments and in the presence of toxic smoke and gases. It also provides automatic positive pressure and can be quickly fitted with NBC filter canisters for stand-by use in areas of low toxicity. Switching from ambient to compressed air is effected by clapping a hand over the front of the NBC filter canister; the pneumatic system is actuated by the wearer's breathing impulses.

Operator air supply is switched automatically on the first inhalation while another safety device prevents accidental turn-off. All cylinders are charged to 300 bar and cannot be over-filled. A warning whistle sounds to indicate when only 12 minutes of compressed air remain. If required the system can then be connected to an auxiliary air supply using quick-coupling connectors that can be employed safely even in a highly toxic environment.

An anatomically-contoured backplate is fixed to a stainless steel frame with adjustable shoulder and waist straps. The low profile twin-cylinder configuration ensures the apparatus is fitted close to the body and reduces the 'capsizing moment' of the wearer when crossing rough terrain. Non-standard harnesses for special applications are available. A one-piece NBC protection garment worn with the Spiromatic has space at the back to completely enclose the equipment.

Regulator peak flow capacity is 2000 litres per minute between 50 and 300 bar which permits two-man use, or one man plus a resuscitator or an air-powered tool. Cylinders are available in a range of sizes in a single cylinder or low-profile twin configuration.

A radio transceiver interface known as Savox can be fitted to the face mask in seconds without tools. Savox is intrinsically safe, lightweight and water resistant.

In an emergency, Spiromatic breathing apparatus can safely be used underwater but if diving is anticipated the Interspiro Divator Mark 2 range matches Spiromatic performance standards.

For use in a wide range of applications, a trolley- or sled-mounted auxiliary air supply unit is available to permit two-man operation up to 60 metres away from the unit. An air compressor unit for re-charging cylinders in the field is also available.

STATUS

In production. In service with Greece, Iceland, Netherlands, Norway, United Kingdom (Army and Royal Air Force) and the US Navy. The equipment is also used by numerous police and fire-fighting forces.

MANUFACTURER

Interspiro Limited, Halesfield 19, Telford, Shropshire TF7 4QT, United Kingdom.

Side view of a Spiromatic 324 breathing apparatus showing the low profile twin-cylinder configuration, with the wearer doffing the face mask

Spiromatic breathing apparatus worn by an EOD operative and with a NBC filter canister fitted to extend the duration of the wearer's compressed air in low toxicity areas or during decontamination; the one-piece NBC protective garment is partially removed

An EOD operative in action wearing an Interspiro Spiromatic high performance breathing apparatus

Medimech Binary Auto-Injector

DESCRIPTION

The Medimech binary auto-injector is used for the self-administration of nerve and chemical agent antidotes. The auto-injector has a distinctive appearance which means that it cannot be easily confused with any other similar object, having an olive drab body, a yellow cap at one end and a removeable blue 'C' clip at the other. Instructions for using the auto-injector are provided on the body of the device.

The auto-injectors are issued in individual plasticised pvc sleeves in strips of three. The auto-injector is removed from its sleeve (which can subsequently be used as a casualty tag), the blue 'C' clip is removed and a red pin on top of the body is pressed until it stops. The user then places the yellow cap against the injection point, removes the red pin and presses a green cap on top of the auto-injector body. A needle then injects the antidote through the user's clothing and into the body. The device is held in place for five seconds and is then removed and discarded.

The injector contains solutions of atropine sulphate BP and pralidoxime chloride USP or obidoxime chloride EP, in separate compartments. The resultant solution is colourless, or almost so. Each solution is contained in a cartridge assembly which is a sealed sterile container made of a non glass material and housed within the injector body. Tamper indicating seals are provided.

The length of the auto-injector is 150 mm and the diameter is 16 mm. The weight of each auto-injector is approximately 30 grammes. Shelf life is five years at normal temperatures and the auto-injector was designed for use in a temperature range from −45 to +50°C.

Medimech Limited also produce a single solution auto-injector for atropine or morphine self-administration. These auto-injectors follow the same general lines as the binary auto-injector but are 95 mm long and 16 mm in diameter. Weight is approximately 16 grammes. A morphine auto-injector is also produced for use by civilian authorities.

STATUS

In production.

MANUFACTURER

Medimech Limited, Unit 34, Riverside Estate, Sir Thomas Longley Road, Frindsbury, Rochester, Kent ME2 4DP, United Kingdom.

Stellar NBC Decontamination Water Purification Equipment

DESCRIPTION
Stella-Meta Filters make equipment for precoat filtration, reverse osmosis, carbon filtration, chlorination and chemical dosing to remove turbid, ionic, organic, chemical, nuclear and bacterial contaminants from water supplies. Three of the available equipment sets (AB3/NBC1, 8M3/NBC4 and FSD/NBC3) have outputs of purified water from 4500 to 13 000 litres per hour from fresh water, or from 1250 to 4100 litres per hour from NBC-contaminated fresh water. The units are available either as integral self-contained, or add-on units, compatible with non-NBC Stellar water purification sets. The company has been assessed by the Ministry of Defence to quality assurance DEF STAN 05-21/AQAP-1.

The company has produced a demonstration/pilot equipment integral water treatment unit known as the NBC5. This equipment can remove NBC contaminants from any water supply source by combining several existing Stellar components.

STATUS
In production.

MANUFACTURER
Stella-Meta Filters, Laverstoke Mill, Whitchurch, Hampshire RG28 7NR, United Kingdom.

Stella-Meta Filters NBC4A trailer for treatment of NBC contaminated water

NBC5 decontamination water treatment unit developed by Stella-Meta Filters for use in NBC environment

Charcoal Cloth Field Battle Dressing

DESCRIPTION
The Charcoal Cloth Field Battle Dressing was developed using a combination of modern wound care materials and technology to address the immediate treatment of battlefield wounds. The dressing contains activated charcoal cloth for infection control and for the adsorbtion of chemical agents from contaminated wounds. As the dressing contains a chemical agent-proof barrier it can also be used to patch areas of NBC protective clothing that may have been breached.

The field dressing provides firm, even compression around a wound to stem blood flow without impairing circulation, a high level of blood adsorbency, good infection control and strike-through prevention.

The dressing and its chemical agent-proof packaging were designed to be as simple to use as possible. Instruction for use are visible through the transparant package and the instructions remain intact once the package has been opened. Once opened the wound-facing side of the dressing is easily distinguishable from the outer cover material and the cohesive bandage will remain in a roll. A long 'tail' is provided to facilitate self-bandaging. The outer material can be produced according to customer requirements but is usually a heavy duty lightweight cotton in a camouflage colour.

The field dressing is available in two sizes, 200 × 100 mm and 200 × 200 mm.

STATUS
In production.

Charcoal Cloth field battle dressing

MANUFACTURER
Charcoal Cloth Limited, East Wing, Bridgewater Lodge, 160 Bridge Road, Maidenhead, Berkshire SL6 8DG, United Kingdom.

UNITED STATES OF AMERICA

PEG-Enzyme Therapeutic Process

DESCRIPTION
During January 1986 the US Army Medical Research and Development Command (USAMRDC) headquartered at Fort Detrick, Maryland, awarded a contract to ENZON Inc to develop effective enzyme-based antidotes to military chemical agents, including Soman (GD) and the trichothecene mycotoxins.

The development work carried out by ENZON Inc will concentrate on the PEG-enzyme process. With this process special enzymes that hydrolise (ie neutralise by reacting with water) are used to detoxify toxic substances such as nerve agents both inside and outside the body. The body would normally reject the enzymes involved but after attaching strands of a polyethylene glycol (PEG) polymer modified enzymes are disguised from the actions of the body's immunogenic determinants, ie the body will not set its immune system into action to destroy the enzyme. The enzymes involved are called PEG-enzymes and produce no allergic reactions. The catalytic action of the enzymes involved remain unchanged by the PEG modification.

PEG-enzymes can be introduced into the body by auto-injection devices or can be spread onto exposed areas of the body using a special cream. Other uses are the spraying or other application of PEG-enzymes in solution to detoxify large areas of contaminated terrain or equipment.

The PEG-enzyme process may also be used for the treatment of leukemia, immune deficiency and other diseases.

STATUS
Development.

MANUFACTURER
ENZON Inc., 300 C. Corporate Court, South Plainfield, New Jersey 07080, USA.

Mask, Chemical-biological: Headwound, ABC-M18

DESCRIPTION
This head mask is intended for casualties who are unable to wear a normal protective mask and is essentially a large hood made from a permeable filter material which will allow air to pass through for breathing and allow exhaled air and moisture to be diffused. The mask is secured at the neck by a drawstring and a flexible rubber collar. There are two 203 × 76 mm plastic windows. The head-wound mask can provide protection against chemical and biological agents for at least an hour and is also effective against radioactive dust. The mask is issued folded flat in a vinyl plastic carrying bag.

STATUS
Production complete. Declared obsolete but may be held in reserve by the US Army.

Vesicant Agent Protective Ointment, M5

DESCRIPTION
This ointment is issued to all field personnel for protection against blister agents and for neutralising such agents once used. It is primarily intended for use on the skin, but can also be used for decontaminating personal weapons and equipment. The ointment is white and the main constituent is chloromide S-330. It is issued in tubes 95 mm long and 19 mm in diameter.

STATUS
In service with the US Army, Navy and Air Force.

M272 Water Testing Kit, Chemical Agent

DESCRIPTION
The M272 Water Testing Kit, Chemical Agent replaces the AN-M2 water testing kit (see next entry) and the ABC-M30A1 refill kit. The M272 water testing kit can detect hazardous levels of nerve, mustard, lewisite and blood agents in treated or untreated water sources.

The kit contains a heat resistant plastic test container with a rubber stopper and connector; chemical agent detector tubes banded with blue or red; chemical agent test reagents; nerve agent test tickets; an instruction card; training simulants; a thermometer; waterproof matches and striking strip; a tube holder that fits in the lid, a clip and extra rubber connectors.

To test for nerve agent the white patch on a test ticket is wetted by a water sample. It is placed in a clip for an indicated time and then pressed against the white patch opposite on the ticket. If the patch turns blue there are no nerve agents present.

Tests for other agents involve filling the test container with a water sample. A reagent is added and after breaking off the ends of either the blue- or red-banded glass tubes, the tubes are inserted in the rubber connector. Heating the test container base will result in a colour change in the glass tubes which can then be compared with a sample colour chart on the instruction card to determine which agents, if any, are present.

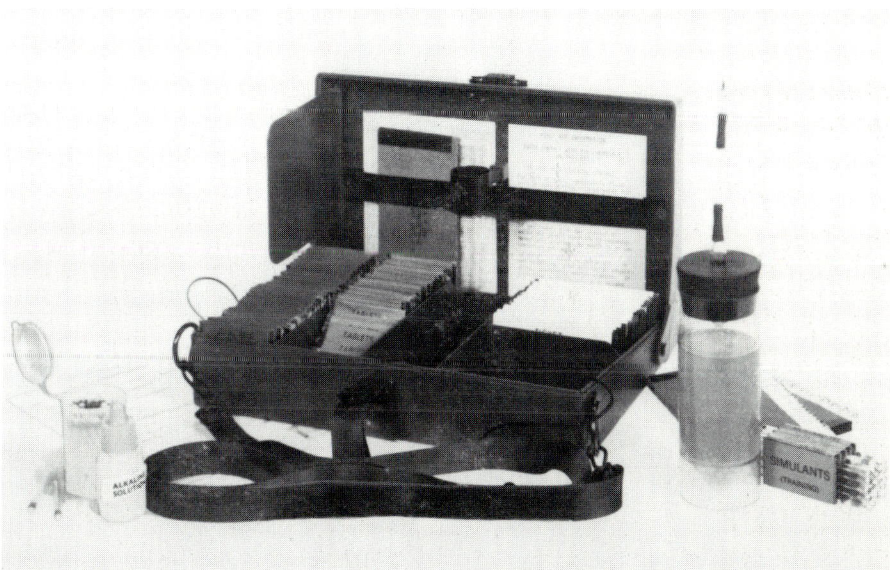

M272 Water Testing Kit, Chemical Agent

STATUS
In production.

MANUFACTURER
Anachemia Canada, 500 Second Avenue, PO Box 147, Lachine (Montreal), Quebec H8S 4A7, Canada.

Water Testing Kit, Chemical Agents, AN-M2

DESCRIPTION
The AN-M2 kit is used to test water samples in the field for the presence of various chemical warfare agents. It can be used only with unchlorinated water samples. The kit is contained in a plastic case measuring approximately 152 × 64 × 89 mm, and consists of 12 reagent containers, a metal scoop, glass containers and cleaning materials. The kits are issued to the various units in boxes of 24 kits, with each box weighing 22.7 kg.

STATUS
In service with the US Army, Navy and Air Force.

KT Corporation M10 Micro-Climate Cooling Unit (MCCU)

DESCRIPTION
The M10 Micro-Climate Cooling Unit is a personal cooling device designed to alleviate the problems of heat stress and dehydration caused by the prolonged wearing of NBC protective clothing and respirators for extended periods of time, especially in hot climates. The M10 MCCU is a small, lightweight, thermo-electric cooling module designed for comfort plus ease of operation and maintenance.

The M10 MCCU circulates a chilled water/glycol coolant through a vest worn under NBC protective clothing, thereby cooling an individual. Depending on the ambient conditions the closed loop system will draw over 200 watts from an individual, allowing him to continue to operate for extended periods of time, even outside a collectively-cooled environment. Power for operation can be taken from the dc battery or alternator in a vehicle or aircraft by connecting a single power cord.

Typical cooling figures are 240 watts at 35°C ambient temperature and 200 watts at 41°C. Power requirements are 500 (±20) watts provided from a 24 to 28-volt source. A lightweight, high-energy battery capable of supply power to the M10 for up to eight hours is under development.

The M10 MCCU module was designed to be worn attached to load-carrying equipment or carried in a respirator bag. In an aircraft or vehicle the M10 can be stowed at a prepared installation point or worn with little or no discomfort or inconvenience. The M10 unit links to the cooling vest via a leak-tight quick-connector. Two levels of cooling are available, Low and High. An additional modification will provide forced air to the user's respirator.

Several types of cooling vest are under development for use with the M10. One of these is an air/liquid hybrid vest capable of being used by crewmen on a vehicle equipped for air cooling. The crewman can use the M10 for dismounted operations with this system. Another design incorporates the cooling system into an armoured spall-proof vest for armour crewmen.

The M10 MCCU has undergone evaluation at the US Army Natick Research, Development and Engineering Center at Natick, Massachusetts. It has been selected to provide individual cooling to crew members of two different developmental armoured vehicles.

The M10 MCCU weighs 3.74 kg, is 310 mm high, 211 mm wide and 127 mm deep. The system can be decontaminated by full immersion in standard US Army decontamination agents.

STATUS
Final development.

MANUFACTURER
KT Corporation, 300 Long Beach Boulevard, Stratford, Connecticut 06497, USA.

CHEMICAL TRAINING, MAINTENANCE AND OTHER EQUIPMENT

CANADA

Hands Fireworks Chemical Simulators

DESCRIPTION
The Canadian Armed Forces are evaluating two types of chemical simulator that simulate a chemical warfare attack by projecting clouds of chemical powder or liquids without using any form of mortar. The simulators are in two basic forms. One is the Groundburst Chemical Simulator and the other the Airburst Chemical Simulator.

Both types of simulator can be fired from the ground or from the top of a vehicle. Each consists of a canister which acts as the projector. If required the canister may be held steady by a simple spiked holder into which the canister is clipped once the spike has been driven into the ground. The groundburst version projects a cloud of chemical powder or liquid 6 to 10 metres into the air where the wind will carry the cloud. The airburst version projects a cloud of liquid to a height of approximately 60 metres.

Both types of simulator are packed in plastic foam boxes containing 18 canisters and spike holders. The boxes measure 570 x 390 x 180 mm.

SPECIFICATIONS

Type	**Groundburst**	**Airburst**
Weight	350 g	400 g
Weight of filling	225 g	165 g
Height of canister	143 mm	143 mm
Diameter of canister	73 mm	73 mm
Height of burst	6–10 m	61 m

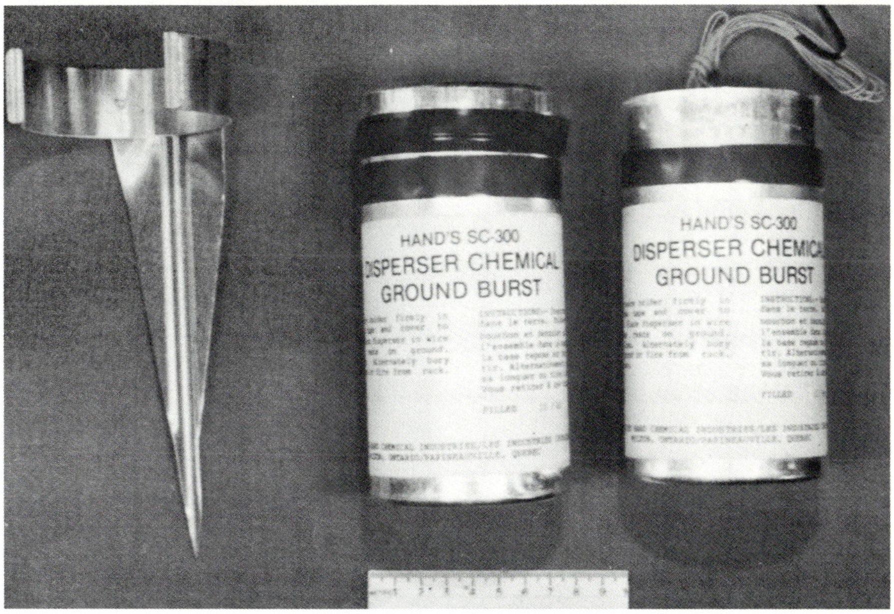

SC-300 groundburst chemical simulator with one ground spike holder

STATUS
Under evaluation by the Canadian Armed Forces.

MANUFACTURER
Hands Fireworks Inc, 221 Nipissing Road, Milton, Ontario, Canada L3Y 1R3.

Chemical Agent Simulant Projectile

DESCRIPTION
The chemical agent simulant projectile is designed to disperse a chemical agent simulant that will cause M8, M9 and 3-Way chemical agent detector papers to respond in the same manner as if they were exposed to actual chemical agents. The projectile has an ovoid two-part body with vestigial stabiliser fins at the rear. Before use the projectile is loaded with a bursting charge and filled with the chemical simulant. The projectile is then loaded into a disposable launch tube which contains a launch charge. After launching the projectile rises to a height of approximately 50 metres where the bursting charge detonates and scatters the chemical simulant in aerosol form over the training area.

STATUS
In production. In service with the Canadian Armed Forces.

MANUFACTURER
Anachemia Canada Inc, Case postale 147, Lachine (Québec), Canada H8S 4A7.

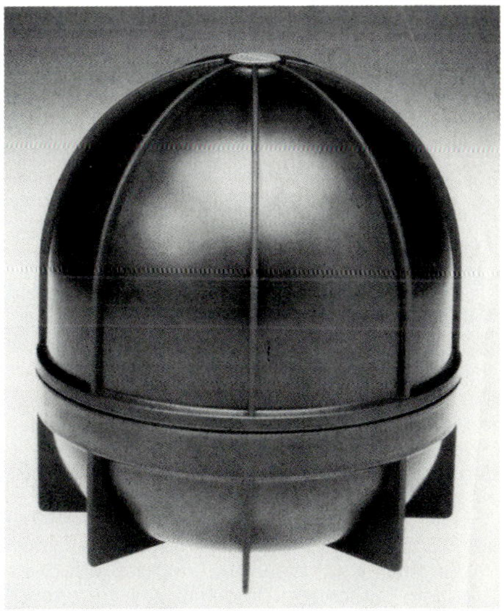

Chemical agent simulant projectile

Mark 2 Portable Leak Tester

DESCRIPTION
The Mark 2 portable leak tester is designed to measure with precision the leakage of a NBC respirator face piece. The tester consists of a measuring unit coupled to a test head onto which the respirator to be tested is fitted (the filter canister is removed and replaced by a metal bung for the leak test operation). The measuring unit is contained in a dust and moisture-proof case which can also carry the test head and the tubing required to connect the test head to the measuring unit. When in use the tester displays the differential pressure and the leak rate on a digital readout. If required, green and red accept/reject lights can be supplied to simplify the testing procedure. A selectable dual-inlet port allows the unit to serve two test heads for greater testing capacity. A testing cycle lasts 45 seconds.

Weight of the Mark 2 portable leak tester is 30 kg and dimensions are 300 × 530 × 640 mm. The external power source may be 110 volts, 60 Hz, or 220 volts, 50 Hz. Power consumption is 2 amps.

STATUS
In production. In service with the Canadian Armed Forces.

MANUFACTURER
Canadian Arsenals Limited, 2 Place Félix Martin, Montreal, Canada H2Z 2Z2.

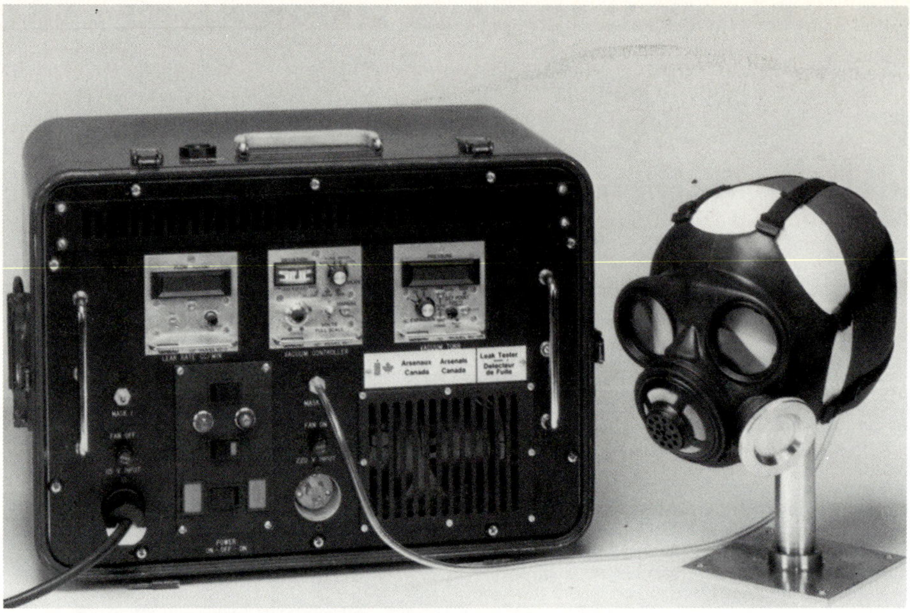

Mark 2 portable leak tester in use with a Gas Mask CML-BIO C3

FRANCE

Lacroix SIMULTITOX Chemical Attack Simulator

DESCRIPTION
The Lacroix SIMULTITOX chemical attack simulator consists of a six-barrelled 154 mm mortar that fires smoke or chemical training agent grenades to simulate a chemical warfare attack. The mortar may be set up on the ground or on a light vehicle trailer and can be used to fire a single salvo of six or four grenades or three salvos each of two grenades. The polystyrene grenades can be fired to a range of between 125 to 240 metres and can cover an area of one to five hectares, depending on wind conditions, with various forms of training agent. The launcher weighs 45 kg and takes three minutes to prepare for use. Each grenade weighs 0.3 kg (empty) and has a diameter of 154 mm. The filling may be a training agent such as 1.5 litres of water with a washable dye in solution, smoke, tear gas or a CS derivative.

STATUS
In production. In service with the French Army and some other armed forces.

MANUFACTURER
Étienne Lacroix, BP 213, 31601 Muret Cedex, France.

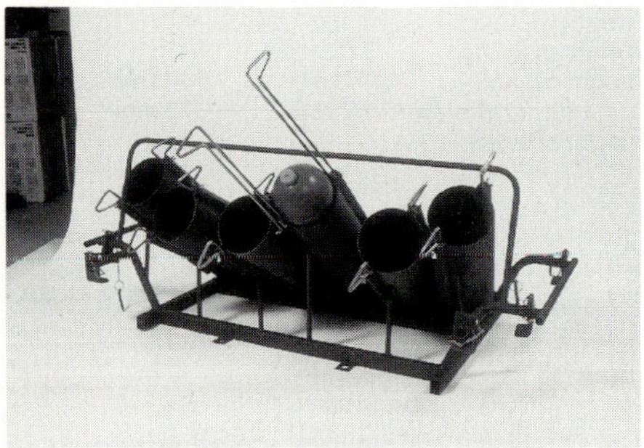

Lacroix SIMULTITOX chemical attack simulator

Lacroix ASATOX Chemical Attack Warning Device

DESCRIPTION
The Lacroix ASATOX chemical attack warning device is a hand-held pyrotechnic launcher intended to produce audible and visual warning signals to alert military units in the field of the approach of a chemical agent attack.

In use the rocket container is held so that one hand can turn the main body to arm the launching system and the other hand can pull the firing grip at the base of the unit. This ignites a propellant charge to launch an assembly to a height of approximately 200 metres. At that height a parachute is deployed to suspend visual signals involving a yellow-red-yellow flare pattern and an audible signal warning module. The visual signals remain ignited for 20 seconds while the audible signal lasts 10 seconds.

The complete unit weighs 0.69 kg. Length is 347 mm and the cartridge diameter is 47.5 mm.

STATUS
In production. In service with the French Army.

MANUFACTURER
Étienne Lacroix, BP 213, 31601 Muret Cedex, France.

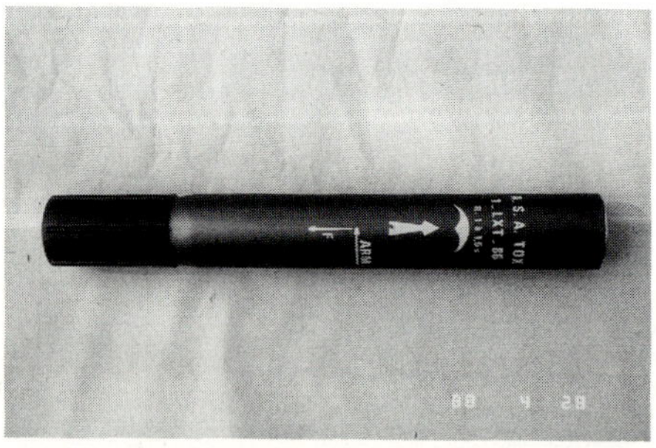

ASATOX chemical warning device

NBC Training Suits

DESCRIPTION
These suits are designed for training purposes only and provide no NBC protection whatsoever. They are intended to replace normal NBC clothing during training and exercises and reduce training costs by resembling full NBC clothing, thus allowing NBC alarm drills to be practised fully. The complete suit comprises a jacket and trousers that can be worn over normal battledress, with the jacket having an integral hood. The material used is conventional cloth through which air can permeate freely. The suit is available in three sizes and is reusable.

STATUS
In production.

MANUFACTURER
Manufactures de Vêtements Paul Boyé, 53 quai de Bosc, 34202 Sête, France.

GERMANY, EAST

WSpT Special Chemical Technology Workshop

DESCRIPTION

The WSpT special chemical technology workshop is one element in the KRCW combined radiological/chemical workshop unit (the other components being the WKCA, the WPSA and the Chemical Spares Store). The WSpT workshop is carried on a Ural-375 (6 × 6) 4000 kg truck which normally tows a single-axle trailer carrying a 6-6376 diesel-engined electrical generator.

The workshop is contained in a LAK 11 detachable box body unit which weighs 4305 kg when fully equipped and with a removeable canvas awning. The box body is 4.3 metres long, 2.5 metres wide and 2 metres high. It is mounted on a standard ISO 10-foot container bed and contains a small lathe, a generator, heating and ventilation plant, a spark plug tester and cleaner, a compressor and welding apparatus. This equipment, along with more specialised tools and machines, is used to service and maintain chemical decontamination apparatus and in particular the ARS-12 and ARS-14 truck-mounted equipments, along with the various DD series of mass decontamination equipments. The repairs and maintenance involved is classed as either routine or medium-grade, and the WSpT can operate for periods of up to ten hours in the field with the fuels and other supplies carried.

Combat weight of the WSpT is 12 875 kg. Dimensions are 7.46 × 2.69 × 3.35 metres. It has an operating range of 650 km and can travel at speeds up to 75 km/h.

STATUS

Production probably complete. In service with the East German armed forces.

WSpT special chemical technology workshop

MANUFACTURER

East German state factories.

Chemical Training Kit, Model CAS-60

DESCRIPTION

This kit is used in the classroom to train chemical reconnaissance personnel and consists mainly of a number of sample bottles containing various chemical substances. Trainees can thus be taught to identify certain substances either by smell or by appearance. Included are various types of agent, decontaminants, solvents, smoke generating compounds and incendiaries. A small case containing first-aid supplies is included in the kit.

The Model CAS-60 (CAS – 'Chemische Ausbildungssatz') is contained in a wooden box measuring 750 x 600 x 900 mm.

STATUS

Probably no longer in quantity production but easily produced.

MANUFACTURER

State factories and army schools.

NORWAY

Radiac RAS 100 Simulated Fallout Radiation Training Set

DESCRIPTION

The Radiac RAS 100 Simulated Fallout Radiation Training Set was developed in a joint programme between the manufacturers AME, the Norwegian National Defence Research Institute (FFI) and the Norwegian Army Material Command (HFK). The Radiac RAS 100 provides realistic training for NBC personnel by using the same equipment working exactly as it would during or after a nuclear attack or in the aftermath of a nuclear power incident.

The complete set consists of a control unit/transmitter in a water-tight case, receivers in the shape of simulated sensor units, 'hot spot' transmitters, helmets with receiving antennae, and a directional transmitting antenna. A typical training set could consist of one control unit/transmitter with a directional transmitting antenna, three to five 'hot spot' transmitters, five to eight sensors and the same number of helmets with receiving antennae.

The training system simulates a real situation in an operational area after a nuclear explosion. The control unit/transmitter and the directional antenna represent the ground zero of an explosion. Before operating the following parameters must be set: decay rate; simulation speed; time after explosion; and time for the accumulation of fallout. The simulation system may be operated in real time but this can be increased for training purposes in nine steps.

The 'hot spot' transmitters simulate increased radioactivity and are placed at points where fallout concentrations are likely to occur. Signals from the directional and 'hot spot' transmitters are detected by the antennae carried on the top of the trainees' helmets and are sent to the simulated sensor instruments by cable. Sensor instruments used with the Radiac RAS 100 include a Norwegian radiac detector, the NM 70, and the system is used by Norwegian civil defence organisations who are issued with the PDRM 82 portable dose rate meter. The programming of the simulation equipment allows breaks for instruction, briefings, etc.

The Radiac RAS 100 training system can operate off a 24-volt dc supply or 110 and 220-volt ac mains supplies. The control unit/transmitter operates on a frequency of 29.5 MHz. Its weight is 3.7 kg (excluding the antenna) and it measures 300 × 115 × 230 mm.

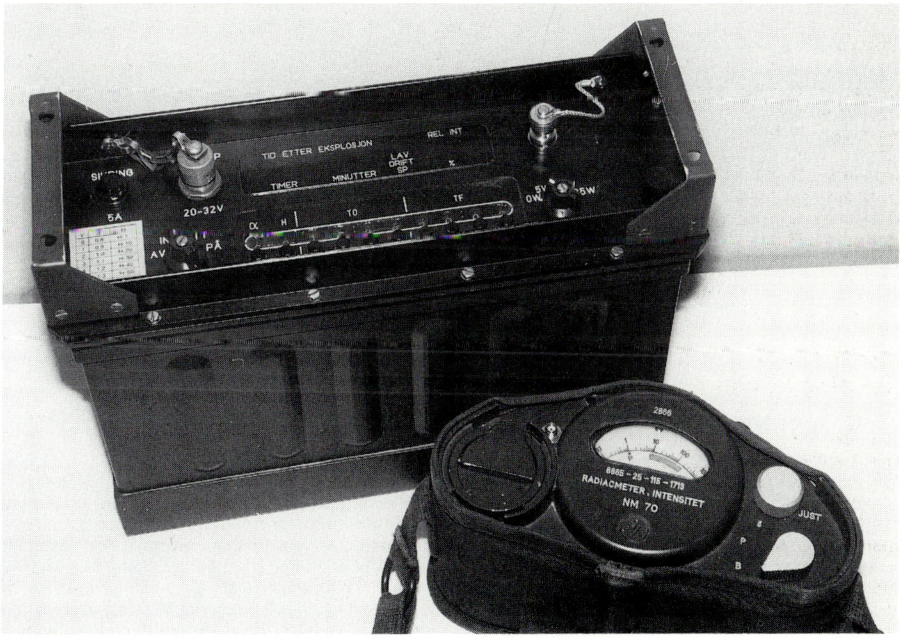

Radiac RAS 100 simulated fallout radiation training set showing the control unit/transmitter and a simulated radiac meter (the NM 70) sensor unit

STATUS

In production. In service with the Norwegian armed forces and civil defence organisations.

MANUFACTURER

AME as micro electronics, PO Box 83, N-3191 Horten, Norway.

POLAND

Portable Gas Training Chamber

DESCRIPTION
This portable device is used for training recruits and NCOs with their chemical protection equipment. It consists of a suspended metal cover with an attached shroud which is lowered over an individual until it covers to the knees. Inside the metal cover is a device for releasing tear gas agent which, once scattered over the masked trainee, enables him to test the efficiency and fit of his face mask, followed by personal decontamination drills. The device can be hung from a specially-erected metal or timber framework, and in the field can be suspended from a tree branch.

STATUS
Production complete. In service with the Polish forces.

MANUFACTURER
Probably state factories but could be easily improvised in the field by any unit.

UNION OF SOVIET SOCIALIST REPUBLICS

Protective Mask and Canister Leakage Tester, Model PGP

DESCRIPTION
This equipment is used to test most types of Warsaw Pact protective masks and filter canisters for leakage. All the components are contained in a metal box and include two rubber bulb-type air pumps, a manometer, a one-minute sand glass timer, rubber tubing and plugs, valves and various fittings. With both the masks and the filter canisters the units are sealed and pumped up to pressure. The pressure is measured on the simple manometer and has to be maintained for one minute without leakage; the time is determined by the sand glass timer.

STATUS
Production probably complete. In service with the Soviet and Warsaw Pact forces.

MANUFACTURER
State factories.

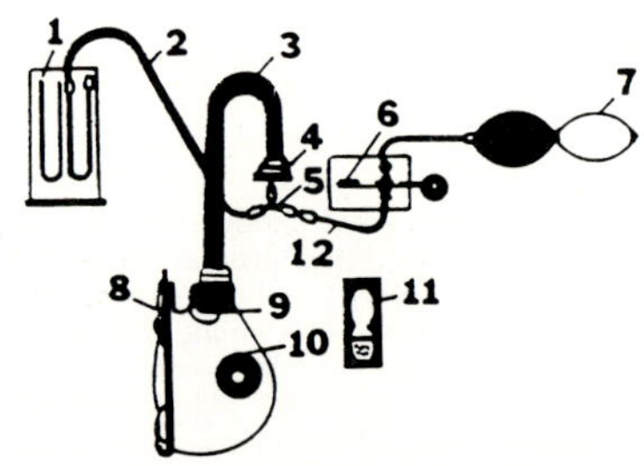

Model PGP protective mask and canister leakage tester mask testing **(1)** *pressure gauge* **(2)** *rubber tubing* **(3)** *protective mask nose* **(4)** *connector* **(5)** *T-tube* **(6)** *wire spring clamp* **(7)** *rubber bulb* **(8)** *face-piece clamp* **(9)** *rubber plug* **(10)** *protective mask face-piece* **(11)** *1-minute sand glass timer* **(12)** *rubber tubing*

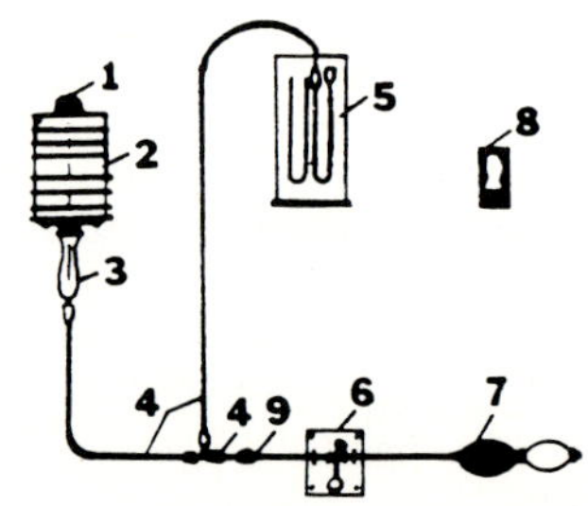

Model PGP protective mask and canister leakage tester canister testing **(1)** *rubber plug* **(2)** *canister* **(3)** *connecting tube* **(4)** *rubber tubing* **(5)** *pressure gauge* **(6)** *wire spring clamp* **(7)** *rubber bulb* **(8)** *1-minute sand glass timer* **(9)** *connector*

Mobile Chemical Workshop PPKhM-1M

DESCRIPTION
Based on the chassis of the GAZ-66 (4 × 4) 2000 kg truck, this mobile workshop is used for field repair, maintenance, calibration and servicing of chemical and radiological equipments. The workshop is contained within a specially-equipped box body and additional tentage and equipment are carried in a towed IAPZ-738 single-axle trailer. The box body has four work spaces although more can be provided within the carried tents. Wherever possible, electrical power is taken from an external source but an AB-2T230 portable generator can supply power for lighting and the various items of test and repair equipment. A petrol heater and an air compressor are also provided.

STATUS
Production status uncertain. In service with the Soviet armed forces.

MANUFACTURER
State factories.

UNITED KINGDOM

NBC Operational Simulator

DESCRIPTION
The NBC Operational Simulator was developed to meet a British armed forces requirement (MER(LA)4/18/86) for equipment to simulate, by safe means, the presence of radioactive and chemical contamination, and the detection and measuring of ionising radiation and chemical agent vapours. The prototype simulator was developed by the nuclear research establishment at Harwell, advised by the Chemical Defence Establishment (CDE) at Porton, Wiltshire, and the Defence NBC Centre at Winterbourne Gunner, Wiltshire.

A computer at an exercise headquarters is used to prepare an exercise strike plan and to display NBC contamination hazards. The strike information is transmitted to instruments in the field which mimic operational NBC equipments. These instruments are used for NBC surveying and monitoring operations, providing realistic hazard information at unit level. Thus NBC reporting can be included in unit and formation exercises, allowing an entire command chain to be realistically exercised.

The NBC Operational Simulator has a central control computer which displays a map of an exercise area. The operator enters an NBC strike plan together with current meteorological conditions into the computer, which then displays the resultant spread of contamination on the computer screen. The strike plan can be instantly modified in response to the display until the desired effects are achieved for the field exercise concerned. Once the plan has been finalised, the strike information is transmitted via a Base Transmitter (which can be stationary or mobile) to simulator-dedicated field instruments.

The field instruments mimic operational NBC equipments. The user enters his grid reference location into the instrument using a keyboard on the instrument. A built-in processor then calculates the local level of contamination by using the information received from the Base Transmitter. Once a strike plan has been received and stored, the subsequent variation of contamination levels with time is automatically followed and the user grid reference can be altered as necessary. The instruments can be used for surveying and monitoring to provide NBC reports for commanders in real time. The simulator can also be used for command post exercises and for training NBC cell controllers and operators.

Instruments used with the NBC Operational Simulator to date include the Chemical Agent Monitor (CAM) and the Personal Dose Rate Meter; a version of the PDRM 82 portable dose rate meter is under development for use with the simulator and it is planned to simulate other instruments such as NAIAD and the AFV Meter as development progresses.

STATUS
Prototype.

CONTACT ADDRESS
LSOR 4, Ministry of Defence, Neville House, Page Street, London SW1P 4LS, United Kingdom.

Simulator Projectile Airburst Liquid Chemical L1A2

DESCRIPTION
The Simulator Projectile Airburst Liquid Chemical Agent L1A2 (SPAL) was designed to simulate chemical warfare spray and liquid-filled airburst weapons, by projecting a container filled with Chemical Agent Training Mixture (CATM) into the air and bursting it at a height of not less than 12 metres. The bursting of the container spreads the CATM in the form of droplets over an area approximately 15 metres wide and 50 metres long and downwind from the point of burst. The droplets are large enough to produce a visible colour on chemical agent detector papers.

The SPAL consists of a cardboard mortar-type barrel, a polythene bottle filled with CATM, and an electrically-initiated actuator for projecting and bursting the bottle.

STATUS
In production. In service with the British Armed Forces.

MANUFACTURERS
Astra Holding plc, Astra House, 43 Old Dover Road, Canterbury, Kent CT1 3DE, United Kingdom.
Brocks Pyrotechnics, Sanquhar, Dumfriesshire DG4 6JP, United Kingdom.

Portable Non-destructive Filter Test Kit

DESCRIPTION

During the late 1950s the Chemical Defence Establishment (CDE) at Porton Down, Wiltshire, developed a simple and reliable procedure for testing the integrity and effectiveness of installed chemical agent filters. Known as the Non-destructive Test the procedure consists of the rapid injection of a known amount of test agent, an organo-halogen compound, into the air stream passing through a filter and measuring with a sensitive halogen detector the amount in the effluent air stream.

The sensitivity of the halogen detector, developed by the CDE enables filter penetrations down to 0.00001 per cent to be measured using a small amount of test agent. Only 1 ml of l-bromutane test agent is needed to verify a filter of 170 cubic metres per hour capacity. This is less than 0.1 per cent of the filter capacity and the test is therefore non-destructive.

The Portable Non-destructive Filter Test Kit is manufactured by Hi-Tech Scientific Limited under an exclusive licence agreement with the CDE. The kit contains everything necessary for the testing of installed gas filters of capacities up to 1700 cubic metres per hour.

All the items in the kit are carried in a reinforced carrying case. Items included are a HGD-20 halogen detector, a clean air tube, a calibration tube and one metre of sample hose. Also included is a HGD-40 multi-range spray gun, 200 ml of l-bromobutane, and 40 carbon dioxide cartridges (20 × 28 g, 10 × 16 g, 10 × 8 g). Also included are operator and maintenance manuals.

The HGD-20 halogen detector is a robust and portable instrument with a direct reading digital display. It is intended to be used by semi-skilled technicians. Normally it is powered from a 240- or 110-volt ac supply but it can be deployed in the field for up to 5 hours running time by using the optional HGD-21 portable power source. The HGD-21 will also run off a 12-volt vehicle battery.

The HGD-40 multi-range spray gun is hand held and powered by carbon dioxide propellant enabling up to 10 ml of test agent to be dispensed. Three sizes of pressurised carbon dioxide cartridge can be fitted into the gun barrel. When the trigger is squeezed a piercing assembly releases the propellant which disperses the test agent as an aerosol within seconds.

The complete kit weighs 3.1 kg and measures 250 × 150 × 260 mm.

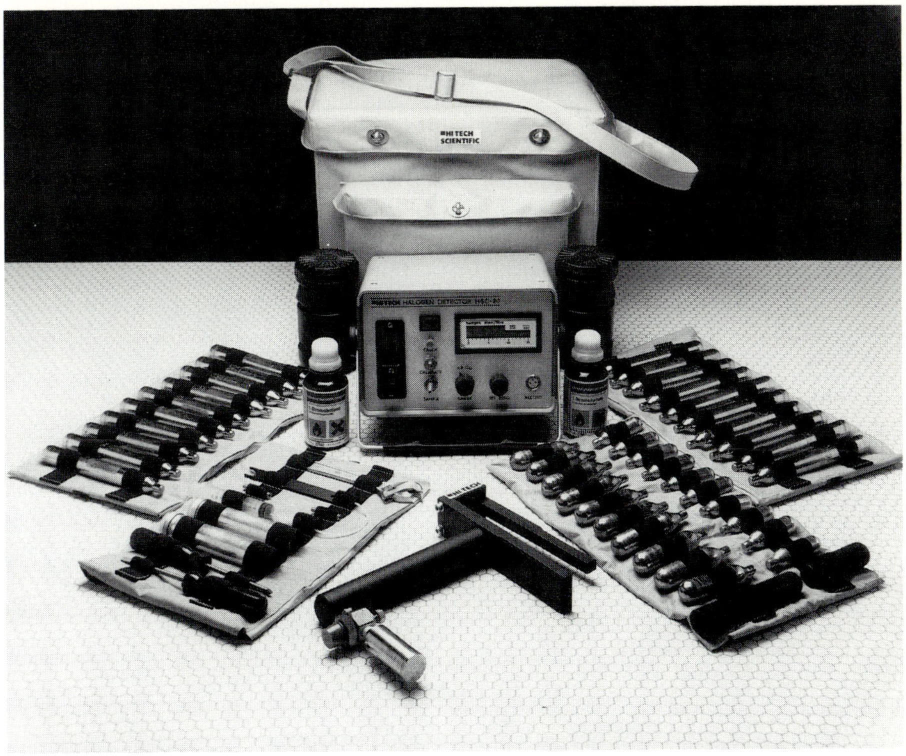

Complete portable non-destructive filter test kit

STATUS

In production. In service with the British Armed Forces.

MANUFACTURER

Instrument Division, Hi-Tech Scientific Limited, Brunel Road, Salisbury, Wiltshire SP2 7PU, United Kingdom.

Moore's (Wallisdown) Limited On Site Filter Testing Equipment

DESCRIPTION

This on site filter testing equipment is used to test the efficiency of air filters and filter installations used in a wide variety of applications. The test equipment operates by introducing a controlled amount of sodium chloride (salt) into the air flow in front of the filter concerned, and measuring the amount of sodium chloride left in the air after passing through the filter, thus determining the filter's efficiency. A portable version of this test equipment has been developed to cover ship-borne systems.

The testing system employs two items of test equipment, a portable sodium chloride detector and a thermal generator.

The portable sodium chloride detector is a compact portable flame photometer with its own built-in sampling fan. It features an LCD panel meter and automatic sampling of airflow rate through a feedback system to the fan motor which is positioned downstream from the flame. The sampled air flows through a closed system so that, if necessary, it may be returned to a contaminated duct. In this application the sampling fan is required only to move the air around the closed loop, allowing systems at low pressure to be tested.

The Light Sensitive Detector is a fast acting photo conductive diode. The linear response to sodium chloride concentrations is utilised for display on the digital panel meter. Where flame emission is non-linear (at high concentrations above 2 milligrammes per cubic metre) it is necessary to refer to calibration graphs. An outlet socket is available as standard for a chart recorder.

The portable sodium chloride detector is powered by a 110/220-volt 50/60 Hertz power supply but can operate from an optional rechargeable lead acid battery pack. The unit uses a one-litre hydrogen bottle with a consumption of 500 millilitres per minute. The weight of the detector is 17 kg which increases to 22 kg when the battery pack is used. Dimensions are 480 × 200 × 480 mm.

The thermal generator is available in two models,

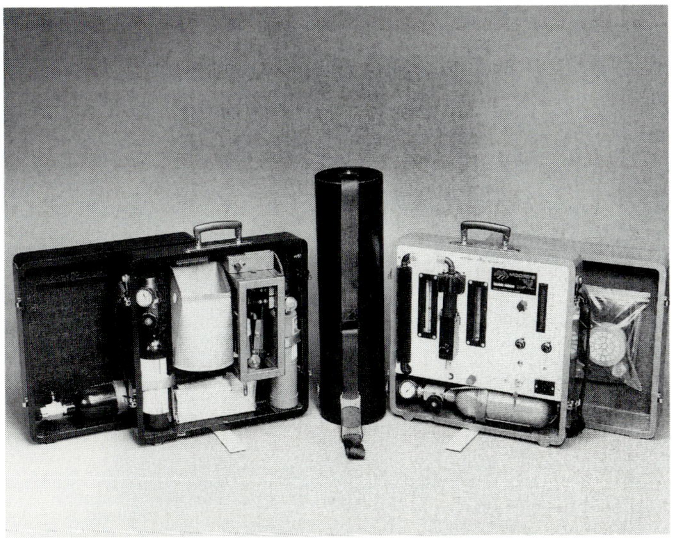

Moore's(Wallisdown) Limited on site filter testing equipment with (left) thermal generator in its carrying case complete with associated hoses, and (right) the portable sodium chloride detector

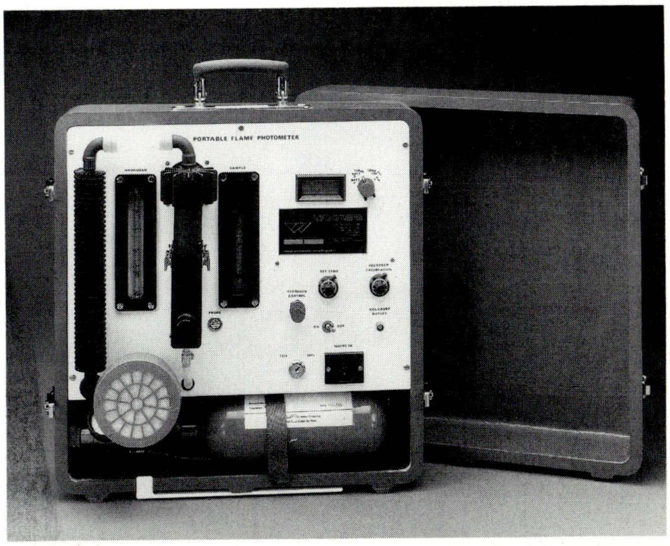

Portable sodium chloride detector used with the on site filter testing equipment

standard and mini. Both produce an aerosol by burning a 'stick' of sodium chloride in an oxy-propane flame at a constant controlled rate. The sodium chloride evaporates to re-condense in the form of a dry sub-micron aerosol.

The standard thermal generator weighs 6.4 kg and measures 400 × 180 × 270 mm. It requires oxygen and propane gas supplies and can be powered by eight 1.5-volt AA batteries. The aerosol range is variable up to 6 grammes per minute.

The mini model of thermal generator weighs 3.2 kg and measures 245 × 120 × 150 mm. It also requires oxygen and propane gas supplies and is powered by eight 1.5-volt AA batteries. The aerosol range is variable up to 0.5 grammes per minute.

STATUS
In production.

MANUFACTURER
Moore's (Wallisdown) Limited, Wallisdown Road, Bournemouth BH11 8QN, United Kingdom.

Moore's (Wallisdown) Limited Sodium Chloride Bench Test Rig

DESCRIPTION
This bench test rig is used for the production and development testing of filters, filter cansiters and respirators for personal protection. It uses the sodium chloride (salt) aerosol test principle in which a sodium chloride aerosol is introduced to one side of a filter and the amount detected on the other side is used to determine the filter's efficiency.

The rig provides a pass/fail system for production testing. A change-over valve enables two test chambers to be used if required and the rig can be used by semi-skilled operators. Respirator filters and materials may be tested at flow rates ranging from 15 litres a minute to 95 litres a minute with a minimum detectable penetration of 0.00005 per cent. The unit can also be used for respirator face seal checks.

The test rig weighs 90 kg and measures 1448 × 535 × 560 mm.

STATUS
In production.

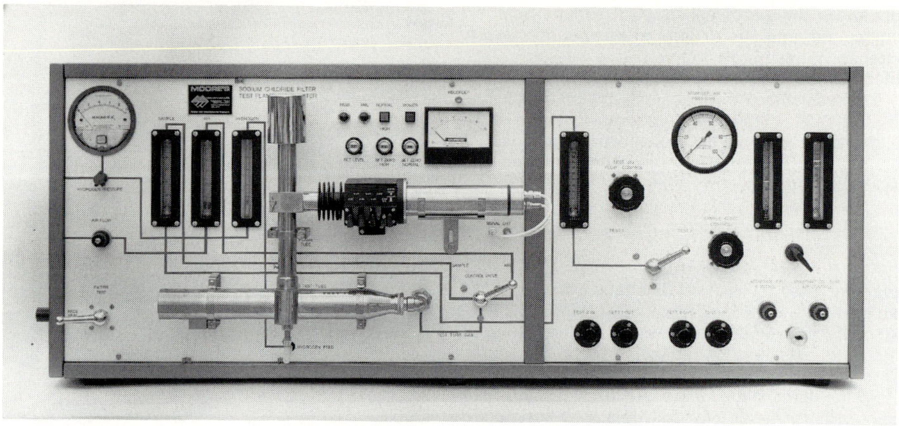

Sodium chloride aerosol test bench rig

MANUFACTURER
Moore's (Wallisdown) Limited, Wallisdown Road, Bournemouth BH11 8QN, United Kingdom.

UNITED STATES OF AMERICA

Simulated Area Weapons Effects – Nuclear, Biological, Chemical I and II (SAWE-NBC I and II)

DESCRIPTION
The Project Manager for Training Devices, US Army Materiel Command, is developing a series of devices and simulants that will simulate the effects of NBC weapons to provide a realistic NBC environment in which to train.

The Simulated Area Weapons Effects – Nuclear, Biological, Chemical I (SAWE-NBC I) includes the Casualty Assessment System (CAS), the Persistent Chemical Agent Simulant (PCAS) and the Chemical Agent Decontamination Simulant (CADS). These will provide real-time casualty assessment during force-on-force engagement simulation exercises. The CAS consists of an electronic package located inside the standard protective mask. The PCAS and CADS are simulants which replicate various chemical and biological agents on the battlefield and provide the opportunity to exercise proper chemical and biological avoidance, detection and decontamination procedures.

SAWE-NBC II will consist of biological and nuclear training device systems that will provide the means to assess casualties in real time during force-on-force engagement exercises in a simulated biological and/or nuclear warfare environment. The systems will activate casualty assessment devices worn by participants in exercises and will be MILES inter-operable.

The research and development stages for SAWE-NBC I and II are scheduled to commence during fiscal year 1989, with production commencing during fiscal year 1990. Production should be completed by the end of fiscal year 1995.

Associated with SAWE-NBC I and II is SAWE-RF (RF – radio frequency). This is a projected training system that will simulate accurately and in real time the effects of persistent NBC agent contamination, as well as the effects of indirect fire and land mines, in force-on-force training exercises. The system will make use of Global Positioning System technology to provide accurate position location. Competitive development is scheduled to commence by the end of fiscal year 1990, to be completed by the end of fiscal year 1991. Production will be undertaken in two phases, the first during fiscal years 1992 and 1993, and the second during fiscal years 1994 and 1995.

All three versions of SAWE will be fielded at the US Army's three Combat Training Centers (CTCs); at Fort Irwin, California, Fort Chaffee, Arkansas, and Hohenfels, West Germany.

STATUS
Early development.

Detector Unit, Simulator Automatic Chemical Agent Alarm, XM81

DEVELOPMENT
The XM81 simulator has been designed to be used with the M8 Automatic Chemical Alarm System (ACAS) and the exploratory development programme was successfully completed in December 1980. A 21-month engineering development programme was initiated in March 1981 and initial plans called for the system to be type-classified in early fiscal year 1985 with a subsequent Initial Operational Capability (IOC) in fiscal year 1987.

DESCRIPTION
The XM81 system consists of a hand-held transmitter unit and four receiver modules which are fitted to the M43 detector units between the upper and lower assemblies. The transmitter can send four distinct coded radio signals to any or all four of the receiver modules which then activate the M43 detector units. The transmitter has a range of approximately 1000 metres and the frequency used is 141.925 Mhz; a frequency chosen so that it will not interfere with other electronic systems, and vice versa.

When not in use the XM81 system components are

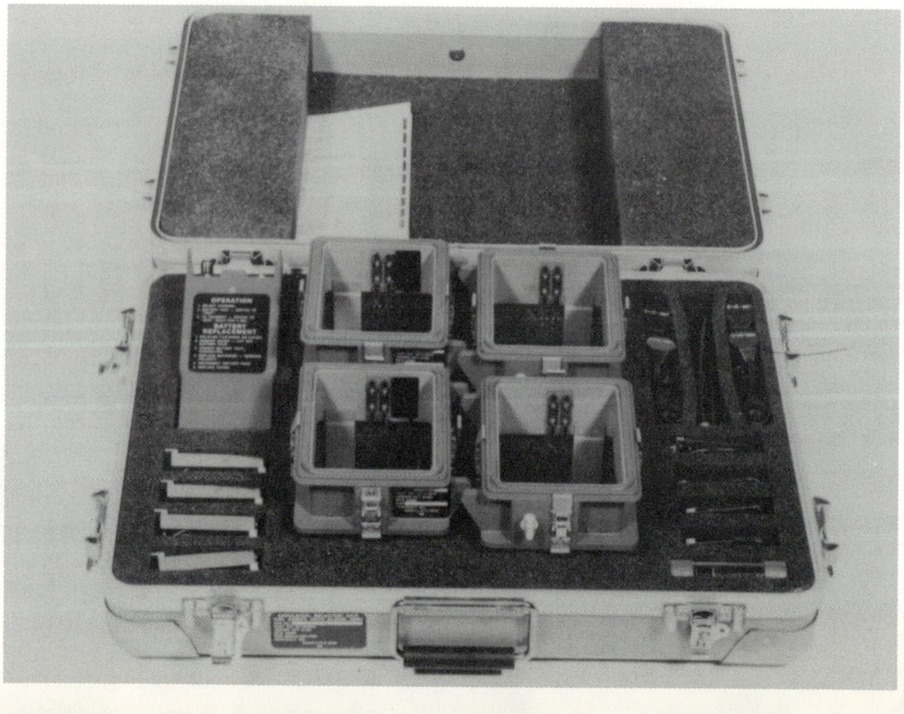

XM81 simulator set in carrying case (US Army)

kept in a metal carrying case which contains four receivers, one transmitter unit and associated cables and spares.

SPECIFICATIONS
Transmitter unit
Height: 184 mm
Width: 114 mm
Depth: 57 mm
Weight: 1.14 kg
Receiver module
Height: 152.4 mm
Width: 178 mm
Depth: 140 mm
Weight: 0.91 kg
System in case
Height: 241 mm
Width: 813 mm
Depth: 406 mm
Weight: 15.66 kg
System range: approx 1000 m
Frequency: 141.925 Mhz

STATUS
Initial production.

DEVELOPMENT AGENCY
Chemical Research, Development and Engineering Center, Aberdeen Proving Ground, Maryland, USA (now closed).

XM81 receiver in M43 detector module

XM81 transmitter module

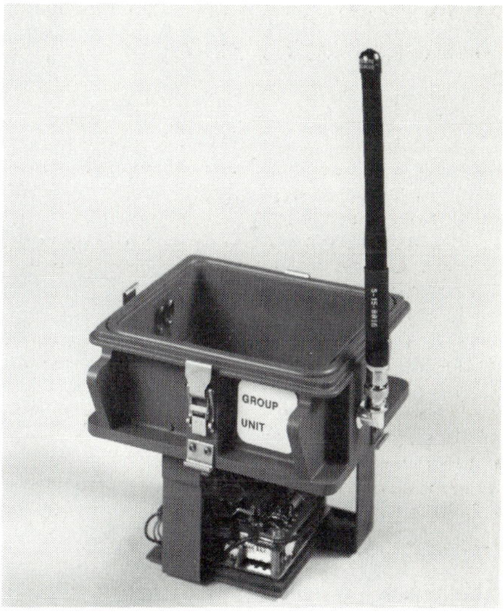

XM81 receiver module

TDA-99 Field Mask Leakage Tester

DESCRIPTION
The TDA-99 field mask leakage tester utilises the basic design of the US Army's M14 Protective Mask Leakage Tester combined with state-of-the art solid state technologies. The unit may be used in the field to test the integrity of any military protective face mask using a universal clamping system on a high impact urethane test head. Apart from the test head, other components include a vacuum pump, a light scattering chamber, a flowmeter, a digital penetration indicator, a solid state amplifier, an aerosol generator and control components.

The TDA-99 takes about 10 minutes to prepare for use and operates off a mains electrical supply. The test agent employed is a corn oil aerosol with an approximate mass median size of 0.57 micrometres. The tester uses a forward light detection system with a sensitivity of 0.001 per cent to detect significant leaks in lens, inhalation-exhalation, voicemitter and drinking tube assemblies and the facepiece itself. A typical test takes approximately three to five minutes for a complete analysis.

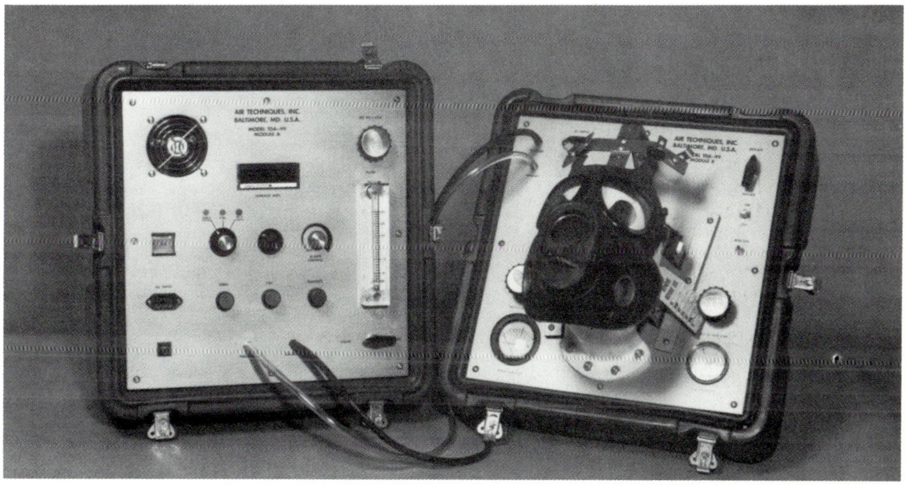
TDA-99 field mask leakage tester

STATUS
In production.

MARKETING AGENCY
Tradewinds Limited, 307-F Maple Avenue West, Vienna, Virginia 22180, USA.

MANUFACTURER
Air Techniques Inc, 1801 Whitehead Road, Baltimore, Maryland 21207, USA.

AudioPack Portable Sound System

DESCRIPTION
The AudioPack portable sound system is a version of a lightweight, self-contained commercial loudspeaker and microphone system adapted to use a microphone that can be fitted inside standard M17-series and other NBC masks. The AudioPack may be carried slung on the small of the back to allow free use of both hands, and when the system is used it is possible for a masked user to produce speech that can be understood from over 80 metres away.

The AudioPack unit weighs 2.72 kg, is 381 mm long and 178 mm in diameter. Power is derived from a rechargeable battery that can provide up to 12 hours of use. The frequency range is from 60 Hertz to 12 kiloHertz and audio output is 25 watts. Waterproof 8 ohm speakers are used.

STATUS
In production. In service with the US Army and some other armed forces.

MANUFACTURER
AudioPack Sound Systems, 10011 Walford Avenue, Cleveland, Ohio 44102, USA.

AudioPack portable sound system in use

Earmark Loudmouth Voice Amplifier

DESCRIPTION
The Earmark Loudmouth voice amplifier is a lightweight unit designed for troops wearing NBC protective equipment or face-obscuring devices. With the audio power to amplify voice commands from 65 dB to over 100 dB the Loudmouth has many applications in an NBC environment or in areas of high background noise. The unit is normally used with a throat microphone but it is possible to fit a small microphone to the exterior of a voicemitter on an NBC respirator. The Loudmouth amplifier unit can be worn anywhere on a soldier's uniform requipment and more than one unit can be worn and used to direct speech in more than one direction.

The case used for the Loudmouth is thick-walled polycarbonate resin and is fully waterproofed. Power is provided by a single 9-volt alkaline cell with a 12-hour duty cycle.

The Loudmouth weighs 345 grammes, is 132 mm high, 97 mm wide and 38 mm deep.

STATUS
In production.

MARKETING AGENCY
KT Corporation, 300 Long Beach Boulevard, Stratford, Connecticut 06497, USA.

Earmark Loudmouth voice amplifier

Earmark Loudmouth voice amplifier in use

WARSAW PACT

Protective Equipment Workbench, Model SKhM-M

DESCRIPTION
This item of equipment is a compartmented workbench specially equipped for the maintenance of protective clothing, protective masks and filter canisters, and breathing gear. It contains special and other tools, instruments, some spare parts and various materials for the servicing, testing and repairing of all the above.

STATUS
In service with the Soviet and Warsaw Pact forces.

MANUFACTURER
State factories.

YUGOSLAVIA

KODŠ M2, M3 and M4 Chemical Agent Training Smoke Pots

DESCRIPTION

The KODŠ M2, M3 and M4 chemical agent smoke pots are used as training aids during training and tactical exercises when a simulated chemical agent attack is required.

The KODŠ M2 and M3 resemble cylindrical hand grenades and may be thrown or placed on the ground for use. To prepare the smoke pots a screw top is removed from one end of the cylidrical body to expose a cord handle. This is then pulled and the smoke pot is thrown or placed on the ground. The cord operates a friction fuze to initiate smoke production and the simulated chemical agent cloud issues through three small holes close to the fuze. The simulated agent used with the M2 is a substance known as HAF which is not dangerous but provides an unmistakeable odour and a degree of tear action. The M3 uses CS and may be ignited electrically. A cloud of simulated agent will be produced for 8 to 10 minutes.

The M4 smoke pot is a much larger cylindrical device 240 mm high and 180 mm in diameter. It weighs 5.2 kg and contains 4 kg of CS to produce a cloud of simulated agent for up to 18 minutes. Electrical or friction ignition may be used.

SPECIFICATIONS
(M2 and M3)
Weight: 0.7 kg
Weight of smoke production material: 0.45 kg
Diameter: 80 mm
Height: 130 mm
Smoke emission time: 8-10 min
Shelf life: 3 years

STATUS

In production. In service with the Yugoslav armed forces.

MANUFACTURER

Federal Directorate of Supply and Procurement (SDPR), 9 Nemanjina Street, 11001 Belgrade, Yugoslavia.

Addenda

PROTECTIVE MASKS AND CLOTHING

BEM 4GP General Purpose Respirator

The BEM 4GP general purpose respirator was developed by ENGICOM NV in close collaboration with the Belgian Armed Forces and various university research centres. It has been developed to the pre-production stage in answer to a Belgian Armed Forces requirement with the intention of producing the respirator in three sizes. It is anticipated that production will commence during 1990 with initial production concentrating on the medium-sized version.

The BEM 4GP uses a facepiece that is stated to be easy to decontaminate and is held in position by a head harness system. The vision device is a one-piece panoramic panel made of coated polycarbonate which is impact and scratch resistant and which can be supplied as an easily replaceable spare part. Ordinary spectacles may be worn with the vision panel. In front of the user's mouth is a monoblock assembly which combines the filter canister interface, the inhalation and exhalation systems, a communication device and a drinking system. There is also a drainage point.

The BEM 4GP general purpose respirator

The filter canister interface is unusual in that the canister position can be readily altered to suit the wearer. The canister can be moved through 180 degrees (centre, left or right) to allow the user to use weapons or other equipment in various positions. The interface has a locking system and there is no break in protection as the filter canister position is altered.

The speech diaphragm is a straight-through device that is compatible with most types of communication equipment. A microphone that fits into the speech system is under development. The drinking system uses a 'drinking straw' arrangement that is guided by a protective sheath. This system allows the intake of high viscosity liquids and prevents any pollution of the interior of the respirator.

The respirator is carried in a bag that also contains other parts such as spectacles. Available as spare parts are the vision panel, the monoblock assembly, the speech diaphragm and the inhalation and expiration valves.

STATUS
Pre-production.

MANUFACTURER
ENGICOM SA NV, Binnensteenweg 172, B-2530 Boechout, Belgium.
Telephone: 32.3.455.41.77. Telex: 33808.

Gas Mask Set Type 70 M

The Hungarian Gas Mask Set Type 70 M is intended to provide protection for the individual soldier against NBC warfare agents and has three main parts, a face mask, a filter canister (described as a filter insert) and a fabric carrier bag.

The face mask is of the helmet type and uses a rubber body stretched over the head to provide the necessary protective seal around the face. Four sizes are available. There are two small flat eyepieces with anti-condensation inserts placed over the inside of the eyepieces. The filter canister is located on the left-hand side of the face mask and does not provide protection against carbon monoxide. A chin-mounted exhalation valve directs spent air out of the mask while a speech unit is mounted in front of the mouth area.

The fabric carrying bag is used to carry the face mask and filter and is also used to carry accessories. These include a box containing three pairs of anti-condensation inserts, a separate box containing four sets of spares inhalation and exhalation valves, and a piece of anti-condensation soap.

STATUS
In production. In service with the Hungarian armed forces.

AGENCY
TECHNIKA Foreign Trade Company, Budapest-Hungary, X., Salgótarjáni u.20, 1475 Budapest, POB 125, Hungary.

Face mask and filter for Gas Mask Set Type 70 M

Iraqi Military Protective Mask and Protective Clothing

Few details of this protective respirator are available. It is stated to protect the wearer against all toxic gases likely to be encountered during warfare and is provided with a CF4 filter canister. There are two eyepieces that appear to be polycarbonate and there is a drinking system that connects with a standard water bottle. The mask is said to provide protection against Mustard-type agents in aerosol form for a minimum of six hours.

The mask weighs 1.02 kg and is supplied together with a fabric carrying satchel. Storage life is two years.

The protective suit is a two-piece garment with an integral hood. It provides protection against Mustard agent aerosols for a minimum of four hours and a minimum of six hours against Mustard agent vapours. Storage life is two years.

The protective outfit is completed by rubber gloves 0.8 mm ($\pm$0.1 mm) thick and rubber protective overshoes. Both gloves and overshoes are supplied in three sizes. The gloves weigh 280 grammes per pair and the overshoes 0.81 to 1.03 kg per pair.

STATUS
In production. In service with the Iraqi armed forces.

Iraqi military protective mask

MANUFACTURER
Iraqi state factories.

Iraqi military protective suit

SGE 1000 NBC Respirator

The SGE 1000 NBC respirator was developed as a private venture and is of a novel and unusual design in which the usual rubber-based facepiece is replaced by a stiff transparent structure which acts as a load-bearing unit to support the head harness, valves, filters, regulators and other components. The advantages claimed by the new arrangement are a substantially reduced weight, an unlimited field of vision, the weight is held by a rigid structure, the impact-resistant facepiece provides eye and face protection against shell splinters and impacts, and that the wearer can be readily identified visually.

On the SGE 1000 the entire body of the mask is made of a strong, rigid and transparent polycarbonate. The facepiece is coated with polysiloxane resin to make it scratch and chemical resistant and the material is strong enough to provide eye and face protection against violent impacts and missiles such as projectile fragments. The main part of the facepiece consists of a large set-back flat surface in front of the eyes. The flat area permits the use of the respirator with optical and weapon sighting devices and a protective shield which filters ultra-violet and infra-red rays can be clipped (snapped) in and out of place without having to remove the mask. If required the non-optical transparent areas of the facepiece can be etched or painted to reduce reflectivity.

Due to the use of the load-bearing facepiece the rubber face seal can be made much lighter than usual and it has to act as a seal only. The seal is made of a very soft bromobutyl rubber with a supple bellows-type structure to combine a perfect facial seal with wearer comfort, even after prolonged use. (When the mask is worn with a protective hood the seal can be made of platinum-catalysed medical grade silicone rubber). The bellows structure allows it to adjust easily to all shapes of face so it is not necessary to stock a wide range of mask sizes. The face seal has a seat into which the edge of the load-bearing facepiece is inserted and the coupling of these two parts is ensured by a fixing rim.

The mask has two side fittings for filter canisters with NATO standard threads. Either can be sealed with a screw-on cap or a sound-transmitting membrane. Each filter fitting consists of an external pipe union, an internal ring nut, a one-way membrane, a flat sealing gasket and an O-ring seal on the facepiece. The material used is mainly black polyarilamide with the O-rings and gaskets being bromobutyl rubber.

The speech device consists of an open circular membrane which also acts as an exhalation valve. This device is fitted in the centre, around the feeding device (see below), and allows sound waves to be released together with exhaled air. Externally the exhalation valve/speech device is covered by a stiff dome-shaped element which protects the membrane and maintains a sufficient volume of uncontaminated air to be in constant contact with the membrane. An auxiliary exhalation and drainage valve is fitted to the chin area of the mask to increase the exhalation flow rate needed during high work rates and to drain away any liquids which may accumulate inside the mask.

A liquid feeding device is fitted to the centre of the speech and exhalation unit and consists of an automatic gas-tight valve made of bromobutyl rubber and incorporated into the mask. The valve is self-cleaning and can be connected by a bayonet fitting to a liquid food container. On rotating the container, after fitting it to the mask, a nozzle is released which penetrates the valve to reach a folded flexible tube inside the oronasal unit, straightening it and bringing it into contact with the wearer's mouth. This system allows easy refilling of the container in contaminated environments and utilises a tank equipped with the equivalent fittings to those on the mask.

The oronasal unit of the SGE 1000 is made of soft medical grade silicone rubber which adheres softly and evenly to the facial contours and provides a sufficiently rigid structure to ensure effective sealing during inhalation. Two inhalation valves are fitted laterally to the unit, the lower part of which is connected to the auxiliary drainage unit.

The fixing rim connects the head straps to the mask and also locks the face seal and the incorporated hood, if used. The pull exerted on the fixing points of the head harness is thus transferred and distributed evenly around the whole circumference of the mask.

The SGE 1000 can use corrective lenses are mounted on a frame which snaps into the inside of the mask. It is possible to fit a resuscitation system directly to the oronasal mask via the liquid feeding valve. This system is independent of the normal filter fitting.

The SGE 1000 can be provided in a version with an integral NBC hood manufactured from Saratoga cloth. The hood is fitted with a double overlapping collar which reaches to the shoulders. This version of the SGE 1000 is fitted with a silicone rubber facial seal.

As an option the SGE 1000 can be fitted with an automatic pre-filter inserted between the filter canister and the mask to allow safe canister replacement. The pre-filter slides inside its support and is constantly pushed against the main filter by a spring. When the main filter is fitted the pre-filter is bypassed but when the main filter is removed the bypass closes automatically and the pre-filter takes over until the new filter is inserted.

STATUS
Ready for production.

MANUFACTURER
SGE Società Generale Elastomeri SpA, Via Arvigo 6, 16010 Sant'Olcese (GE), Italy.
Telephone: (010) 407686, 711523. Telex: 286375 SGE I.

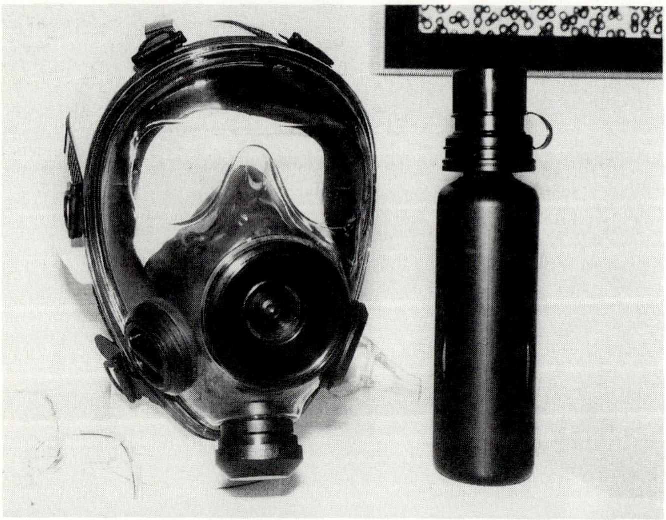

SGE 1000 NBC respirator with liquid food container

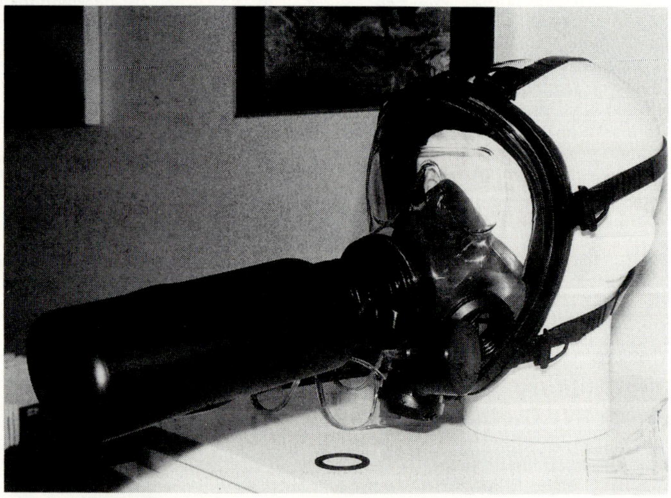

SGE 1000 NBC respirator with liquid food container fitted to feeding device on mask

Romanian NBC Respirator with Sound Amplifier

Few details are available regarding this NBC respirator which is described by it's producers, ROMTEHNICA, as a gas mask. It is held in position on the head by rubber straps, has two polycarbonate eyepieces and a chin-mounted filter canister, and is stated to be effective for up to six hours against chemical warfare agents in the form of droplets, aerosols and vapours. The respirator is intended to be worn for time periods of up to 12 hours.

The respirator is fitted with a drinking device which can be connected to a standard water bottle, and can also be connected to a sound amplifier unit via a cable from a cheek-mounted internal microphone. The sound amplifier unit is intended for use by commanders and weighs 0.5 kg.

The respirator and sound amplifier unit can be carried in a special carrying satchel weighing 0.5 kg. The respirator facepiece weighs 0.6 kg and the filter canister 0.45 kg.

STATUS
In production. In service with the Romanian armed forces.

MANUFACTURER
ROMTEHNICA, 5-7 Drumul Taberei St., Bucharest, Romania.
Telephone: 33 10 61.

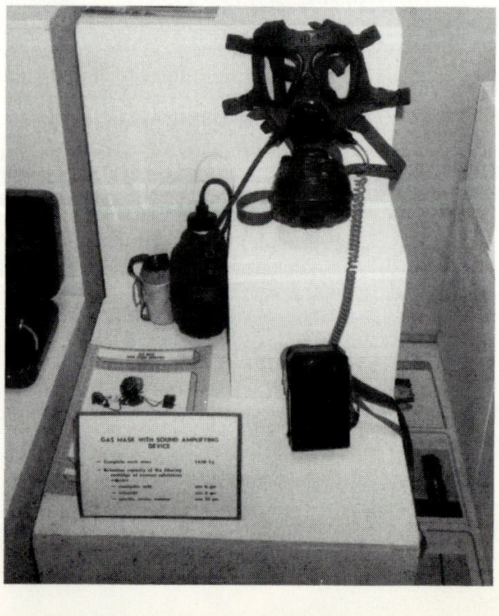

Romanian NBC respirator with sound amplifier and fitted with drinking device (C F Foss)

New Pac Disposable C-Cover S/89

In 1985 the Swedish Defence Material Administration (FMV) invited the Swedish plastics industry to take part in a development competition to design a disposable plastic cover for chemical warfare protection following FMV requirements. In close co-operation with FMV and following troop trials, the New Pac-designed C-Cover was accepted in December 1987. In February 1988 FMV ordered 79 000 C-Covers for the Swedish Air Force. Deliveries of the first 5000 units to FMV were made by December 1988 with final deliveries during June 1989. A subsequent order for a further 500 000 units has been made by FMV for delivery during 1989 and 1990.

A complete C-Cover S-89 consists of two packages, a body cover and a pair of foot covers, both tightly rolled and capable of being carried inside the face cavity of a stowed respirator or in a uniform pocket. When required the rolled body cover package is taken and the disposable stretch film cover is removed. The body cover is then unrolled to its full length and the entry fold is found. The thumbs are then placed into the centre fold and the user assumes a crouching position. The body cover is then placed over the head and shoulders. At this point the disposable outer bag falls to the ground and the cover's patented zig-zag folds unravel downwards to cover the entire body. The entire donning process takes 10 seconds and can be accomplished even when the wearer is lying on the ground. There is space on the back of the body cover to accommodate a back pack, and two rubber bands and a plastic belt are stored inside one of the end-sealed sleeves.

The protection process is continued by placing the respirator onto the head through a perforated membrane. Before moving out the foot covers are put into place. They are long enough to protect the entire length of the leg and fastening bands are used to tie them to a combat waist belt or webbing. The foot covers have soles with extra reinforcement around the toe and heel and two pairs of integrated shoe strings are used to fix the soles around the boots. The sole material can withstand at least 5 kilometres of walking across rough ground.

The C-Cover S/89 barrier material consists of five layers of co-extruded polyethylene/polyamide blown film 0.05 mm thick. The foot cover sole material consists of two layers of non-woven polypropylene fibre fabrics with a weight of 180 grammes per square metre.

The weight of a complete C-Cover S/89 is 560 grammes. The body cover is packed in a cylinder 65 mm in diameter and 150 mm long. The foot covers package measures 150 × 150 × 60 mm.

STATUS
In production. In service with the Swedish Air Force and on order for the Swedish Army and Marines.

MANUFACTURER
New Pac Systems AB, PO Box 15, S-564 01 Bankeryd, Sweden.
Telephone: 46 36 710 40. Telex: 70 412 LMARK S.

Swedish troops wearing the New Pac disposable C-Cover S/89

Combat Suit 90

Following a period of investigative research the Swedish Armed Forces issued a specification for an NBC protective outfit to be known as the Combat Suit 90. The specifications for the Combat Suit 90 were that it should consist of a jacket, trousers and overboots; it should be capable of being worn in place of the standard Combat Uniform 90; it should be capable of being donned and removed within five minutes; it should be capable of remaining in use in a non-combat area for 28 days; it should be available in four sizes; when packed it should not weigh more than 2 kg plus the footwear; the outer material should be self-extinguishing in the case of fire; it should be camouflaged; it should provide at least 12 hours protection against chemical warfare agents; the shelf life should be 10 years and 20 years if possible; it should be rain resistant; and the operational temperature range was given as −20 to +20°C. There was no nuclear flash protection demand and the combat suit was not required to be washable as washable training suits would be provided.

After evaluation a firm order was placed by FMV for 19 000 suits with delivery of the first 9000 during 1989 and the remaining 10 000 during 1990. An option exists for a further 28 000 suits with the latest option exercise date being 31 December 1990.

For the firm order 9500 suits are being produced by Ekman Safety AB at Avesta in Sweden (the garments are manufactured by Eagle Military Gear Overseas Limited of Tel-Aviv, Israel), and 9500 suits by K Stormar Konfeksjonfabrik A/S of Dokka, Norway. Outer fabric sub-contractors are Borås Wafveri AB of Borås, Sweden, and inner fabric sub-contractors are Helsawerke of Gefrees, West Germany. The overboots are supplied by Bata Schuh AG of Möhlin, Switzerland.

In addition to the provision of the Combat Suit 90 it is known that there is a Swedish Armed Forces requirement for an NBC respirator to replace the existing Skyddsmask 51 which was based on the American M9. Several designs are known to be under consideration including the British Avon S10, modified by Trelleborg AB, and designs from the Swedish concerns Forsheda and Horda.

SM90 NBC Protection Mask

Following technical and other tests carried out by the Swiss Defence Technology and Procurement Agency, the Swiss Armed Forces decided to procure the SM90 NBC protection mask. The SM90 was selected due to its sealing qualities for various head shapes, its high level of wearer comfort and low breathing resistance, and for its wide field of vision. Other factors involved were the use of insertable corrective lenses for spectacle wearers, the positive guidance of intake air to prevent eye-piece fogging, a good communication capability with the provision of a speaking aid device, and the provision of a drinking device.

The SM90 uses a bromobutyl facepiece with twin polycarbonate eyepieces. Corrective lenses can be inserted into the eyepiece mounting frames. The filter canister is located on the chin with a speech assistance device located in front of the mouth. A drinking device is inserted into the speech assembly and can be used with a standard water bottle. A head net arrangement held in place by six straps supports the respirator when worn.

When not in use the SM90 is carried in a fabric case which also holds spares and accessories.

STATUS
In production for the Swiss Armed Forces.

MANUFACTURER
Huber + Suhner AG, Rubber Products Division, CH-8330 Pfäfflikon/ZH, Switzerland.

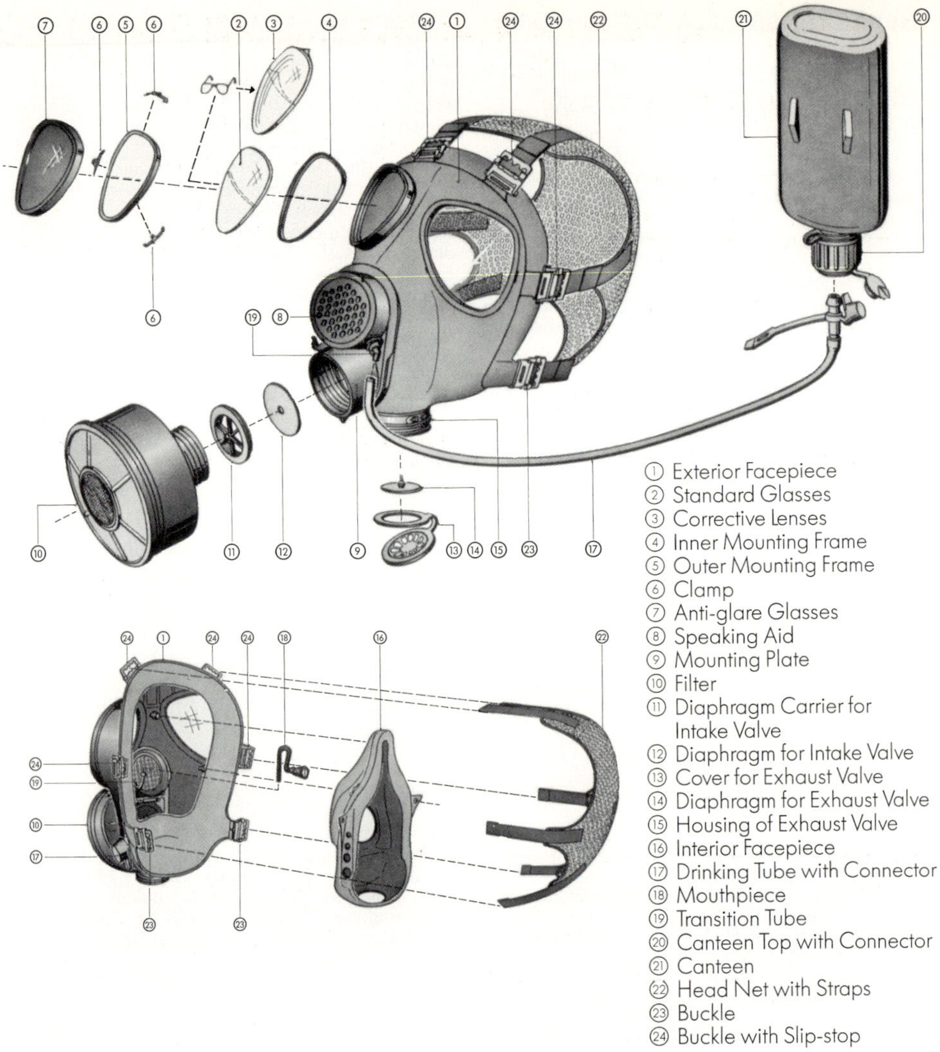

Component parts of the SM90 protective mask

The SM90 NBC protective mask together with its carrying case

DECONTAMINATION KITS AND EQUIPMENT

DEW 3-litre Decontamination Device

The portable DEW 3-litre decontamination device disseminates C1 (DS2) decontaminant in a controlled spray to remove chemical warfare agents from the surface of any military equipment. The device is reusable and can be easily cleaned and purged.

The tank on the device holds 3 litres of usable decontaminant and, once pressurised, disseminates a nozzle-controlled stream in a fan spray pattern directable to between 1 and 3 metres. An integral hand pump and quick-filler valve for external air sources such as a vehicle compressor pump ensure easy pressurisation. The device is provided with a pressure relief valve, a vehicle mounting bracket and a pump handle which also acts as a carrying handle. The device is corrosion resistant to C1 decontaminant and is adaptable to other chemicals. It has been shown to be easy to fill, charge and operate even when NBC protective clothing is worn. An operating instruction plate is attached. No special tools are required for disassembly and an optional spare part kit is available. The operational temperature range is from −45 to +50°C.

STATUS
In production for the Canadian Armed Forces (20 000 units).

MANUFACTURER
DEW Engineering and Development Limited, 3429 Hawthorne Road, Ottawa, Ontario, Canada K1G 4G2. Telephone: (613) 523 8150. Telex: 053 3167.

DEW 3-litre decontamination device

DEW 20-litre Decontamination Device

The DEW 20-litre decontamination device dispenses C8 decontaminant from standard steel or plastic jerrycans. The portable device is supplied in a fabric bag as a kit that contains an integral pump, a quick fill valve, a pressure gauge, a safety relief valve, a quick-disconnect hose, a wand extension, a scraper pole, scrubbing brushes, basic tools and a spares package.

The device is reusable and has a usable capacity of 18.5 litres. The tank can be initially pressurised within one minute using the hand pump and the contents can then be dispersed over an eight minute period with two re-pressurisations. The device uses a controllable spray with a spray distance of from 1 to 3 metres. The device has been shown to be easy to fill, pressurise and operate when wearing full NBC protective clothing and the device is easily cleaned and purged.

The entire kit weighs 8 kg and fits into a 180 × 180 × 600 mm storage area.

STATUS
In production.

MANUFACTURER
DEW Engineering and Development Limited, 3429 Hawthorne Road, Ottawa, Ontario, Canada K1G 4G2. Telephone: (613) 523 8150. Telex: 053 3167.

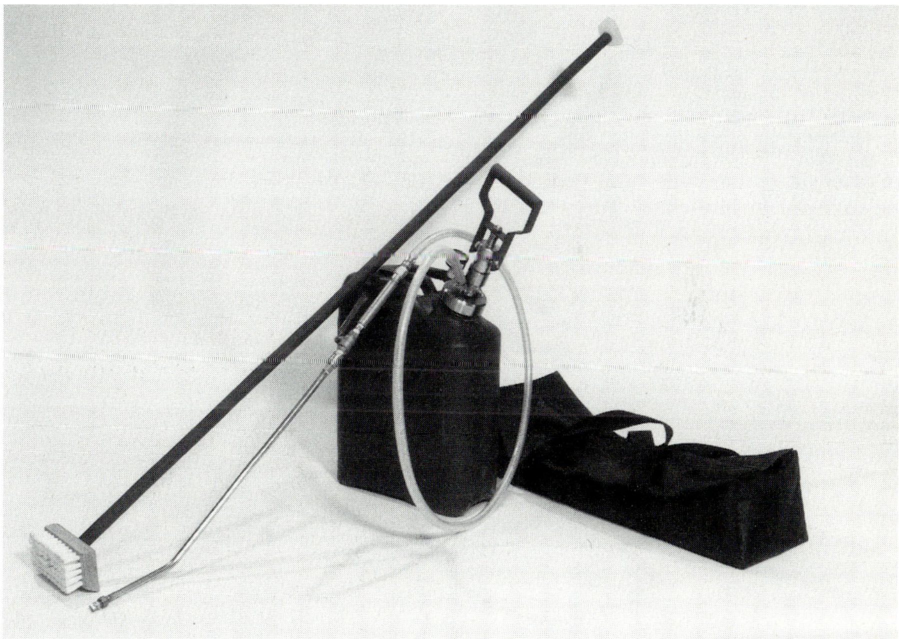

DEW 20-litre decontamination device

ACMAT-Kärcher Mobile Decontamination System

In order to determine the future NBC decontamination requirements of the French civil defence authorities, Kärcher France and the military vehicle manufacturers ACMAT have combined to produce a prototype of a mobile NBC decontamination system that is heli-portable. Normally the system would be carried on an ACMAT TPK 4.30 or 4.35 (4 × 4) truck.

The system consists of a number of standard Kärcher decontamination modules mounted in a steel carrying frame manufactured by ACMAT. The frame, which is fitted with carrying and locking lugs on each corner, contains Kärcher MPS, MPDS and DADS decontamination systems (see entries in *Decontamination kits and equipment* section for details of these systems), a water tank, two hose reels, and an accessories box containing spray lances and other items. Future examples will probably provide storage space for decontamination chemicals.

The weight of the complete system with a full water tank is approximately 2000 kg.

STATUS
Prototype.

MANUFACTURERS
ACMAT, (Ateliers de Construction Mécanique de l'Atlantique, Le Point de Jour, 44600 Saint-Nazaire, France.
Telephone: 40.22.33.71. Telex: 700913 F.

ACMAT-Kärcher mobile decontamination system

Kärcher France, 5 avenue des Coquelicots, ZA des Petits-Carreaux, 94385 Bonneuil/Marne Cedex, France.
Telephone: 43.77.12.70. Telex: 213071 F.

INTER-CB Decontamination Apparatus

The INTER-CB decontamination apparatus is a portable and rugged handpump device for the spraying of NBC decontamination solutions. It has an internal volume of 1.5 litres which is sufficient to decontaminate an area of about 10 square metres. About 20 operations of the handpump handle are sufficient to empty the apparatus. It is claimed that no maintenance of the apparatus is necessary. Training decontamination solutions can be used with this device.

STATUS
In production.

MANUFACTURER
INTER-CB, F H Schneider, Postfach 47, CH-3000 Bern 15, Switzerland.
Telephone: 031 52 44 34. Telex: 912 353.

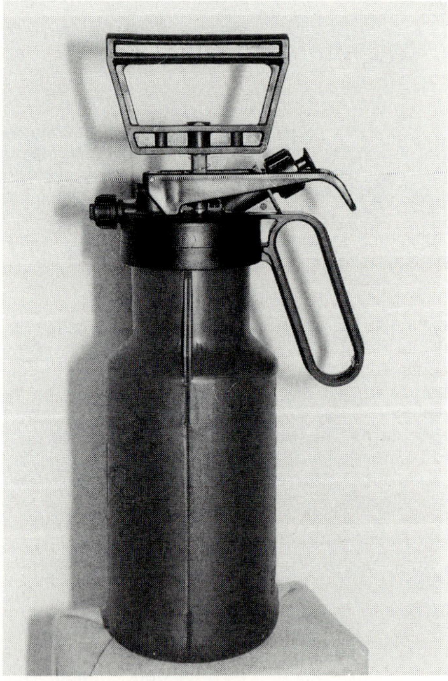

INTER-CB decontamination apparatus

CHEMICAL AND BIOLOGICAL DETECTION AND IDENTIFICATION EQUIPMENT

Honeywell ELAC Chemical Agent Detection System A2

The West German Armed Forces have used the American M43A1 nerve agent detection and alarm system since 1981 but a need was recognised for a vapour alarm system that would also be able to detect nerve, blister, blood and choking agents. Development of a miniature modular system that would initially detect nerve agents only was initiated by the West German Armed Forces Procurement Agency. The next step was the development of a combined blister, blood and choking agent alarm. By combining these two modules the A2 vapour detector and alarm system was produced. The A2 system is currently at its final development stage with the different agent sensing systems undergoing extensive testing at the Munster Armed Forces NBC Defence Research Establishment. By late 1989 a number of pre-production A2 detectors will be made available to the West German Armed Forces for both technical evaluation and extensive user trials.

The A2 detector is a small and lightweight microprocessor-based portable chemical agent vapour detector and alarm for both personal and point detector use. It includes three different sensors to detect nerve, blister, blood and choking agents at low concentrations and complying to NATO requirements.

The A2 detector and alarm is contained in a sealed aluminium case with a 11.4-volt 4 Ah lithium battery pack attached to the bottom of the case to provide 72 hours of continuous unattended operation. If required a remote control and alarm unit can monitor and power the A2 detector unit remotely by means of two-core field telephone cable. For portable and vehicle-borne use the A2 is contained in a special carrying case which holds spares and accessories. For fixed installation use an additional central control unit is available in which up to 12 remote control and alarms can be integrated to monitor an array of the same number of remotely deployed A2 detectors. This unit consists of two operating consoles including a personal computer; all units are installed in a 19-inch (483 mm) rack.

The electronic system consists of different individual modules which are placed on a mother board allowing easy replacement of individual circuits. A personal computer-controlled test set will be available for maintenance, trouble shooting and repair at depot level. A built-in test and a special confidence test set will offer test and training capabilities at organisational levels.

Within the A2 detector unit, nerve and blister agent detection is performed by two different ionisation cell systems using americium 241 as an ionising source to produce the agent cluster ions to be detected. In the nerve agent ionisation cell the selection between the heavier agent clusters and the lighter air ions is provided by a dynamic grid which is modulated with a transversal electrical field, while across the length of the cell a continuous longitudinal electrical field is applied. The positive nerve agent ion clusters exhibit a high permeation rate through the modulated grid towards a collecting electrode.

For blister agent detection a negatively biased M43A1 type baffle cell is used as these agents predominantly form negative ion clusters. For the purpose of both agent enrichment and humidity reduction, a silicone rubber membrane is placed between the sample air and the blister agent sensing cell. Since blister agent detection is considerably affected by sample air humidity the A2 blister agent ionisation cell is arranged into a closed internal sensing loop where humidity is kept low by an expendable desiccant module lasting 72 hours. Blister agents permeate into the internal sensing loop through the silicone rubber membrane which separates the sensing flow from the external sample air flow. To improve the signal-to-noise ratio an additional reference baffle cell is placed in the internal sensing loop between the desiccant module and the blister agent sensing cell. Due to a molecular sieve in the desiccant module the reference cell is always provided with clean dry air when the A2 detector is running, resulting in a well defined clean air reference signal to compare with the signal exhibited by the blister agent sensing cell.

Blood and choking agents are detected by a commercially-available Bayer electro-chemical cell placed close to the sample air outlet. When not required this expendable cell can be easily removed and stored separately; its shelf life is approximately five years. When the cell (a two-electrode amperometric cell) is activated it lasts for six to twelve months.

Detector and alarm unit (left) and remote control and alarm unit (right) of Honeywell ELAC Chemical Agent Detection System A2

The remote control and alarm unit associated with the A2 detector can be linked to the detector and alarm unit by a two-core field cable up to 1000 metres long. The remote control and alarm unit, usually referred to as a remote unit, provides both remote alarm and control functions. All status information of the detector and alarm unit, usually referred to as the alarm unit, is displayed on the control panel of the remote unit. Power for the remote unit comes either from its own 11.2-volt 4 Ah batteries or from vehicle batteries.

The A2 detector and alarm unit and the remote control and alarm unit both weigh approximately 1.5 kg and measure 207 × 125 × 52 mm.

Using the technology involved in the A2 programme an add-on module for the M43A1 detector is under development. This will enable an otherwise unmodified M43A1 to detect blister, blood and choking agents in addition to its usual nerve agent detection capability. The add-on module will be capable of use as a stand-alone device or in connection with the M43A1.

STATUS
Final development – see text.

MANUFACTURER
Honeywell-ETEC-Nautik GmbH, PO Box 2520, D-Kiel 1, Federal Republic of Germany.
Telephone: (0431) 883 0. Telex: 292 521 hone d.

Chemical Agent Sensor GVJ-2

The Chemical Agent Sensor GVJ-2 is a portable hand-held instrument able to detect nerve and blister chemical warfare agents and various toxic industrial materials. The device automatically monitors the air for the presence of agents with air analysis being carried out in an ionisation cell under the control of a microprocessor. Ion mobility spectroscopy techniques are employed.

The GVJ-2 has a sensitivity of 1 $\times$ 10-5 mg/litre for GB, GD and VX, and 2 $\times$ 10-3 mg/litre for HD. The response time in both instances is five seconds.

The GVJ-2 has only two operator controls, an on/off push button and a mode change button. The two operating modes employed are a single measurement mode in which only one sample of air is pumped into the system, analysed and the results are then displayed, while in the periodic measurement mode the same sequence is repeated at pre-selected time periods. Results are displayed on a liquid crystal display panel using an array of eight blocks. For instance a heavy concentration of nerve agent will result in all eight blocks being displayed. Lesser concentrations will produce fewer blocks. The display can also be used to provide a battery low indication (BAT.LOW.) and an indication that the usual self test after switching on has resulted in a component failure (TEST.FAIL).

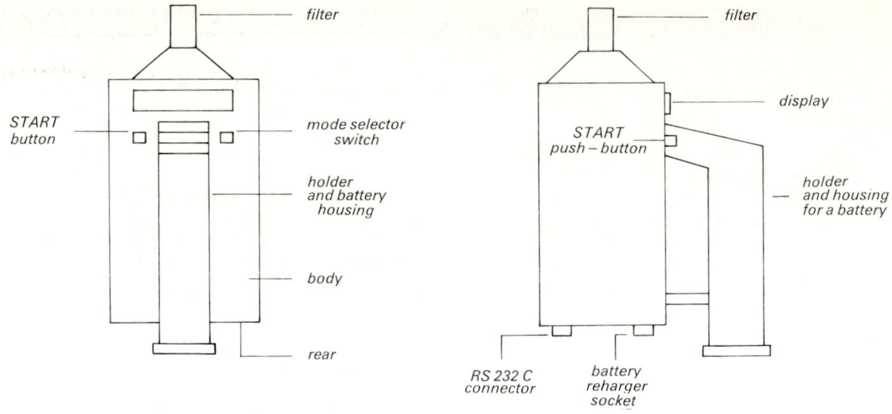

Outline drawings of Chemical Agent Sensor GVJ-2 with top view (left) and side view (right)

The GVJ-2 is 400 mm long over its longest point and weighs approximately 1.8 kg with four NiCd batteries. The batteries provide at least six hours of continuous use and (typically) ten hours of more normal use. Its operating temperature range is from −25 to +55°C with a relative humidity of up to 95 per cent.

STATUS
Under development with a view to production commencing during 1990.

MANUFACTURER
TECHNIKA Foreign Trade Company, Budapest-Hungary, X., Salgótarjáni u.20, H-1475 Budapest, POB 125, Hungary.

Nerve Agent Alarm ASTN-2

The Nerve Agent Alarm ASTN-2 automatically provides optical and acoustic signals following the detection of nerve agents in the surrounding atmosphere. The signals are provided only when a pre-determined nerve agent concentration has been exceeded. Warning signals can be provided from a remote location via a cable if required, and the device can be mounted in a vehicle in which instance an anti-vibration base is supplied.

The ASTN-2 employs a sampling air flow of approximately one litre per minute and an alarm can be provided within a maximum of 20 seconds from detection. The normal autonomous operating cycle is 24 hours. Power is supplied from a battery or from a vehicle power supply between 12 and 24 volts. Maximum power consumption under normal climatic conditions is 0.8 amp in the sampling mode and 1 amp in the alarm mode.

Total weight of the system is approximately 8 kg of which 2 kg is the anti-vibration base. Dimensions, with the base, are 245 $\times$ 210 $\times$ 225 mm.

STATUS
In production.

MANUFACTURER
ROMTEHNICA, 5-7 Drumul Taberei St., Bucharest, Romania.
Telephone: 33 10 61.

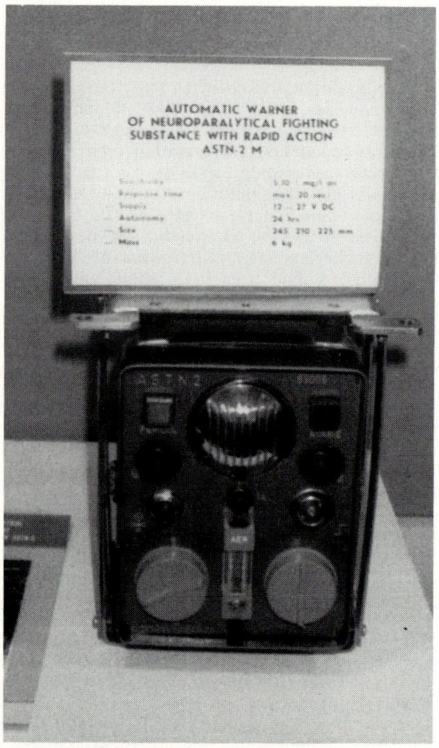

Nerve agent alarm ASTN-2 (C F Foss)

MEDICAL AND LIFE SUPPORT EQUIPMENT

Casualty Respirator Hood, NBC Mark 2

The Casualty Respirator Hood, NBC Mark 2A was developed to meet a requirement from a NATO country for the protection of casualties unable to wear a standard service respirator. The hood is powered by a separate blower unit, fitted with standard NBC filter canisters, that supplies a constant flow of filtered air into the hood. The hood can be fitted over NBC clothing and can be fitted to a person who is unconscious and unable to co-operate.

The Casualty Respirator Hood Mark 2A covers the casualty's face and head with a transparent hood which is strong and highly resistant to heat, flame and penetration by chemical agents. The hood is maintained under positive pressure to give protection levels equivalent to or better than standard service respirators. A constant flushing flow of filtered air through the hood prevents misting and any build-up of carbon dioxide. The airflow is ducted to the top of the hood from where it flows down over the head and face and escapes through exhalation valves at the bottom. The neck seal may be opened to administer oxygen direct to a casualty whilst maintaining overpressure in the hood.

The filtered air supply is provided via a hose from a separate Racal blower unit which may be belt mounted. The blower incorporates a motor fan and a lithium battery pack which can power the system continuously for six hours or longer.

The equipment weighs less than 2 kg, including the battery pack and filter canisters.

STATUS
Final prototypes. Ready for production.

MANUFACTURER
Racal Panorama Limited, Beresford Avenue, Wembley, Middlesex HA0 1QJ, United Kingdom.
Telephone: 01 900 2827. Telex: 922101.

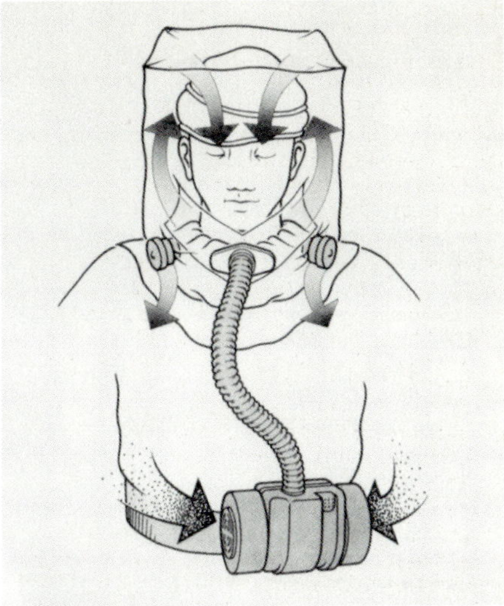

General outline drawing showing the air flow route in the Casualty Respirator Hood, NBC Mark 2A

CHEMICAL TRAINING, MAINTENANCE AND OTHER EQUIPMENT

Norabel Chemical Attack Simulator

The Norabel Chemical Attack Simulator consists of a chemical exercise grenade, a launcher, and a liquid chemical agent simulant. The grenade is fired from the launcher and, at a pre-determined burst point, a 50-metre long cloud of simulant droplets and aerosols is formed to cover an area of from 1000 to 1500 square metres.

In use the liquid chemical agent simulant, which can cause colour changes on chemical agent detection papers, is poured into a plastic container that forms the body of the practice grenade. A cover is then pushed into the filled body and an overflow pipe device is actuated to ensure the correct volume of contents is enclosed. A delay fuze is placed into a housing in the lid and the practice grenade is then loaded into the launcher. The launcher, which is secured to the ground by a spade-type frame, is a simple steel pipe from which the grenade is fired electrically using a black powder charge. The firing charge also initiates the grenade fuze which subsequently detonates to scatter the contents of the grenade.

STATUS
In production. In service with the Swedish Armed Forces.

MANUFACTURER
Norabel, Box 803, S-713 25 Nora, Sweden.
Telephone: 46 587 115 50. Telex: 73451 norabel s.

Two Norabel Chemical Attack Simulator launchers ready for use together with their practice grenades (T J Gander)

Index

Manufacturer's Index